World War II Monuments of Luxembourg: Dedicated to US Forces

80th Anniversary Edition of the Battle of the Bulge

World War II Monuments of Luxembourg: Dedicated to US Forces

80th Anniversary Edition of the Battle of the Bulge

Note about the cover photo headstone. Since this photo was taken, the unknown remains have been identified as those of Private John P. Cooper, Company B, 778th Tank Battalion. At the request of the family, PVT Cooper's remains have been disinterred and returned to the United States. However, the headstone remains. The US Defense POW/MIA Accounting Agency (DPAA) continues to identify unknown remains from all wars in which there are missing Prisoners of War (POWs) or Missing in Action (MIA) personnel.

Robert O. Walton

Published by:
Hang Time Publishing Ltd. Co.
c/o P. M. Politano
15 Palomino Court
Charleston, SC 29407

Walton, R.O. (2024). *World War II Monuments of Luxembourg: Dedicated to US Forces, 2nd edition*. Hang Time Publishing.

To correspond with the authors, please send an email to: waltonlux@pt.lu

ISBN 979-8-218-40031-6

Dedication

To the men and women of Luxembourg who lost their lives during World War II and to the valiant US and Allied Forces who fought and died to liberate Europe from the Axis powers – let you never be forgotten.

-Robert O. Walton

Contents

From the Ambassador of the United States of America to the Grand Duchy of Luxembourg

The second edition of *World War II Monuments of Luxembourg* is a publication that stands out as a testament to the shared history between the United States and Luxembourg and serves as a bridge connecting our past to the present and future at an extraordinarily pivotal time.

In 2024, we mark the 80th anniversary of the Battle of the Bulge, a turning point in the war where American soldiers displayed extraordinary valor and resilience. For that reason, the publication of this book gains an even deeper significance. It was in the harsh winter of 1944 that Luxembourg experienced firsthand the commitment and bravery of American forces, who fought tirelessly against the oppression of tyranny. The monuments and remembrance sites detailed in these pages are far more than mere stone and metal; they are living symbols of gratitude, honoring those who fought for Luxembourg's freedom. On behalf of the United States, I am forever grateful to Luxembourg for preserving and continuing to honor these important memories with grace and reverence.

We also mark the 75th anniversary of the North Atlantic Treaty Organization (NATO) in 2024. NATO is an alliance born out of the lessons of World War II. It is a cornerstone of transatlantic security. The relationship between the United States and Luxembourg, cemented on the battlefields of the past, continues to thrive within the framework of NATO. Luxembourg's liberation in WWII and the subsequent establishment of NATO are intertwined narratives of resilience and unity, narratives that this book eloquently captures. This book, in its own special way, reminds us that the peace and stability we enjoy today were hard-won and must be actively preserved.

In today's world, as we face the resurgence of conflict and war, even on the European continent, we are starkly reminded that the fight for democracy and freedom is a perpetual endeavor. It requires our unwavering commitment and constant vigilance. The lessons we have garnered from World War II are invaluable, and we must ensure they are never forgotten, continually guiding our actions and decisions.

Bob Walton's work through these pages truly serves as a reminder of our collective journey from the shadows of war into the light of peace. As we look to the future, new generations must understand these lessons. This book serves as an educational tool, ensuring that the stories of courage and camaraderie, and of liberation and freedom, are passed down. It is through understanding our shared history that we can best appreciate the freedoms we now enjoy and recognize the importance of working together to safeguard them.

May this book inspire you as it has inspired me. May it enlighten you about our shared past. May it also inspire a dedication to the principles of liberty and cooperation. Together, let us uphold the values for which so many have valiantly fought and for which too many have made the ultimate sacrifice.

Tom Barrett

United States Ambassador to Luxembourg

Foreword

As we reach the 80th anniversary of the battles to free Luxembourg from German occupation during World War II, it seems fitting to update my original text on the *Monuments of Luxembourg Dedicated to US Forces.* The 1st edition was published in 2021, however I missed a few monuments and new ones have been dedicated. In this edition, I have added the missing monuments, updated some of the photos, fixed errors in the previous edition and added the many information panels that have been erected to describe the events in the area of various monuments. I hope that you find this text a valuable resource. These monuments have been erected to remember the events, bravery and heroism of the US Forces to liberate Luxembourg. At this point in time, almost all of the men who fought have died, either during the war, of old age, or somewhere in between. These monuments are a continuing memory of their actions that will remain long after the last soldier dies.

A special thanks to my family and friends who took the time to accompany me as I searched for the many monuments in Luxembourg. For this edition my wife Petra, and Terry Farrelly joined me on several excursions. In addition, I would like to pay special thanks to Philippe Victor with the National Military Museum in Diekirch, Tom Scholtes with CEBA, Roland Gaul retired director of the National Military Museum in Diekirch, Fern Barbel with Centre de services norTIC, Marcel Scheidweiler, Erny Kohn, Christian Pettinger, Frank Rockenbrod and Yves Rasqui for helping me track down some of the harder to find monuments and for reviewing and contributing to the final manuscript. I would also like to thank Volker Teuschler with cube-werbung.de for providing the print proofs of the information panels along with the Ghost Army Legacy Project. Hopefully, these will reproduce better and make the information panels in the book more readable.

Finally, a very special thanks to Tom Scholtes, Erny Kohn, Marcel Scheidweiler, Fern Barbel, and Roland Gaul, who were instrumental in erecting many of the monuments over the years.

Robert O. Walton, PhD
Major, US Army Reserves (retired)

If you know of any monument to US Forces in Luxembourg that is not documented in this book or any incorrect information, please contact me so that I can update the text. Emails can be sent to: waltonlux@pt.lu

Introduction

On the morning of May 10, 1940, the Grand Duchy of Luxembourg was invaded by German forces. While resistance was mounted, it was futile. By nightfall German forces had occupied most of the country and members of the Grand-Ducal family had escaped. Grand Duchess Charlotte finally moved to London in 1941. The Royal family eventually settled in Canada for the duration of the war while the Grand Duchess Charlotte stayed in London and became an important symbol of national unity. The Grand Duchess' husband, Prince Félix, served in the Northern Command of the British Army, her son, Hereditary Prince Jean, joined the British Irish Guards and took part in several battles.

The Germans considered Luxembourg a Germanic country with similar language and culture, so at first was placed under military administration and later annexed directly into Germany under the Gauleiter Gustav Simon. Active and passive resistance continued throughout the war, which Simon and his Nazi hordes always brutally put down. In all, 5,700 Luxembourgers died during World War II, which was about 2% of the population at that time.

In September 1944 Allied forces started to push the German Army behind the *Westwall* fortifications, and by early December this had been mostly accomplished. In Luxembourg the Westwall ran along the eastern border of the country, east of the Our, Sauer and Mosel rivers. South of Luxembourg Patton's Third Army was preparing to move east into Germany. However, at 0530 hours on the morning of December 16, 1944, German forces started to move westward through northern Luxembourg and parts of Belgium in a surprise offensive. The battle became known officially as the "Battle of the Ardennes" by the US Army, and "Wacht am Rhein" by the German military, but most of us know the offensive as the "Battle of the Bulge." On 18 December, Bradley requested that Patton prepare to move north to counterattack the German forces. The Battle of the Bulge was the bloodiest engagement of US Forces during WWII, with over 19,000 Americans dead, almost 48,000 wounded, and over 23,000 captured or missing. The official end of the Battle of the Bulge was January 25, 1945, and by January 28, 1945, the first Allied offensive entered Germany.

After German forces finally departed Luxembourg in late January 1945, US troops also quickly departed, pushing German forces further into Germany. Finally, after 5 years of occupation, Luxembourg was again a free country.

General Tactical Situation in the area of Luxembourg on December 15, 1944

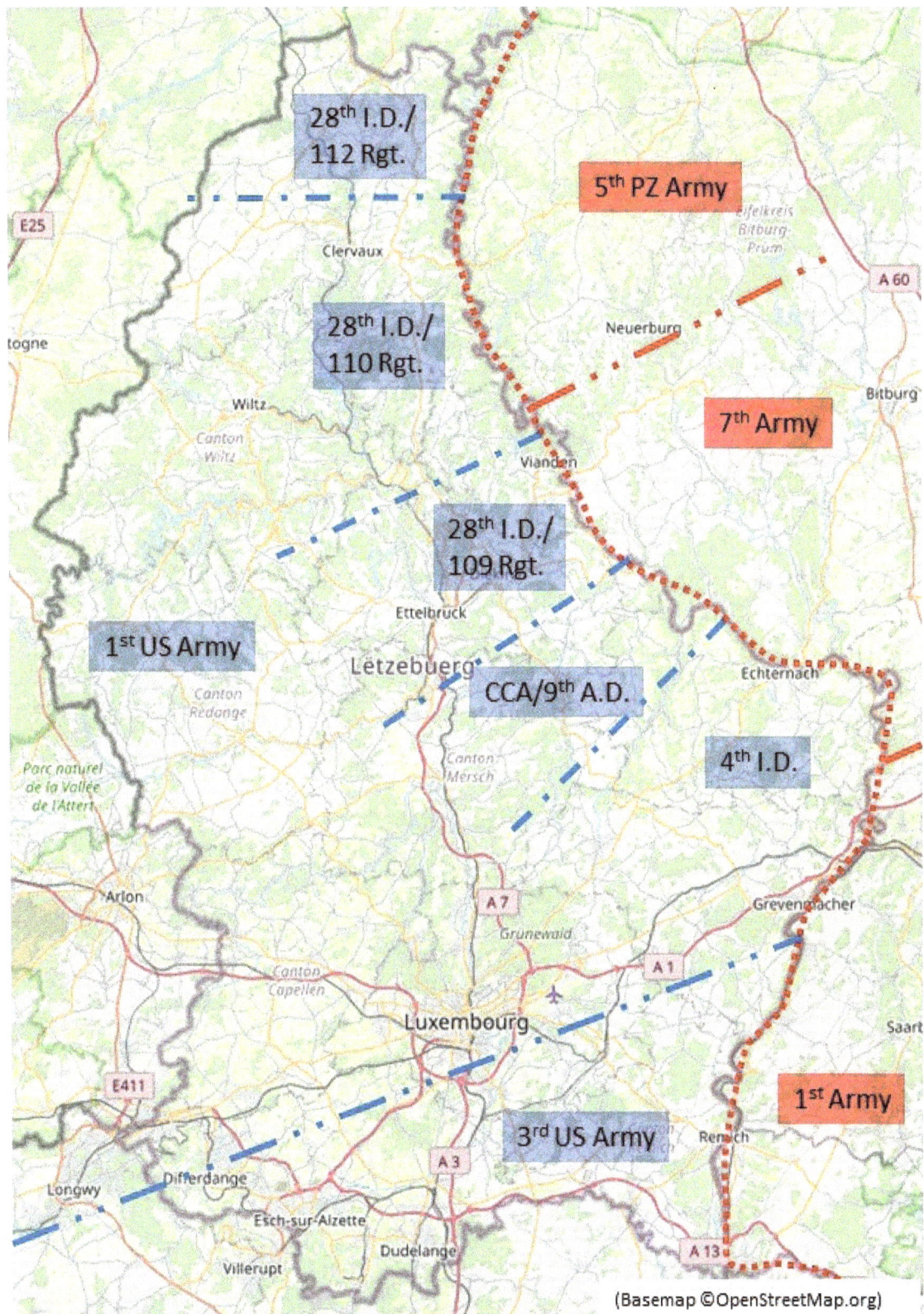

(Basemap ©OpenStreetMap.org)

Major American Units in the Luxembourg Area of Operations September 1944 – February 1945

The first American forces enter Luxembourg near the town of Pétange on September 9, 1944, and with the liberation of Vianden on February 12, 1945, marked the end of combat operations in Luxembourg. This is a list of the major units (Division and higher) that were stationed in Luxembourg during this time. The patch or emblem has been included since many of the monuments have the unit emblem on them. Also note that during this time some units were reassigned.

Unit	Commander	Nickname	Emblem
12th Army Group	LTG Omar N. Bradley		
94th Infantry Division	MG Harry J. Malony	Neufcata	
1st US Army	LTG Courtney H. Hodges	Doughboys	
V Corps	MG Leonard T. Gerow	Victory	
9th Infantry Division	MG Louis A. Craig	Old Reliables	
5th Armored Division	MG Lunsford E. Oliver	Victory	
VII Corps	MG J. Lawton Collins	Jayhawk	
83rd Infantry Division	MG Robert C. Macon	Ohio	

8th Infantry Division	MG Donald A. Stroh/MG William G. Weaver	Golden Arrow
Third US Army	LTG George S. Patton Jr.	Patton's Own
III Corps	MG John Milikin	Phantom
4th Armored Division	MG Hugh J. Gaffey	Breakthrough
6th Armored Division	MG Robert W. Grow	Super Sixth
11th Armored Division	MG Edward H. Brooks	Thunderbolt
26th Infantry Division	MG Willard S. Paul	Yankee
35th Infantry Division	MG Paul W. Baade	Santa Fe
90th Infantry Division	MG James A. Van Fleet	Tough 'Ombres

VIII Corps

(*under First US Army until December 20, 1944)   MG Troy H. Middleton

9th Armored Division   MG John W. Leonard   Phantom

17th Airborne Division   MG William M. Miley   Golden Talon

28th Infantry Division   MG Norman D. Cota   Keystone

87th Infantry Division   MG Frank L. Culin Jr.   Golden Acorn

XII Corps   MG Manton S. Eddy

4th Infantry Division   MG Raymond O. Barton   Ivy

5th Infantry Division   MG Stafford L. Irwin   Red Diamond

10th Armored Division   MG William H. H. Morris Jr.   Tiger

80th Infantry Division   MG Horace L. McBride   Blue Ridge

76th Infantry Division   MG William R. Schmidt   Liberty Bell

XX Corps   MG Walton H. Walker

Monuments

The following pages provide location, photos and background on the World War II monuments of Luxembourg dedicated to US Forces. Most of these monuments are dedicated to US forces who liberated Luxembourg in the fall and winter of 1944-45. However, there are a few monuments from US Forces to the Luxembourg people and have been included. There are many more monuments that are not included, that are dedicated to the Luxembourgers who were deported to concentration camps, conscripted into the Germany military, joined resistance groups or simply ran afoul of the German military government and were murdered. In addition, there are a few monuments to the Royal Airforce (RAF) that are dedicated to aircraft crash sites and crew that are not included in this text.

The address, and latitude and longitude listed for each of the monuments are based on the data provided by Apple maps, in-car GPS, or openstreetmap.org. The monuments are listed by the nearest town. Many of the monuments are either in the town center, at the church or on a major road leading into or out of a village, or at a river crossing point. However, some are located in the woods or in remote areas. The latitude and longitude data would be a better option for finding some of the monuments but will only get you close. Just be prepared for some "hunting" in some cases.

Copies of the information panels have been included in this edition. Unfortunately, some of the text is hard to read due to the size of the panels and the required reduction to fit into the book. However, the QR codes on the panels can be scanned to find more information about the site.

Asselborn (Domaine du Moulin d'Asselborn)

A small personal memorial to the US Forces in the hotel's inner courtyard. There are also two more monuments just to the north of the hotel. The first is about the historical site of the mill with one panel on the Battle of the Bulge. The other is dedicated to a future of peace for all.

Op der Millen 1

9940 Asselborn

Latitude: 50.0880

Longitude: 5.9645

Beaufort

In appreciation to citizens of Beaufort and the Luxembourg survivors of the German concentration camp at Buchenwald from grateful friends in the 6th Armored Division.

Rue de la Libération

6315 Beaufort

Latitude: 49.833126

Longitude: 6.297766

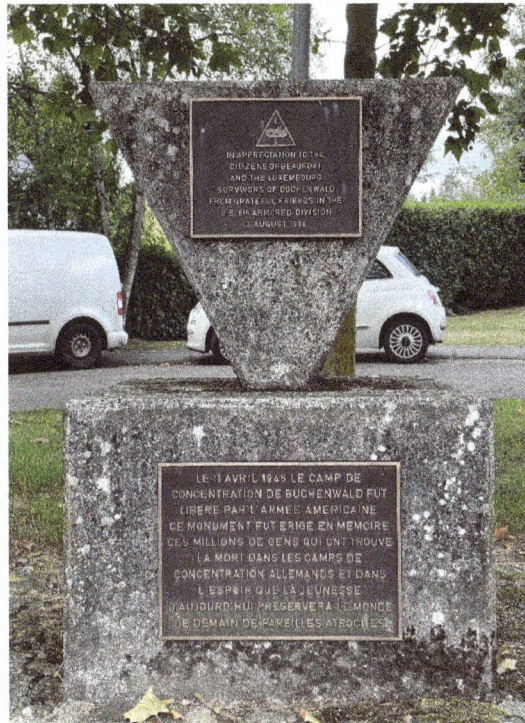

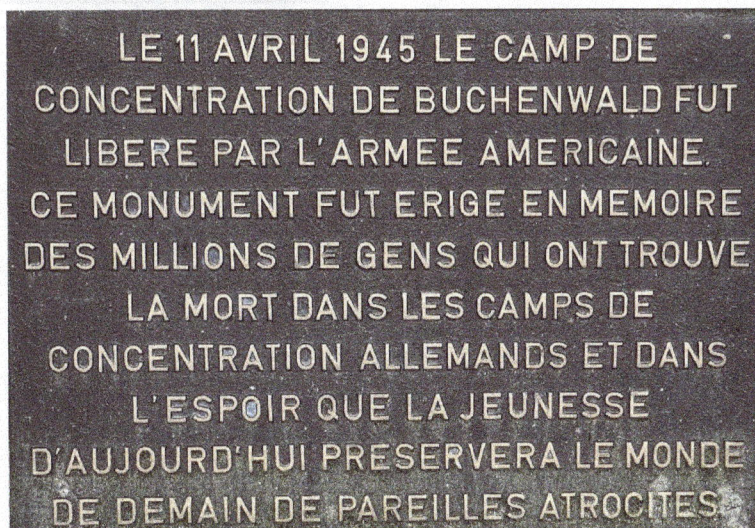

Inscription reads: On April 11, 1945 the Buchenwald concentration camp was liberated by the American Army. This monument was established in memory of the millions of people who died in Germany's concentration camps and in the hope that the youth of today will keep the world of tomorrow free from a similar atrocity.

Beaufort

In honor to the 5th and 9th Armored Divisions and the 9th Infantry Division. Monument is located on side of church in Beaufort.

Rue de l'Eglise 5

6315 Beaufort

Located by the church in Beaufort

Latitude: 49.833749

Longitude: 6.290736

There is a mistake on the plaque. The 9th Inf Div. was never in Beaufort, but the 5th Inf Div "Red Diamond" was, as shown in the upper left-hand corner of the plaque.

Beaufort-Dillingen

Located on CR 364 between Beaufort and Dillingen there are two identical information charts located about 100m apart on either side of the road that covers the 60th Armored Infantry Battalion's, 9th Armored Division, defending of this area on December 16-21, 1944.

CR364

6340 Beaufort

Latitude: 49.84621

Longitude: 6.30553

The deadly woods
The 60th Armored Infantry Battalion, CCA/9th U.S. Armored Division
defending Beaufort; December 16-21,1944
(Close-up on companies "A" and "B" in the "Kuesselt" and "Millebach" woods)
(compiled by Roland Gaul, MNHM/Diekirch)

On **December 10, 1944**, the 60th Armored Infantry Battalion (AIB) of Combat Command A of the 9th U.S. Armored Division relieved units of the 28th U.S. Infantry Division to assume responsibility of the defense of a sector extending from the heights above REISDORF to the fringing high ground east of BEAUFORT, overlooking the Sûre (Sauer) river and the opposed German border area. The 60th AIB, commanded by **Lt.Col. Kenneth W. COLLINS**, initially deployed its "A" and "C" companies to this defense line task, whereas "B" company, the battalion command post and the service company remained quartered in BEAUFORT. For immediate artillery support, the 60th AIB could call on the nearby 9th Armored Division's own 3rd Armored Field Artillery Battalion (AFB) (**Lt.Col. George RUHLEN**), positioned in and around HALLER On its right flank towards the "MULLERTAL", the 60th AIB had as its neighbor units of the 4th U.S. Infantry division on the defense line running east.

In the early morning of **December 16, 1944**, the German surprise attack in this sector began with a huge artillery- and rocket projector barrage falling down on the front-line companies' dugouts and observation posts around BEAUFORT. Although most communication lines – except radio networks - were disrupted, it became quickly obvious that this was more than a strong local attack!

At first daylight, strong enemy infantry patrols, after having crossed the Sûre river at DILLINGEN, began attacking the frontline positions of companies "C", commanded by **Captain Roger A. SHINN** (left) and "A" (**Captain John W. SCHALLES**) on the right end of the 60th AIB's defense sector. The well-dug in individual rifle platoons inflicted heavy losses to the first wave of attackers and artillery fire was called in on the advancing German troops spreading out in the dense forests. A straight enemy infantry attack coming from the road to DILLINGEN hit the center of company "A" around 10 a.m. and succeeded in dislocating the first platoon from its original position, forcing it to withdraw and began exploiting to the left and right.

(NB: It was later learned that those German assaults in the center were staged by elements of the 988th "Grenadier-Regiment" of the 276th "Volksgrenadierdivision", commanded by Major General Kurt MOEHRING (**), whereas the division's other two regiments, the 986th and the 987th were to attack along the "Enz blanche" (on the right) and "Ernz noire" (left) flanks of the German sector of attack. The 276th "Volksgrenadierdivision" had received the order to capture the high ground across the Sûre, destroy the American artillery positions around HALLER and form a south western extension in direction of LAROCHETTE.)

"B" company,, which so far had been held in reserve, was immediately committed to seal the developing threatening gap. "B" company commanded by **Captain Floyd D. HARDER**, attacked thru company "A"'s lines, cut across the German advance elements and succeeded in closing the gap and reestablishing contact with companies "C" and A". Although a squad of company "C"'s own 3rd platoon had been killed or captured, the remainder of the "C" and "A"'s platoons resisted fiercely leaving no place to the enemy. Due to the 3rd AFB's accurate artillery fire from HALLER directed by forward observers in company "A" and "C" sectors on the German crossing points, the enemy's attack was slowed down in the afternoon on that first day of the "Bulge".

However, during the night to December 17, 1944, as fighting went on in the darkness in the "Kuesselt" woods, the Germans came thru the extension of the "MULLERTHAL" draw, turned west at HALLERBACH and started attacking behind the 60th AIB's front-line company's positions to cut them off. During the day, more and more Germans appeared in the rear areas and the situation for BEAUFORT became very alarming. The 9th Armored Division commander, **Major General John W. LEONARD** released a reserve force consisting of tanks, tank destroyers, anti-aircraft artillery and armored reconnaissance elements to be transferred immediately in relief support of the endangered CCA units. **Col. Thomas L. HARROLD**, CCA commander ordered one troop of the 89th Reconnaissance Squadron to Beaufort to hold the town, as the 60th AIB's Headquarters were evacuated from "Hôtel Belvédère" in Beaufort along with the HQ- and service companies to SAVELBORN.

During the day of **December 17, 1944**, the enemy pressure grew so strong that the elements of 89th Reconnaissance Squadron had to withdraw from BEAUFORT filling with German troops and it was decided to try relieve the hard pressed companies "A", "B", and "C" with two task forces (Task Force HALL and PHILBECK) in the morning of **December 18**. As ammunition, food and medical supplies in the front-line company sectors were running short, the company commanders of "A" and "C" rushed back for supplies. A half-track was loaded for each of them and they managed to reach their companies and also have a number of wounded evacuated. "B" company could not be re-supplied or relieved. To worsen things, radio contact with the three companies was lost in the morning of December 18, 1944 and a further attempt to relieve them during the entire day with tanks from the direction of SAVELBORN failed again due to strong individual German antitank elements. In the late afternoon a forward artillery observer of the 3rd AFB got his radio working again to connect to 60th AIB's headquarters and Lt.Col. COLLINS got the order to withdraw for all front-line companies out to company "C" via the artillery observer **Lt. Ira CRAVENS**'s radio. Runners penetrating the German lines got the word out to the cut-off companies "B" and "A" to do the same and withdraw back along the ERMSDORF-SAVELBORN road During that move, "C" company commander, Captain SHINN, was captured, but Captains SHALLES and HARDER managed to pull out the exhausted platoons of their companies "A" and "B" by the following day. From the morning of **December 19, 1944** until December 26, the 60th AIB held a new defense line from SAVELBORN to ERMSDORF, when relieved by units of the 5th U.S. Infantry Division, which also liberated BEAUFORT on **December 27, 1944**

During the fighting that occurred in and around BEAUFORT and adjacent areas during the **December 16-22, 1944** time frame, the 60th AIB suffered the following casualties:

1 officer and 23 enlisted men killed
1 officer and 60 enlisted men wounded
5 officers and 132 enlisted men missing in action or taken prisoner

*(**) **Major General Kurt MOEHRING**, commander of the 276th Volksgrenadier-Division, was killed in the early afternoon of December 18, 1944 at the site named "Krewinkel/Kapell" on the road leading from Beaufort to Grundhof in his staff car together with his driver and two guards. The vehicle was most likely hit by machine gun bursts from a rear outpost of company "A". He was succeeded by Major General Hugo DEMPWOLFF*

Berdorf (Fortress Hotel)

Monument for "Fortress Hotel" where F Company, 12th Infantry Regiment, 4th Infantry Division resisted German attackers and held the Hotel du Parc in Berdorf from 16 to 20 December, 1944, during the opening days of the Battle of the Bulge. The original Hotel de Parc has been replaced by a senior residence.

7 An der Ruetsbech

6552 Berdorf

Latitude: 49.82408

Longitude: 6.34718

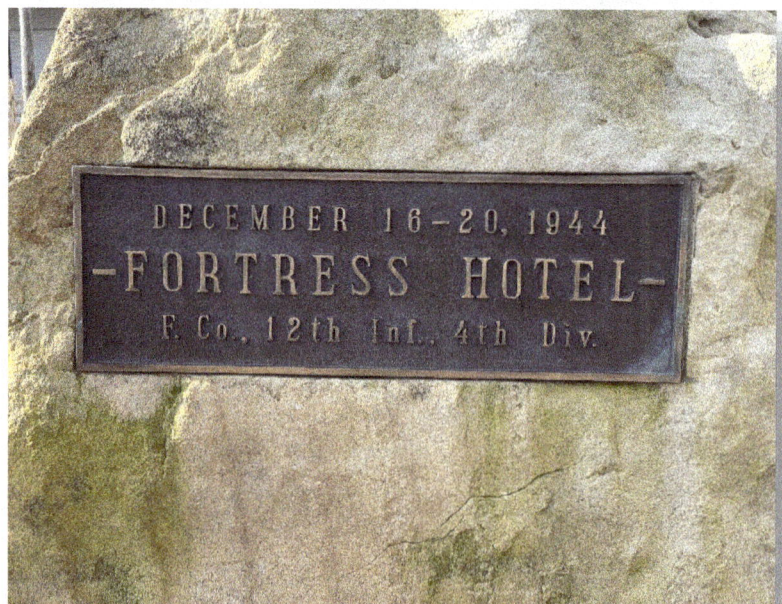

Berdorf

In honor to the 10th Armored Division, Battle of the Bulge, 1944-1945.

Rue de Consdorf

6551 Berdorf

Latitude: 49.8131

Longitude: 6.3412

Berlé

In honor of the 90th US Infantry Division which liberated Berlé on January 9, 1945. Monument is located next to the church in Berlé. Nearby is also a map of the liberation of Berlé.

Duerfstrooss 25

9636 Winseler

Located by the church in Berlé

Latitude: 49.9515

Longitude: 5.8551

IN HONOR OF
THE
90TH US INFANTRY DIVISION
WHICH LIBERATED BERLE
ON JANUARY 9, 1945

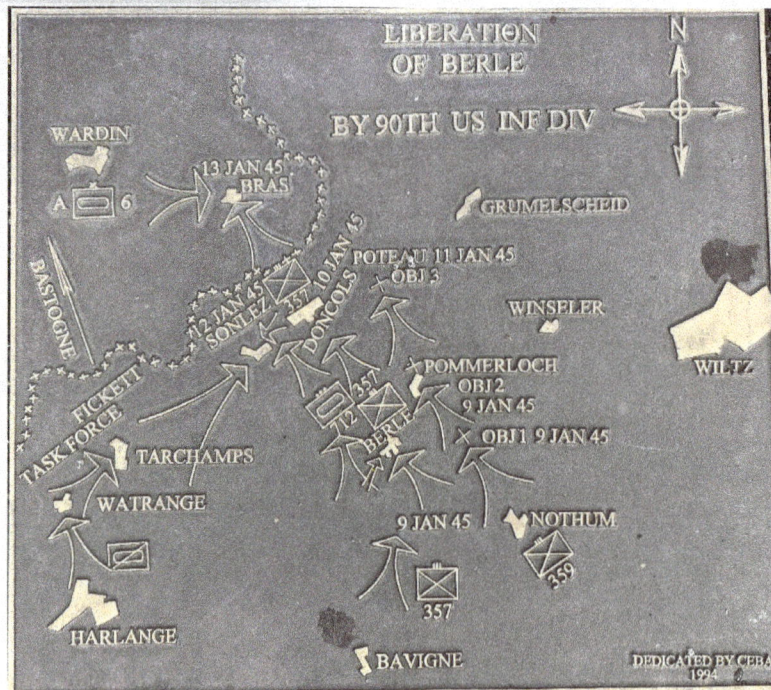

LIBERATION OF BERLE BY 90TH US INF DIV

Berlé

Two plaques at this location, both dedicated to the 90th US Infantry Division, one on the monument and one on the dairy building next to the monument (seen in the background of the photo below).

Duerfstrooss 11

9636 Winseler

Latitude: 49.9507

Longitude: 5.8545

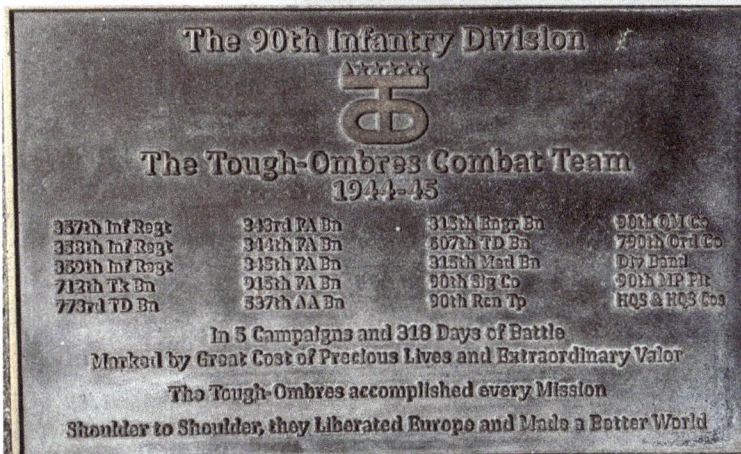

The 90th Infantry Division

The Tough-Ombres Combat Team
1944-45

357th Inf Regt	343rd FA Bn	315th Engr Bn	90th QM Co
358th Inf Regt	344th FA Bn	607th TD Bn	790th Ord Co
359th Inf Regt	345th FA Bn	315th Med Bn	Div Band
712th Tk Bn	915th FA Bn	90th Sig Co	90th MP Plt
773rd TD Bn	537th AA Bn	90th Rcn Tp	HQS & HQS Cos

In 5 Campaigns and 318 Days of Battle
Marked by Great Cost of Precious Lives and Extraordinary Valor

The Tough-Ombres accomplished every Mission

Shoulder to Shoulder, they Liberated Europe and Made a Better World

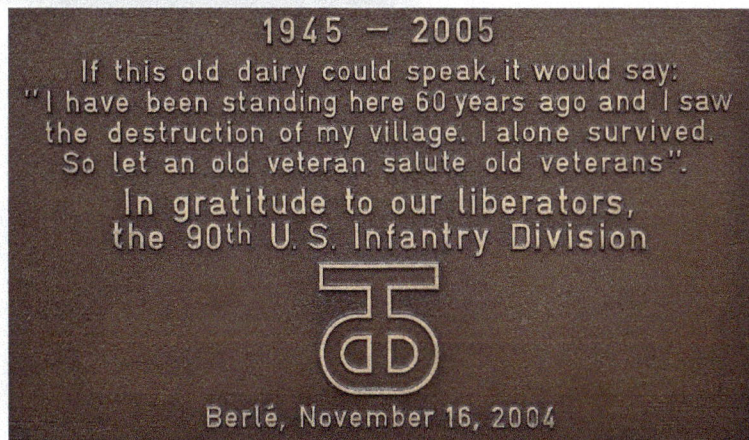

1945 — 2005
If this old dairy could speak, it would say:
"I have been standing here 60 years ago and I saw
the destruction of my village. I alone survived.
So let an old veteran salute old veterans".
In gratitude to our liberators,
the 90th U.S. Infantry Division

Berlé, November 16, 2004

Bertrange

Dedicated to the 5th US Armored Division that fought at this location on September 9, 1944 to restore the liberty of Luxembourg. Monument is located at pull off on E44 near Bertrange.

E44

8059 Bertrange

Latitude: 49.5963

Longitude: 6.0436

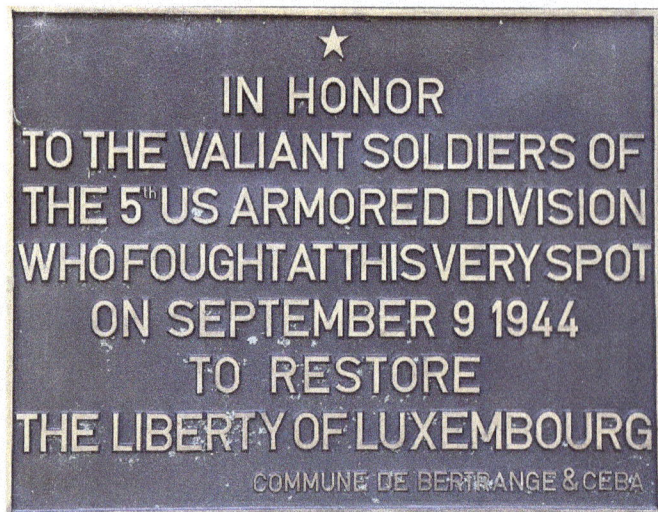

IN HONOR
TO THE VALIANT SOLDIERS OF
THE 5th US ARMORED DIVISION
WHO FOUGHT AT THIS VERY SPOT
ON SEPTEMBER 9 1944
TO RESTORE
THE LIBERTY OF LUXEMBOURG
COMMUNE DE BERTRANGE & CEBA

Bettborn (Schankegriecht)

In memory of the valiant soldiers of the 26th US Infantry Division who fought in this area in December 1944.

Rue Principale

8612 Bettborn

On N12 between Bettborn

and Grosbous

Latitude: 49.8109

Longitude: 5.9464

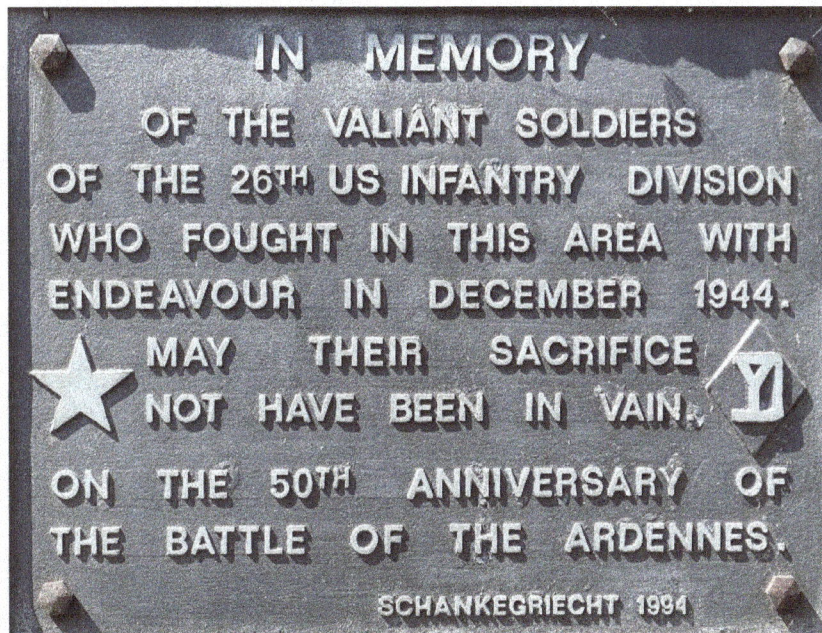

Bettembourg

Monument lays out the major events in Bettembourg during World War II in a series of steppingstones. Two of these stones are related to the US. The first remembers 27 citizens killed in an American bombing raid on May 11, 1944, and the second is dedicated to General J. H. Polk (LTC during WW II) and his troops that liberated Bettembourg on September 11, 1944.

Parc Municipal

3260 Bettembourg

Latitude: 49.5150

Longitude: 6.1033

11. Mee 1944

D'Amerikaner
bombardéiere
Beetebuerg.
Et gi 27
Doudeger.

11. September 1944

Den U.S.
Generol J.H. Polk
a seng Truppen
befreie
Beetebuerg.

Bettembourg

Monument to the 23rd Headquarters Special Troops "The Ghost Army". The 23rd was a top-secret US army tactical deception unit whose mission was to deceive and mislead German forces on the size and location of Allied forces. The existence of the unit was only declassified in 1996.

Route de Leudelange

Bettembourg

About 1km north of Abweiler

Latitude: 49.5435

Longitude: 6.0809

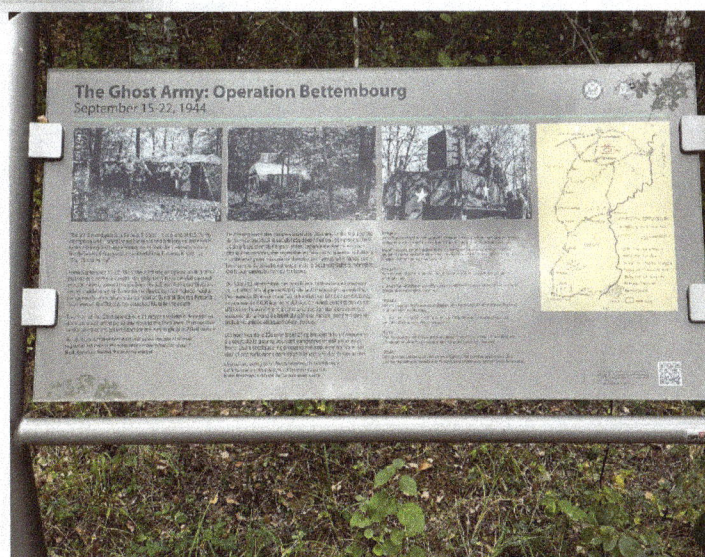

Thanks to ghostarmy.org for providing a high-resolution version of the information board (see next page).

The Ghost Army: Operation Bettembourg

September 15-22, 1944

603E- P - 26 - (20/9/44)
Staff at Bettembourg

The 23rd Headquarters Special Troops – a top-secret U.S. Army deception unit – used inflatable tanks and artillery, sound effects, radio trickery and impersonation to fool the Germans on the battlefields of Europe during World War II. It was known as "The Ghost Army."

From September 15-22, 1944, the 23rd staged Operation BETTEM-BOURG out of these woods. Roughly 400 GIs successfully posed as 8000 heavily armed troops from the U.S. 6th Armored Division. Using a wide array of deception techniques, they helped hold a dangerously undermanned segment of General George Patton's front line, as the Third Army attacked Metz to the south.

The men of the 23rd carried out 21 major battlefield deceptions during the war, often perilously close to the front lines. Their creative tactics saved countless lives and contributed mightily to Allied victory.

"Rarely, if ever, has there been a group of such a few men which had so great an influence on the outcome of a major military campaign."
Mark Kronman, United States Army analyst

Le 23e régiment des troupes spéciales était une unité top-secrète de l'armée américaine spécialisée dans l'illusion : équipée de chars et de pièces d'artillerie gonflables, capable de mettre en place des leurres sonores, des supercheries radio ainsi que des imitations, l'unité avait pour mission de désorienter l'armée allemande sur les champs de bataille européens de la Seconde Guerre mondiale. On la surnommait l'Armée fantôme.

Du 15 au 22 septembre, ces bois furent le théâtre de l'opération BETTEMBOURG. A peine 400 GI de la 23e réussirent à convaincre l'ennemi qu'ils étaient en fait la 6e division blindée américaine, composée de 8 000 hommes. A l'aide de nombreuses techniques d'illusion, ils parvinrent à tenir une section dangereusement exposée de la ligne de front du général Patton, pendant que la troisième armée attaquait Metz, au sud.

Les hommes de la 23e menèrent 21 opérations d'illusion majeures au cours de la guerre, souvent dangereusement proches du front. Leurs tactiques ingénieuses ont sauvé de nombreuses vies et ont fortement contribué à la victoire des forces alliées.

" Jamais un si petit groupe d'homme n'a eu une telle influence sur le résultat d'une campagne militaire d'envergure. "
Mark Kronman, analyste de l'armée américaine

Image 1:
Staff officers set up a phony 6th Armored headquarters on this spot. From here, deception troops ranged many miles up and down the front to dupe the enemy about the division's presence.

Des officiers installent un faux état-major de la 6e division blindée. Les troupes utilisaient ce QG factice comme base pour patrouiller le long de la ligne de front et persuader les Allemands que la 6e division était bel et bien présente.

Image 2:
Several dozen rubber dummies were inflated in the surrounding woods to fool German aerial reconnaissance.

Plusieurs dizaines de véhicules en caoutchouc furent insufflés dans les bois environnants pour tromper la reconnaissance aérienne allemande.

Image 3:
Halftracks equipped with 500-pound speakers played recordings of tanks moving in. The sound could be heard 15 miles away.

Des autochenilles équipées de haut-parleurs de 200 kg diffusaient des bruits de chars, qui pouvaient être entendus 25 km à la ronde.

Right:
This 1944 operation map shows the phony division in red. The early other U.S. unit along the 25-mile section of the front was one squadron from the 3rd Cavalry (Armored).

Droite:
Cette carte des opérations de 1944 montre la fausse division en rouge. La seule autre unité américaine présente sur ce secteur du front long de 40 km était le 3e régiment blindé de cavalerie.

OPERATION BETTEMBOURG
Luxembourg, 15-22 Sept 1944

NOTE: Fictional troops including Patton's 6th Armored Division except for enemy, lie instead of Thionville, and 3rd Cavalry, North of Bettembourg.

Bettendorf

Monument dedicated to the 10th Infantry Regiment, 5th Infantry Division "Red Diamond" and located in the Major General W. M. Breckinridge Square. Plaque to Breckinridge is located on a nearby pillar to the left side of the Regimental monument.

rue de la Gare 8

9353 Bettendorf

Latitude: 49.8724

Longitude: 6.2182

In gratitude to the 10th Inf. Rgt. US ARMY The citizens of BETTENDORF

SQUARE
W.M. BRECKINRIDGE
MAJOR GENERAL
10TH INF. RGT 5TH US INF. DIV
BETTENDORF-RAID
JANUARY 10TH 1945

Bettendorf

Dedicated to 10th Infantry Regiment, 5th Infantry Division, and 4th US Infantry Division.

rue de la Gare 3

9353 Bettendorf

Latitude: 49.8724

Longitude: 6.2197

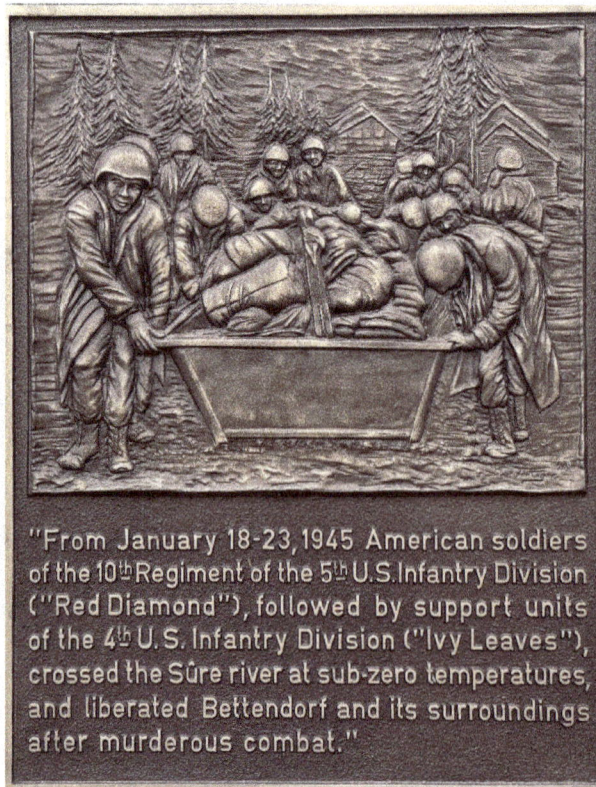

"From January 18-23, 1945 American soldiers of the 10th Regiment of the 5th U.S. Infantry Division ("Red Diamond"), followed by support units of the 4th U.S. Infantry Division ("Ivy Leaves"), crossed the Sûre river at sub-zero temperatures, and liberated Bettendorf and its surroundings after murderous combat."

Bettendorf-Reisdorf-Hoesdorf

Monument is located in the woods between Bettendorf, Reisdorf and Hoesdorf at one of the tour stops along the historical circuit at stop #5 (see hiking trails sections). Dedicated to soldiers of the US 109th Infantry Regiment and the German GR 916th Volksgrenadier Division who fought and died in this area at the beginning of the Battle of the Bulge on December 16, 1944.

Reisdorf

Latitude: 49.8818

Longitude: 6.2483

Inscription reads:

In memory of the soldiers from the 109th Regiment U.S. 28th Infantry Division and the German G.R.91G.352 Volksgrenadier Division who fought and died on this high ground above Hoesdorf on the opening day of the "Battle of the Bulge" 16 December 1941. May their sacrifice never be forgotten.

Bigonville

Two plaques are at the site, the first is dedicated to the 4th Armored Division that liberated Bigonville on December 24, 1944. The second plaque is in remembrance of Lt. Robert M. Lamar, who was hit by flak over Bastogne on January 2, 1945 and belly-landed his P-47D in "Hoch" and survived. The plaque also pays tribute to the men of the 8th and 9th US Army Air Forces.

Rue Principale 20

8814 Rambrouch

Latitude: 49.850156

Longitude: 5.792403

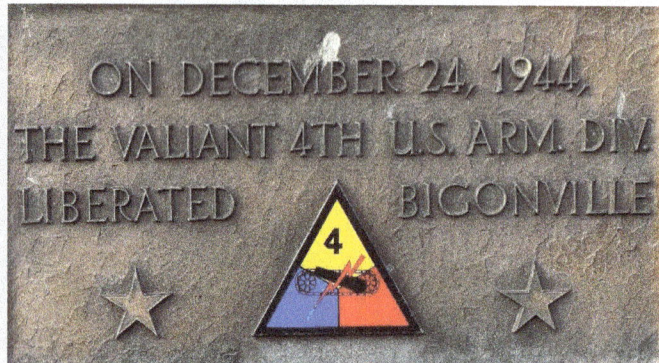

ON DECEMBER 24, 1944,
THE VALIANT 4TH U.S. ARM. DIV
LIBERATED   BIGONVILLE

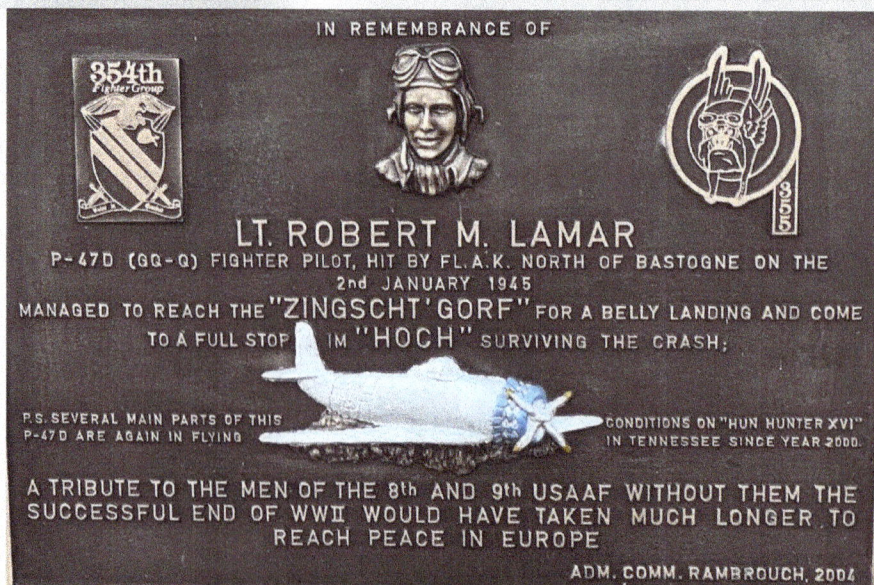

IN REMEMBRANCE OF

LT. ROBERT M. LAMAR
P-47D (GQ-Q) FIGHTER PILOT, HIT BY FL.A.K. NORTH OF BASTOGNE ON THE
2nd JANUARY 1945
MANAGED TO REACH THE "ZINGSCHT'GORF" FOR A BELLY LANDING AND COME
TO A FULL STOP IM "HOCH" SURVIVING THE CRASH;

P.S. SEVERAL MAIN PARTS OF THIS
P-47D ARE AGAIN IN FLYING
CONDITIONS ON "HUN HUNTER XVI"
IN TENNESSEE SINCE YEAR 2000.

A TRIBUTE TO THE MEN OF THE 8th AND 9th USAAF WITHOUT THEM THE
SUCCESSFUL END OF WWII WOULD HAVE TAKEN MUCH LONGER TO
REACH PEACE IN EUROPE

ADM. COMM. RAMBROUCH, 2004

Bissen

Dedicated to those who lost their lives in defense of Luxembourg.

Route de Mersch 6

7780 Bissen

Latitude: 49.7876

Longitude: 6.0663

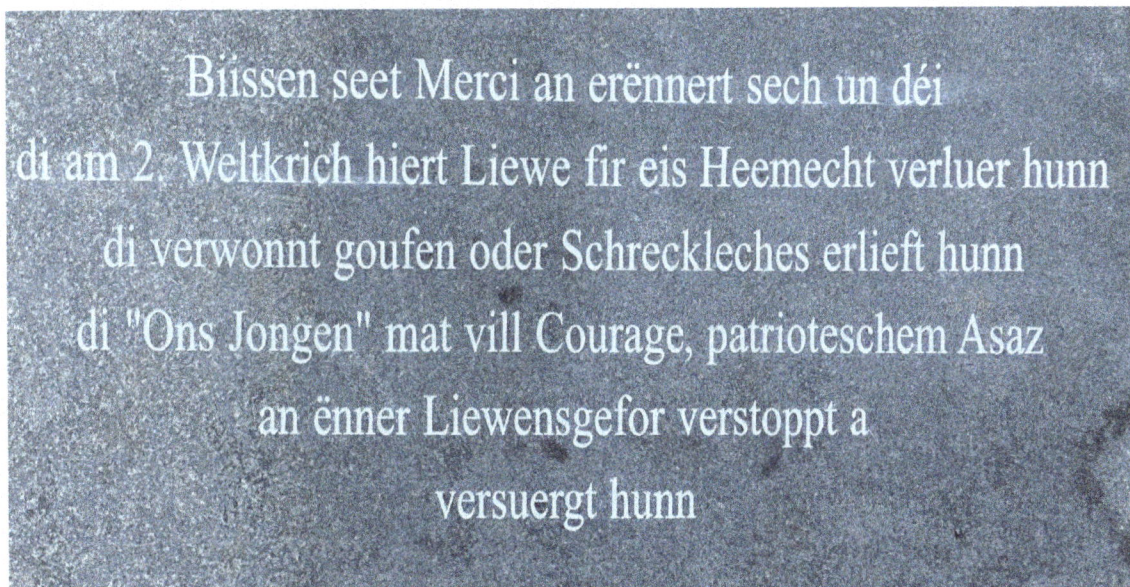

Biissen seet Merci an erënnert sech un déi

di am 2. Weltkrich hiert Liewe fir eis Heemecht verluer hunn

di verwonnt goufen oder Schreckleches erlieft hunn

di "Ons Jongen" mat vill Courage, patrioteschem Asaz

an ënner Liewensgefor verstoppt a

versuergt hunn

Inscription reads: Bissen says thank you and remembers those who lost their lives for our homeland in World War II, those who were wounded and those who suffered terribly while caring for our youth with courage, patriotic commitment and under threat of life.

Bleesbréck

Information panels that outline the events at the "Bleesbréck" crossroads (now a traffic roundabout) from 1940 to 1945. At the beginning of the Battle of the Bulge, the "Bleesbréck" crossroads was an initial defensive position held by the 28th Infantry Division, 109th Regiment before withdrawing and reestablishing a new defensive line behind the Sauer river.

Bleesbrueck 1

9359 Tandel

Located inside the campground

at the intersection of N17 and N19

Latitude: 49.8734

Longitude: 6.1883

"Bleesbréck",
the crossroads of destiny 1940 – 45

"Bleesbréck" 1945

"Bleesbréck" is the local denomination given to a "Y" junction at the east exit of Diekirch / Gilsdorf, leading northeast in direction of Vianden and southeast in direction of Echternach with both roads paralleling the Our and Sauer rivers and the German border areas. It is also here where the "Blees" creek running from the Luxembourg Ardennes converges into the "Sauer" river - hence the name "Bleesbréck", meaning Blees' bridge.

During World War II, because of its vicinity to Germany and the "Westwall" or "Siegfriedline". (German defense line running from the Swiss border up to the Dutch coastline, constructed in the years 1936-1940), the "Bleesbréck" crossroads inevitably became drawn into a number of military operations, especially in winter 1944-45, during the so-called "Battle of the Bulge".

Already on May 10, 1940, the first day of the German armies' invasion of Western Europe, German troops were pouring from their assembly areas in the "Westwall" into neutral Luxembourg territory onward to Diekirch in direction of France across the undestroyed border bridges in the area. As Luxembourg strictly observed its status of neutrality, no resistance was made: German troops destroyed the concrete and steel obstacles and barriers of Luxembourg's territorial border defense and routed through "Bleesbréck" for days with mechanized-, horse-drawn- and infantry units.

Under the German occupation (May 10, 1940 - September 1944) "Bleesbréck" remained an important road junction for crossing into mainland Germany.

After the successful allied landings in Normandy on June 6, 1944 and their progressive retreat from France, defeated German troops again transited through Luxembourg in direction of the "Westwall". For days and days in late August-early September 1944, "Bleesbréck" silently witnessed hundreds of German troops on bicycles, horse carts and trucks with wounded soldiers and damaged equipment, Nazi party officials in their mustard color uniforms and collaborators in confiscated and stolen cars speeding towards mainland Germany pushed by the advancing US troops.

On September 11, 1944 the first US units, primarily reconnaissance troops of the 5th US armored division - soon followed by tanks and dismounted troops - began to show up, cheered at by the just liberated civilians of the villages surrounding "Bleesbréck" as Tandel, Bastendorf, Brandenbourg, Longsdorf, Fouhren, Bettel, Gilsdorf, Bettendorf, Moestroff, Reisdorf, Hoesdorf and Diekirch. The German troops completely withdrew from Luxembourg territory without much fighting and destruction, except in the immediate border line. It was also on September 11, 1944 that the first allied patrol crossed the Our river near Stolzembourg into Germany after routing through "Bleesbréck" - making world history... but the regained freedom and liberty were not for long!

In the early morning hours of December 16, 1944 (the beginning of the Battle of the Bulge), the German artillery well-dug in behind the "Westwall", opened up and a hailstorm of shells and rockets fell on the American positions around "Bleesbréck", then held in this sector by units of the 109th regiment of the 28th US Infantry "Keystone" division. "Bleesbréck" itself was part of the rear-elements' US defensive line and junction point of the 2nd and 3rd battalion's front-line units of the 109th regiment. A battery of 40 mm Bofors guns of the 447th anti-aircraft artillery battalion for both anti-aircraft and ground use defense had previously been dug in around "Bleesbréck" to protect the crossroads in case of enemy attack. As the frontline companies of both battalions soon became engaged in heavy fighting with

Bleesbréck Information Panel (continued)

overwhelming and aggressive German troops of the 352nd "Volksgrenadierdivision" on the heights overlooking the Our river, the reserve of the 109th regiment , its own 1st battalion billeted in Diekirch, was immediately committed and ordered in direction of "Bleesbréck". Their primary task was to reinforce the remainder of US troops in the area and to rescue the encircled US units in Longsdorf and Fouhren. Along with them 4 Sherman tanks of the 707th Tank battalion were sent forth to support the infantry and defend the crossroads as kind of a mobile roadblock. Later that day, the first German advance reconnaissance elements were already spotted at the road Tandel-"Bleesbréck" and beaten back. On December 17, the attempt to rescue the encircled US garrison at Fouhren (E-company / 109th Rgt.) failed – one Sherman tank was knocked out by a German "Panzerfaust", another was damaged and had to retreat. Further, "A"-company/109th Rgt. did not succeed either to progress on the road towards Longsdorf and suffered numerous casualties with another Sherman tank becoming damaged. Heavy fighting continued during the day and the night. On December 18, the situation became very alarming, as the 109th regiment received orders from the 28th division headquarters in Wiltz to withdraw from the Our river sector and establish a new defensive line on the Herrenberg / Diekirch in direction of Ettelbruck. Accordingly, exhausted US troops began retreating from their positions, whereas a last reserve unit supported by 57 mm antitank guns and 2 additional Sherman tanks put up a defense spanning the "Sauer" river bridge at Gilsdorf and "Bleesbréck". That same day, a number of German soldiers were captured at the houses at "Seltz", where they had been hiding. The German pressure increased during the night – heavy fighting raged on the road leading from "Seltz" to "Bleesbréck" and a German vehicle was knocked out there. The morning of December 19, the "Bleesbréck" farmhouse and country inn were found to be occupied

Destroyed Sherman Tank at "Seltz-Bleesbréck"

by German troops. Meanwhile the "Diekirch Miliz", consisting of auxiliary policemen and armed resistance fighters had arrived at "Bleesbréck" to support the American defenders. When it became obvious that the area was almost completely in German hands, a Sherman tank was ordered forward from its position on the Gilsdorf bridge to shell the "Bleesbréck" farmhouse. The approaching tank must have scared the Germans, a number of them ran "hands-up" out of the house and surrendered! The tank put a round in the neighboring building – a firefight developed and under the cover of the tank, the "Miliz" stormed the building. Hand grenades went off, bursts of submachine guns from both sides, another tank round into the building... and out came another sizeable number of German who surrendered. Altogether, more than 50 POWs were made and marched off. Late in the afternoon, all remaining US troops were pulled back to Diekirch, with the civilian population being evacuated during the night thanks to the intervention of Capt. Harry M. Kemp, Executive officer, 3rd Bn / 109th Inf. Rgt. With the road "Bleesbréck" – Diekirch now open, German troops kept coming and progressing until Christmas 1944, when stopped at Bettborn driven out of Ettelbruck and forced back through Diekirch to occupy a defensive line on the north banks of the "Sauer" river.

They were able to hold and defend that sector until January 18, 1945 when the 5th US Infantry division "Red Diamond", part of Gen. Patton's Third Army launched its counter-attack in snowbound terrain at sub-zero temperatures in the Ingeldorf-Bettendorf sector. They crossed the icebound "Sauer" river against weak German retaliation and recaptured "Bleesbréck" and surrounding villages during the ensuing days. By January 25, 1945 the area was totally in American hands – most Germans having retreated while fighting across the "Sauer" and "Our" rivers to reoccupy the "Westwall" bunkers. During the "Rhineland" campaign or full-scale US invasion of mainland Nazi Germany, resulting in the breaching of the "Siegfried line" onward in direction of the Rhine, beginning February 7, 1945, "Bleesbréck" remained an important and well-protected road junction for US supplies to route through in direction of Germany.

(compiled by Roland Gaul – photos: National Museum of Military History, Diekirch)
The National Museum of Military History in Diekirch , located at 10, Bamertal, Tel: (+352) 80 89 08; www.mnhm.lu; email: info@mnhm.lu portrays in great detail the "Battle of the Bulge 44/45 in the area, besides other chapters of contemporary military history

Boulaide

B17 Flying Fortress Bomber crashed on this site on September 28th, 1944 while returning from a bombing mission near Magdeburg, Germany. Crew consisted of: Mayer, W. J. Pilot, Simons, R. A. Navigator, Rogers, O. J. Bombardier, Hicks, D. E. Engineer, Tuescher, E. Radio Operator, Messerich, J. R. turret gunner, Whiters, F. J. Waist Gunner, and Gendron, E. J. Tail Gunner.

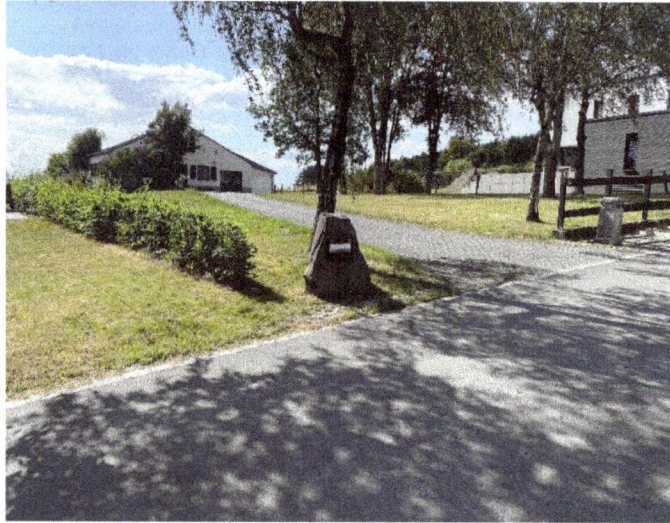

7 rue Haute

9640 Boulaide

Latitude: 49.88840

Longitude: 5.81022

At this place, on September 28th 1944, crashed a B-17 Flying Fortress Bomber which came back from the mission » bombardment of the Ferd Gruson factories near Magdeburg, Germany ».

The crew of eight men parachuted over this ground. The pilot WJ Mayer broke his leg when touching the ground. The engineer DE Hicks jumped out by the bomb trap without a parachute and didn't survive. The bombardier OJ Rogers was injured lightly and the waist gunner FJ Whiters by a 20mm fragment.

The crew was composed of:
WJ Mayer, pilot, RB Gradle, co-pilot, RA Simons, navigator, OJ Rogers, bombardier, DE Hicks(+), engineer, E Tuescher, radio operator, JR Messerich, bull turret gunner, Fj Whiters, waist gunner, EJ Gendron, tail gunner

Ageweit zu Bauschelt, den 26ten September 2009

Boulaide

In memory of the soldiers of the 35th US Infantry Division

Place de la Libération 1

9640 Boulaide

Latitude: 49.8829

Longitude: 5.8089

Inscription reads: In memory of the valiant soldiers of the 35th US Infantry Division 1944-1945

Brandenbourg

Memorial to the 28th US Infantry Division. Monument is located on the backside of several houses in Brandenbourg, behind the former village school (now a restaurant) on the bank of "Blees Creek."

Haaptstrooss 13

9360 Tandel (Brandenbourg)

Latitude: 49.9114

Longitude: 6.1369

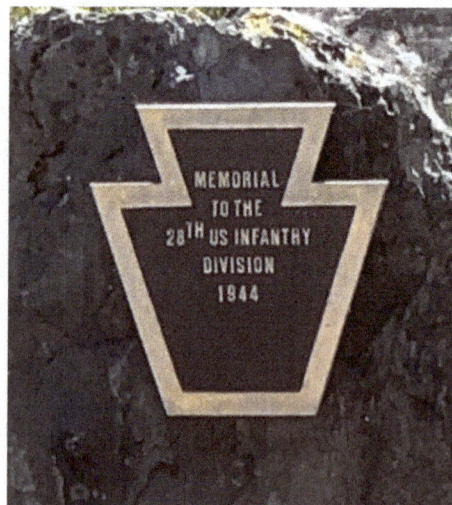

Photo by Felix Vlaminck

Bourscheid

Place Gen. Norman D. Cota, Commander, 28th US Infantry Division.

Um Kraizkapp 4

9140 Bourscheid

Latitude: 49.9099

Longitude: 6.0648

Inscription: A la memoire de nos liberateurs Americains 1944/45, la commune de Bourscheid

In memory of our American Liberators 1944/45, the commune of Bourscheid

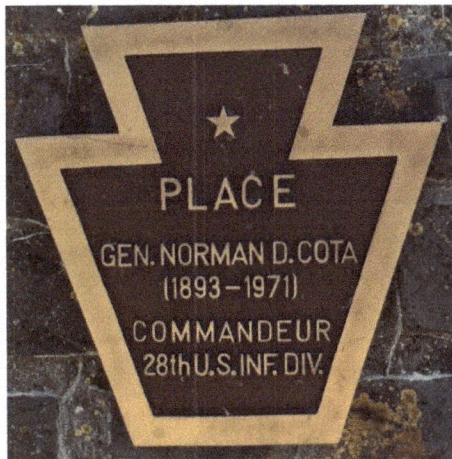

Buschrodt

Located on private property, but viewable from the road, the monument is dedicated to the memory of Pvt Charles J. Ryan, 26th Infantry Division, 3rd US Army, killed in action on December 22, 1944.

Rue Angelsgronn 40

8610 Wahl

Latitude: 49.818814

Longitude: 5.943467

Canach

From the people of Canach to our American liberators who stayed with us from September 16, 1944 to March 24, 1945.

Rue d'Oetrange 1

5411 Lenningen

Latitude: 49.6095

Longitude: 6.3252

MERCI FIR D'FRÄIHEET

MIR KANÉCHER LEIT HALEN ÉIS AMERIKANESCH BEFREIER AN ÉIEREN, DÉI VUM 16.9.1944 BIS DE 24.3.1945 BEI IIS EN HEEM FOND HATEN.

AN DANKBARKEET ERÉNNERE MIR IIS UN SI, DÉI MAM ASATZ VUN HIREM LIEWEN IIS VUM NAZIJOCH BEFREIT AN IIS D'FRÄIHEET ERÉM GIN HUN.

MIR WÄERTEN DAT NI VERGIESSEN A GIN DÉSE MESSAGE UN ÉIS KANNER VIRUN.

D'KANÉCHER LEIT - MÄERZ 1995

Inscription reads: We people of Canach came to love and honor our American liberators who lived with us from September 16, 1944 to March 24, 1945. In gratitude we remember those who gave their lives to liberate us from the Nazis and gave us back our freedom. We will never forget this and will pass this message onto our children. The people of Canach, March 1995

Clervaux

There are many monuments in Clervaux dedicated to US Forces, which are summarized on the following pages. There is also a German PAK 43/41 anti-tank gun on display in the hairpin bend next to the cemetery on Route de Marnach as you exit Clervaux.

Rue Schloff 2

9712 Clervaux

Latitude: 50.0541

Longitude: 6.0299

Inscription

DIR SËDD GESTUARWEN FIR DATT MIR FRÄI LIÄWWEN MCMXL – MCMXLV

You died so that we can live free 1940-1945

A la memoire des enroles de force 1940-1945

In the memory of the forced conscripts 1940-1945

Information Panel 1 Clervaux

REMEMBER US
CLERVAUX

THE BATTLE OF THE BULGE
FROM A CIVILIAN POINT OF VIEW
PART 1

On December 16, 1944 at 5.30 am, a young American soldier on duty on the water tower in Hosingen was sending a routine message to the command post, when he briefly stopped. The whole german front on the other side of the Our had suddenly turned into a string of countless lights. He was still puzzling about this strange phenomenon when the first grenades exploded around him. This young GI had seen nothing else but the muzzle flash of the German artillery. It was the beginning of the Battle of the Bulge.

Driven by painful uncertainty, fear and grief, many families left their homes and farms. Only a few men remained in most of the villages, they took care of the livestock in the stables. They became witnesses to history and – provided they had survived – they gave detailed reports on the fighting in their villages.

Many families loaded all their belongings on carts and hay wagons drawn by horses and oxen, and they fled from the combat operations. Some people found accommodation with family members or helpful people in villages outside the war zone. Other unfortunates were overtaken by combat action. They had to abandon their hopes of escaping the theater of war, and they had no choice but to retrace their steps. Back in the devastated areas, they followed the example of those who had not fled for one reason or another. From then on they had to live in the basements. During these long and arduous weeks they were confined in very small spaces, where they had to brave the freezing cold and to reorganize their daily life. It happened that several families lived in the same basement, crammed like cattle.

Luxembourg civilians in a cellar

Anyone who dared to leave the house risked his life. Shell impacts in the immediate vicinity shook the people in the basements, and they were often on the brink of utter despair. It also happened that Wehrmacht troops requisitioned the basements for themselves and that complete families were driven out of their own homes. People were hit by shrapnel or killed by stray bullets. The despairing population sought solace in prayer, and many vows were taken. A great many of believers promised to erect a statue of Our Lady or to build a chapel in honor of the Consolatrix Afflictorum, if they remained safe and sound. Such a vow is also at the origin of the statue of the Virgin of Fatima which is venerated in Wiltz.

LA BATAILLE DES ARDENNES
D'UN POINT DE VUE CIVIL
1RE PARTIE

Le 16 décembre 1944 à 5.30 h, un jeune soldat américain installé tout en haut du château d'eau de Hosingen est en train de transmettre un message de routine au poste de commandement, lorsqu'il s'arrête brièvement. Le front allemand de l'autre côté de l'Our s'était soudainement transformé en une chaîne composée d'innombrables points lumineux. Les premières grenades éclatent autour de lui alors qu'il est encore en train de réfléchir sur ce drôle de phénomène qui venait de se produire. Ce jeune "GI" n'avait vu rien d'autre que le feu de bouche de l'artillerie allemande.

Poussés par l'incertitude, la peur et le chagrin, beaucoup de familles quittent leurs maisons et leurs fermes. Seuls quelques hommes restent dans la plupart des villages pour s'occuper du bétail dans les étables. Ils vont devenir de véritables témoins de l'époque, et les survivants évoqueront plus tard les combats impitoyables qui se sont déroulés dans leurs villages.

De nombreuses familles chargeaient tous leurs biens sur des charrettes et chariots tirés par des chevaux et des bœufs, et elles prenaient la fuite pour échapper aux opérations de guerre. Certaines personnes réussissaient à se loger auprès de membres de leur famille ou d'habitants compatissants de villages hors de la zone de combat. D'autres malheureux étaient rattrapés par les opérations de guerre. Leur espoir de pouvoir échapper aux combats se trouvant anéanti, ils n'avaient d'autre choix que de retourner dans les zones déjà détruites. Ils y suivaient l'exemple de ceux qui n'avaient pas pris la fuite pour une raison ou une autre. Pendant de longues et difficiles semaines, ils vivaient dans des caves, où il fallait d'abord braver le froid barbare, puis réorganiser la vie quotidienne. La même cave hébergeait souvent plusieurs familles, entassées comme du bétail.

The destroyed village Wiltz

Quiconque osait s'aventurer dehors risquait sa vie. Les impacts de grenades dans le voisinage immédiat faisaient trembler les réfugiés dans les caves : ils étaient souvent au bord du désespoir absolu. Il arrivait également que des soldats de la Wehrmacht réclamaient les caves pour eux-mêmes et chassaient sans scrupules des familles complètes de leurs propres maisons. Bien des habitants étaient touchés par des éclats d'obus ou tués par des projectiles égarés. La population désespérée cherchait du réconfort dans la prière, et bien des croyants faisaient le vœu de faire ériger une statue de la Consolatrice des Affligés voire de lui faire construire une chapelle, au cas où ils sortiraient sains et saufs de la guerre. Un tel vœu est également à l'origine de la statue de la Vierge de Fatima vénérée à Wiltz.

DIE ARDENNENOFFENSIVE
AUS ZIVILER SICHT
TEIL 1

Als ein junger amerikanischer Soldat in der Spitze des Hosinger Wasserturms am 16. Dezember 1944 um 05.30 Uhr seine Routinemeldung an den Gefechtsstand durchgibt, unterbricht er seine Durchsage für einen kurzen Moment. Die gesamte deutsche Front jenseits der Our hatte sich plötzlich in eine Kette unzähliger kleiner Lichtpunkte verwandelt. Während er noch rätselt, was dieser merkwürdige Vorgang wohl bedeute, schlagen rings um ihn herum die ersten Granaten ein. Der junge "GI" hatte nichts anderes als das Aufblitzen der Mündungsfeuer der deutschen Artillerie beobachtet. Die Ardennenoffensive hatte begonnen.

A destroyed farmhouse in the Ardennes

Getrieben von Ungewissheit, Angst und Leid verließen viele Familien ihre Häuser und Bauernhöfe. In den meisten Dörfern blieben einige männliche Dorfbewohner alleine zurück, um das in den Ställen zurückgelassene Vieh zu versorgen. So wurden sie zu wichtigen Zeitzeugen und berichteten später – soweit sie die Kampfhandlungen überlebten – von der Unbarmherzigkeit der Kämpfe in ihren Dörfern.

Viele Familien packten ihr Hab und Gut auf Handwagen und Leiterwagen, von Pferden oder Ochsen gezogen, und flüchteten vor dem Tod und Verderben säenden Beschuss. Einigen gelang es bei Familienangehörigen oder bei hilfsbereiten Bewohnern von Dörfern außerhalb des Kampfgebietes Unterschlupf zu finden. Andere wurden von der Offensive eingeholt. Ihre Hoffnungen, dem Kampfgeschehen noch entrinnen zu können, waren zerstört. Ihnen blieb keine andere Wahl als in die bereits zerstörten Gebiete zurückzukehren. Dort folgten sie dem Beispiel jener die eine Flucht aus den verschiedensten Gründen erst gar nicht angetreten hatten. Fortan lebten sie in den Kellern und es folgte eine lange und beschwerliche Zeit. Auf engstem Raum galt es einerseits der barbarischen Kälte zu trotzen und andererseits das tägliche Leben neu zu organisieren. Nicht selten lebten mehrere Familien im gleichen Keller, zusammengepfercht wie Vieh.

Jeder der sich draußen bewegte befand sich in akuter Lebensgefahr. Nahe Granateinschläge ließen die Kellerinsassen zittern und brachten sie öfters an den Rand der völligen Verzweiflung. Es kam auch vor, dass Wehrmachtstruppen die Keller für sich beanspruchten und ganze Familien rücksichtslos aus ihren eigenen Häusern vertrieben. Viele kamen durch Granathagel und umherirrende Geschosse ums Leben. In ihrer Not suchten die Menschen Trost im Gebet und nicht selten wurden Gelübde abgelegt; dass – falls man den Krieg heil und unbeschadet überleben würde – man der Trösterin der Betrübten ein Gnadenbild oder eine Kapelle errichten würde. So entstand auch das bekannte Gnadenbild der heiligen Jungfrau von Fatima in Wiltz.

Information Panel 2 Clervaux

THE BATTLE OF THE BULGE
FROM A CIVILIAN POINT OF VIEW
PART 2

The German combat troops were quickly followed by units of the German Sicherheitsdienst and the Gestapo, who were looking for "réfractaires" (young men refusing to join the German army), their families and helpers. If they were found, they were deported and killed without further ado. Later on, the bodies of some of these victims were discovered in the surrounding forests, while others are still missing. The cemeteries in the region are the silent witnesses of these barbaric crimes.

Unfortunately we do not have exact figures on civilian casualties in the context of the Battle of the Bulge, but according to reliable estimates these fights brought about 3000 deaths among the civilian population in Luxembourg and Belgium. Unfortunately the end of the hostilities did not go hand in hand with an end of the horror: exploding duds and mines killed or mutilated people, mostly children. The occurrence of epidemics was inevitable, and typhus fever made many victims.

Fitted with the war, the population leaves the villages

At the end of the fighting, many citizens barely escaped with their life. Those who had managed to escape at the beginning of the fighting had to face the total destruction when they returned to their once peaceful villages. Their houses were reduced to ashes and relatives were missing or dead. Many displaced families had not yet been able to come home. And some people did not see their homeland again, as they had died far away from home.

After the end of the war, and the families having gone through a period of mourning, the population attacked the reconstruction of villages and streets. Nowadays, only a few scars remind us the horrible winter days of 1944. Antitank guns, tanks as well as countless memorial stones are found all over the Oesling area and they are always thought-provoking. In the villages affected by the Battle of the Bulge even today, seventy years after the horrible events, each December 16th is marked by commemorative ceremonies.

LA BATAILLE DES ARDENNES
D'UN POINT DE VUE CIVIL
2ᵉ PARTIE

Les troupes de combat allemandes étaient rapidement suivies d'unités du Sicherheitsdienst allemand ainsi que de la Gestapo, qui étaient à la recherche de réfractaires ou de leurs familles et complices. En cas de découverte, les concernés étaient déportés et tués sans autre forme de procès. Les corps de certaines victimes étaient découverts plus tard dans les forêts des alentours, alors que d'autres n'ont jamais été retrouvés. Les cimetières de la région sont les témoins silencieux de ces crimes barbares.

Nous ne disposons malheureusement pas de chiffres exacts sur les pertes civiles dans le contexte de la bataille des Ardennes, mais il ressort d'estimations sérieuses que ces combats ont fait 3000 morts parmi la population civile au Luxembourg et en Belgique. La fin des opérations de guerre ne marquait pas pour autant la fin de l'horreur, car des dizaines de malheureux, souvent des enfants, étaient tués ou mutilés par des obus qui n'avaient pas encore éclaté ou par l'explosion de mines. L'apparition d'épidémies était inévitable, et bien des habitants succombaient aux épidémies de typhus.

À la fin des combats, beaucoup de citoyens n'avaient pu sauver que leur vie. En retournant dans leurs villages jadis paisibles, ceux qui avaient réussi à fuir au début des opérations de guerre se trouvaient confrontés à la destruction totale. Leurs maisons étaient réduites en cendres, bien des proches n'avaient pas survécu. De nombreuses familles déportées n'avaient pas encore pu revenir au pays. Certaines personnes n'ont plus revu leur terre natale, elles sont mortes loin de la patrie.

The village of Berlé lies in ruins

Après la fin des combats, et les familles ayant fait le deuil de leurs disparus, la population s'attaquait à la reconstruction des villages et des rues. De nos jours, on ne trouvera presque plus de traces de l'offensive hivernale de 1944. Les canons antichars et les chars en sont désormais des témoins silencieux, tout comme les innombrables stèles commémoratives que l'on retrouve dans toutes les localités de l'Oesling et qui incitent toujours à la réflexion. Des cérémonies commémoratives se déroulent aujourd'hui encore tous les ans le 16 décembre dans les localités touchées par la bataille des Ardennes.

DIE ARDENNENOFFENSIVE
AUS ZIVILER SICHT
TEIL 2

Sehr rasch folgten den deutschen Kampftruppen sowohl Einheiten des deutschen Sicherheitsdienstes als auch der Geheimen Staatspolizei, welche Ausschau nach Refraktären[1] oder nach deren Familien und Helfern hielten. Beim Antreffen wurden sie kurzerhand verschleppt und getötet. Ihre Leichen wurden später in den umliegenden Wäldern gefunden, andere gelten bis heute als vermisst. Umliegende Friedhöfe sind zu stillen Zeugen dieser barbarischen Verbrechen geworden.

The destroyed train station in Trauweiler

Es bestehen leider keine exakten Zahlen zu den zivilen Verlusten der Ardennenoffensive, seriöse Schätzungen gehen jedoch von rund 3000 getöten Zivilpersonen in Belgien und Luxemburg aus. Doch auch nach dem Ende der Kriegshandlungen hatte das Sterben kein Ende, dutzende Unglückliche, oftmals Kinder, wurden durch explodierende Blindgänger und Minen getötet oder zerstümmelt. Desweiteren konnte ein Ausbrechen von Seuchen nicht verhindert werden. Nicht selten wurden Menschen durch Typhusepidemien hingerafft.

Viele vermochten nach den Kämpfen nichts weiter als das nackte Leben zu retten. Diejenigen, denen die Flucht am Anfang der Schlacht gelungen war fanden bei ihrer Rückkehr ihre einst friedlichen Dörfer komplett zerstört vor. Ihre Häuser waren in Schutt und Asche gelegt worden. Hinzu kamen das Leid und die Trauer um den Verlust von Familienangehörigen. Viele Familien waren auch noch nicht aus der Umsiedlung zurückgekehrt. Einige sollten nie wieder zurückkehren und starben dort, fernab der Heimat.

Nach dem Ende der Kampfhandlungen und nachdem die Familien ihre Toten zu Grabe getragen hatten, begannen die Menschen mit dem Wiederaufbau ihrer Dörfer und Strassen. Heute erinnert fast nichts mehr an die schrecklichen Tage der Winteroffensive von 1944. Zu friedlichen Monumenten umfunktionierte Panzerabwehrkanonen und Panzer, sowie eine schier unübersichtliche Zahl von Gedenksteinen, die in sämtlichen Ortschaften des Oslings zu finden sind, regen immer noch zum Nachdenken an. Selbst heute werden jedes Jahr am 16. Dezember Gedenkfeiern in den damals von der Ardennenschlacht in Mitleidenschaft gezogenen Ortschaften abgehalten.

[1] Junge Männer die sich der Zwangsrekrutierung durch Flucht entzogen hatten.

Clervaux Soldier Remembrance ("GI Monument")

Grand-Rue 16

9710 Clervaux

Latitude: 50.0532

Longitude: 6.0311

IN APPRECIATION TO THE
CITIZENS OF CLERVAUX
AND THE MEMBERS OF CEBA
FROM GRATEFUL FRIENDS IN THE
U.S. 6th ARMORED DIVISION
1 AUGUST 1994

Inscription

TO OUR LIBERATORS
1944 – 1945

SI HUN ÄIS D'FRÄIHEET
ERËMBRUECHT

(They restored our freedom)

Information Panel at Clervaux GI Monument

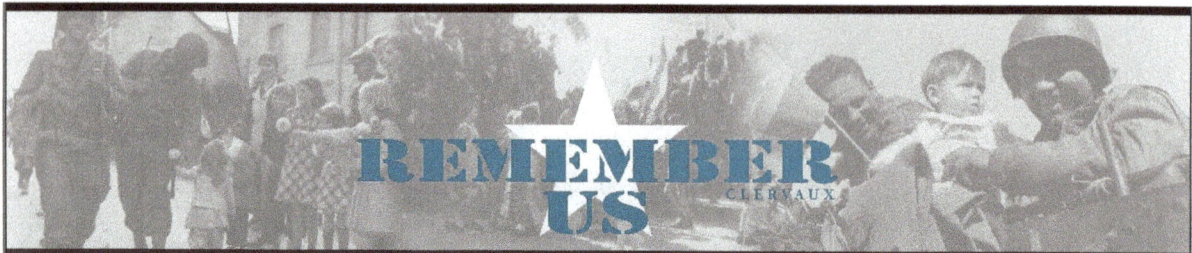

REMEMBER US — CLERVAUX

THE AMERICAN SOLDIER

This bronze statue is dedicated to the 'GI', the american soldier who freed Luxembourg twice from the Nazis, in 1944 and 1945. The monument designed by the Luxembourg sculptor Michel HEINTZ was erected by CEBA on a piece of land made available by the town of Clervaux. The statue was unveiled on September 11, 1983, by the heir to the throne Prince Henri and his wife Princess Maria-Teresa.

The sculpture on the opposite house wall shows the Luxembourg civil population's enthusiastic welcome to the American liberators. This work of art was put up at a later date by the "Caisse d'Epargne de l'Etat".

This 'GI' is supposed to express serenity and impassiveness. Exhausted and weary, he is dreaming of peace and quiet after the fighting. This was the state of mind of the American soldiers stationed at the beginning of the Battle of the Bulge in this supposedly calm front area. After weeks of bloody fighting in the so-called "Hürtgen" forest southwest of Aachen, the 110th regiment of the 28th U.S. Infantry Division had been transferred to the front area along the Our river. As forces proved to be insufficient to establish a continuous front, only the villages along the 'Skyline Drive' - the highway no. 7 crossing the high plateaus between Diekirch and St. Vith (B) - were occupied by American units. The opportunity was used to integrate replacement troops in the units, but also to re-equip soldiers and to conduct tactical field exercises. Soldiers patrolled regularly, day and night, but the general situation was very quiet in this area, the so-called 'Ghostfront'.

'GIs' playing baseball in Consdorf

During the day, most of the soldiers passed the time playing baseball and football, and they enjoyed their regular hot meals. In the evening they attended movie screenings or they sipped a cold beer at the village café, where they contracted friendships that often continued well beyond the war. The USO-Shows - in which the most famous artists performed - were among the highlights of this period.

But it was the calm before the storm! In the early hours of December 16, 1944, the Ghostfront between Echternach (L) and Monschau (D) was exposed to terrible artillery fire. Germany had launched its counterattack in the West.

LE SIMPLE SOLDAT AMÉRICAIN

Cette sculpture en bronze est dédiée au 'GI', au simple soldat américain dont l'intervention a permis de libérer le Luxembourg à deux reprises du joug nazi, en 1944 et en 1945. Le monument conçu par le sculpteur luxembourgeois Michel HEINTZ a été érigé par le CEBA sur un terrain mis à disposition par la commune de Clervaux. La statue fut dévoilée le 11 septembre 1983 par le Grand-Duc Héritier et son épouse la Grande-Duchesse Héritière Maria-Teresa.

La représentation plastique figurant sur la maison d'en face montre la population civile luxembourgeoise réservant un accueil enthousiaste aux libérateurs américains. Cette œuvre a été apposée ultérieurement par la Caisse d'Epargne de l'Etat.

Ce 'GI' est censé exprimer l'impassibilité et de la sérénité. Éreinté et épuisé, il rêve de calme et de repos après les combats. C'était exactement l'état d'âme des soldats américains stationnés au début de la bataille des Ardennes dans ce secteur du front qui était supposé calme. Après des semaines de combats sanglants dans la forêt dite « Hürtgenwald » au sud-ouest d'Aix-la-Chapelle, le 110e régiment de la 28e division d'infanterie américaine avait été transféré vers le secteur du front le long de l'Our. Comme les forces s'avéraient insuffisantes pour assurer la mise en place d'un front continu, seuls les villages le long du 'Skyline Drive', la route nationale 7 qui passe par les hauts plateaux et relie Diekirch et St-Vith (B), étaient occupés par des unités américaines. On profitait du temps disponible pour intégrer des troupes de remplacement dans les unités, mais aussi pour rééquiper les soldats et effectuer des exercices tactiques. Les soldats patrouillaient régulièrement, de jour comme de nuit, et les 'GI' désignaient ce secteur comme 'Ghostfront' (front fantôme).

Pendant la journée, les soldats passaient leur temps à jouer au base-ball et au football américain, et ils prenaient régulièrement des repas chauds. Le soir, ils assistaient à des projections de films ou bien ils dégustaient une bière bien fraîche au café du village. Cela permettait de créer des liens d'amitié qui se poursuivaient souvent bien au-delà de la guerre. Les 'USO - Shows', au cours desquels les artistes les plus renommés de l'époque se produisaient devant les soldats, comptaient parmi les grands moments de cette période.

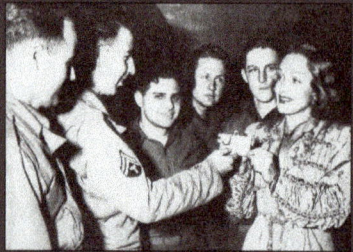

Marlene Dietrich on tour in the Ardennes, December 1944

Mais le calme était trompeur! Aux premières heures du 16 décembre 1944, le front fantôme entre Echternach (L) et Monschau (D) se trouvait soumis à un terrible barrage d'artillerie. L'Allemagne venait de lancer sa contre-attaque à l'ouest.

DER EINFACHE AMERIKANISCHE SOLDAT

Diese Bronzeskulptur ist dem „GI", dem einfachen amerikanischen Soldaten, durch dessen Einsatz Luxemburg 1944 und 1945 zweimal von deutscher Besatzung befreit wurde, gewidmet. Das vom Luxemburger Bildhauer Michel HEINTZ entworfene Denkmal wurde vom CEBA, auf einem, von der Gemeinde Clervaux zur Verfügung gestellten Grundstück, errichtet. Die Statue wurde am 11. September 1983 von Erbgroßherzog Henri und dessen Gemahlin Erbgroßherzogin Maria-Teresa enthüllt.

Die an der gegenüberliegenden Hauswand angebrachte Plastik stellt die, den amerikanischen Befreier frenetisch begrüßende, Zivilbevölkerung dar. Sie wurde nachträglich von der luxemburgischen Staatssparkasse angebracht.

Der dargestellte „GI" soll Ruhe und Gelassenheit ausstrahlen. Abgekämpft und erschöpft sehnt er sich nach Ruhe und Erholung vom Kampfgeschehen. Genau in diesem Gemütszustand befanden sich die amerikanischen Soldaten, die zu Beginn der Ardennenoffensive im vermeintlich ruhigen Frontabschnitt der Ardennen stationiert waren. Nach wochenlangen, blutigen Kämpfen im Hürtgenwald, südwestlich von Aachen, war das 110. Regiment der 28. US-Infanteriedivision an den Frontabschnitt entlang der Our verlegt worden. Zu schwach um eine zusammenhängende Frontlinie aufzustellen, waren nur die Dörfer entlang dem „Skyline Drive", der Nationalstrasse 7 die über den Höhenrücken von Diekirch nach St. Vith (B) führt, von US-Einheiten besetzt. Diese Zeit wurde genutzt um Ersatzmannschaften in die Einheiten einzugliedern, die Truppen neu auszurüsten und taktische Feldübungen durchzuführen. Sowohl nachts als auch tagüber fanden regelmäßige Patrouillengänge statt. Allgemein war die Lage jedoch sehr ruhig, so dass die „GI's" diesen Abschnitt als „Ghostfront" (Geisterfront) bezeichneten.

„GI's" in the Ardennes

Tagsüber vertrieben sich die „GI's" die Zeit mit „Baseball" und „American Football" und genossen regelmäßig warme Mahlzeiten. Abends besuchte man die Vorstellungen im Feldkino oder genoss ein kühles Bier im örtlichen Gasthaus. So entstanden später viele Freundschaften, die über die Zeit hinaus währten. Höhepunkte waren die „USO-Shows" bei denen die berühmtesten Künstler der Zeit auftraten.

Doch die Idylle war trügerisch! Am frühen Morgen des 16. Dezembers 1944 erwachte die Geisterfront zwischen Uchternach (L) und Monschau (B) mit einem mörderischen Artilleriefeuer. Der deutsche Gegenangriff im Westen hatte gerade begonnen.

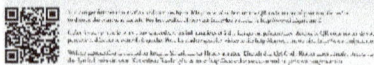

Clervaux Sherman Tank

Monument to Company B, 2nd Tank Battalion, 9th Armored Division.

Montée du Château

9712 Clervaux

Latitude: 50.0543

Longitude: 6.0306

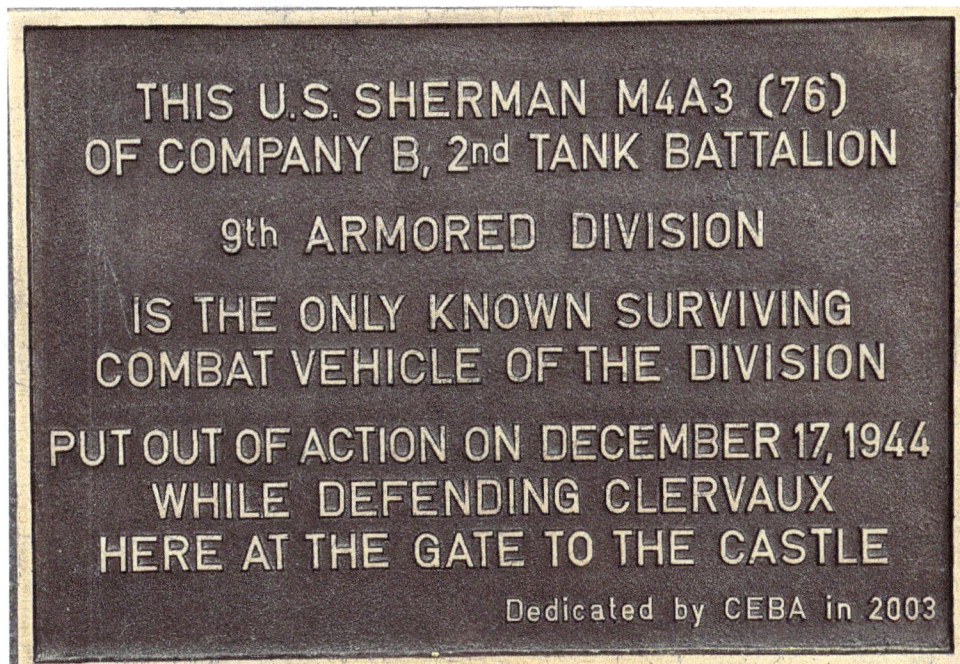

THIS U.S. SHERMAN M4A3 (76)
OF COMPANY B, 2nd TANK BATTALION

9th ARMORED DIVISION

IS THE ONLY KNOWN SURVIVING
COMBAT VEHICLE OF THE DIVISION

PUT OUT OF ACTION ON DECEMBER 17, 1944
WHILE DEFENDING CLERVAUX
HERE AT THE GATE TO THE CASTLE

Dedicated by CEBA in 2003

There is a museum in the Castle next to the tank that is dedicated to the Battle of the Bulge.

The tank was removed in 2019 for refurbishment but is due to be returned by December 2024.

Information Panel at Clervaux Castle

REMEMBER US
CLERVAUX

CLERVAUX IN THE BATTLE OF THE BULGE

The plan of a counterattack through the Ardennes was Adolf Hitler's own idea. He aimed at capturing the harbor of Antwerp, which would have cut the British both from the Americans and from their supplies. Hitler was convinced that this operation would divide the Allies and force them to accept a separate peace in the West. But the attack failed, due to the fierce resistance of the 'GIs'.

As the outnumbered U.S. troops put up an unexpectedly fierce resistance in defending places like Hosingen, Marnach, Reuler, Munshausen and Clervaux, the attacking units of the 5th Panzer Army were unable to take by storm the key road intersections required for a rapid advance.
Clervaux was also the scene of heavy fighting: the units of the 110th Infantry Regiment of the 28th U.S. Division were opposed to the 2nd Armored Division of the Wehrmacht.

The destroyed castle of Clervaux

On the evening of 17 December, Colonel Hurley E. Fuller, commander of the 110th Infantry Regiment, succeeded in escaping at the last minute from his headquarters at Hotel Claravallis, and the organized resistance collapsed completely thereafter. But a group made up of cooks, clerks and some American infantrymen continued to defend the castle in the heart of Clervaux: they prevented German troops from taking the local main road. It was not until the early afternoon of the next day that they had to surrender. Following a German attack with phosphorus grenades the castle caught fire, and the American defenders were running out of ammunition.

At that time, a 'Sherman M4A3 (76)' was used to defend the castle. On December 17 at about 11 am, the tank of the B-Company of the 2nd Armored Battalion of the 9th U.S. Armored Division was positioned in front of the castle gate. The tank had set an ambush behind a medieval house (the so called "Brahaus"). It moved a few meters forwards, fired a shot from it's 76mm main gun onto the column of tanks on the road along the cemetery, and quickly backed up behind the cover of the "Brahaus", again and yet again. Around noon the Sherman was abandoned by it's crew after a grazing shot had bounced off the tank's armour.
The damaged tank stood close to the gable end of the Kratzenberg house until 1956. Using two big recovery vehicles, the Luxembourg army managed to tow the wreck into the castle courtyard (through a hole in the surrounding wall), and damage resulting from combat operations were then repaired on site. Today, the American tank is still at almost the same place, it is the silent witness of the fierce fighting that took place in this region in 1944.

CLERVAUX PENDANT LA BATAILLE DES ARDENNES

C'est Adolf Hitler qui a eu l'idée d'une contre-attaque dans les Ardennes. Il visait la prise du port d'Anvers, ce qui aurait permis de couper les Britanniques à la fois des Américains et de leur ravitaillement. Hitler était convaincu que cette opération permettrait de diviser les alliés et les obligerait à accepter une paix séparée à l'ouest. Mais l'attaque a échoué suite à la résistance acharnée des 'GI'.

Comme les troupes américaines inférieures en nombre défendant des localités comme Hosingen, Marnach, Reuler, Munshausen et Clervaux opposaient une résistance inattendue, les unités attaquantes de la 5e division blindée allemande ne réussissaient pas à s'emparer des points stratégiques essentiels pour une avancée rapide.
Clervaux était également le théâtre de lourds combats: les compagnies du 110e régiment d'infanterie de la 28e division américaine se trouvaient opposées à la 2e division blindée de la Wehrmacht.

Le soir du 17 décembre, le Colonel Hurley E. Fuller, commandant du 110e régiment d'infanterie, réussissait à sortir au tout dernier moment de son quartier général à l'Hôtel Claravallis, et la résistance organisée s'effondrait complètement par la suite. Seul un groupe composé de cuisiniers, de personnel administratif et de quelques fantassins américains continuait à défendre le château au cœur de la localité et empêchait les troupes allemandes d'emprunter la route principale de Clervaux. Ce ne fut qu'en début d'après-midi du jour suivant qu'ils étaient contraints à la capitulation. Le château se trouvait en feu suite à une attaque allemande avec des grenades au phosphore, et les défenseurs américains n'avaient plus de munitions.

À l'époque, un char 'Sherman M4 A3 (76)' était utilisé pour la défense du château. Le 17 décembre vers 11 heures du matin, le char de la compagnie B du 2e bataillon blindé de la 9e division blindée américaine s'était positionné devant le portail du château. Se tenant en embuscade derrière la maison moyenâgeuse dite « Brahaus », il

The 'Sherman' in front of the castle

avançait de quelques mètres pour prendre sous son feu la colonne de blindés passant sur la route longeant le cimetière, puis il reculait en trombe, pour se réfugier derrière les murs du "Brahaus". Ce spectacle se répétait maintes fois, ce n'est que vers midi, que le Sherman était abandonné par son équipage, après avoir été touché par un obus allemand qui n'avait pas réussi à percer le blindage.
Le char endommagé était garé près du pignon de la maison Kratzenberg jusqu'en 1956. C'est à l'aide de deux grosses dépanneuses que l'armée luxembourgeoise réussissait à tirer l'épave dans la cour du château en la faisant passer à travers un trou dans le mur d'enceinte. Les dégâts résultant des opérations de guerre étaient ensuite réparés sur place. Le char américain se trouve toujours quasiment au même endroit, il est le témoin silencieux des combats inhumains qui se sont déroulés en 1944 dans la région.

CLERVAUX IN DER ARDENNENSCHLACHT

Der Plan zum Gegenangriff durch die Ardennen entsprang einer Idee Adolf Hitlers. Ziel war die Einnahme des Hafens von Antwerpen, um so die Briten im Norden des Angriffskeils von den Amerikanern und vom Nachschub abzuschneiden. Hitler war überzeugt, dass er durch dieses Unternehmen die Westalliierten entzweien und zu einem Separatfrieden im Westen zwingen könnte. Doch der erbitterte Widerstand der „GI's" ließ die Offensive scheitern.

Da ab dem Angriffstag, dem 16. Dezember 1944, die zahlenmäßig unterlegenen US-Truppen Ortschaften wie Hosingen, Marnach, Reuler, Munshausen und Clervaux unerwartet zäh verteidigten, gelang es den angreifenden Einheiten der 5. Deutschen Panzerarmee nicht, die für einen raschen Vorstoß wichtigen Verkehrsknotenpunkte einzunehmen. Auch Clervaux war schwer umkämpft: Die amerikanischen Hauptquartiers- und Versorgungskompanien des 110. Infanterieregiments der 28. US-Division, lieferten sich schwere Kämpfe mit der 2. Panzerdivision der Wehrmacht.

Nachdem es am Abend des 17. Dezember Colonel Hurley E. Fuller, dem Kommandeur des 110. Infanterieregimentes, buchstäblich in letzter Sekunde gelungen war, aus seinem Hauptquartier im Hotel Claravallis zu entkommen, brach jeglicher organisierter Widerstand zusammen. Lediglich eine aus Köchen, Schreibkräften und einigen US-Infanteristen zusammengewürfelte Truppe verteidigte weiterhin das Schloss im Ortkern und verhinderte, dass die deutschen Truppen die Hauptstraße durch Clervaux benutzen konnten. Erst am frühen Nachmittag des Folgetages mussten sie kapitulieren. Das Schloss war durch Beschuss mit Phosphorgranaten in Brand geraten und ihnen war die Munition ausgegangen.

Ein „Sherman M4 A3 (76)" wurde seinerzeit zur Schlossverteidigung eingesetzt. Der Panzer der B Kompanie des 2. Panzerbataillons der 9. US Panzerdivision war am 17. Dezember gegen 11 Uhr morgens vor dem Schlosstor in Stellung gegangen. Hinter dem Hausgiebel des mittelalterlichen „Brahaus" in Deckung auf Lauer liegend, fuhr er immer wieder einige Meter vor, um die Panzerkolonne auf der Straße am Friedhof unter Beschuss zu nehmen. Gegen Mittag wurde der „Sherman" seinerseits von einem deutschen Streifschuss getroffen und von der Besatzung aufgegeben.

1945: Grand Duchess Charlotte in Clervaux

Bis 1956 stand der beschädigte Panzer am Giebel des Hauses Kratzenberg. Mit Hilfe von 2 schweren Bergefahrzeugen zog die Luxemburger Armee das Wrack durch ein Loch in der Ringmauer in den Schlosshof und reparierte die Beschädigungen, die noch von den Kampfhandlungen herstammten, vor Ort. Der US-Panzer steht heute noch fast an der gleichen Stelle als stiller Zeuge der unmenschlichen Kämpfe, die 1944 hier in der Gegend stattgefunden haben.

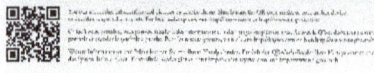

Clervaux

Dedicated to Colonel Hurley E. Fuller, Commanding Officer, 110th Infantry Regiment, 28th Division. Monument is located across the street from the former Hotel Claravallis, his former command post.

Rue de la Gare 4

9707 Clervaux

Latitude: 50.0596

Longitude: 6.0240

Inaugurated on December 15, 2019

Clervaux

In honor to HQ Company, 110th Infantry Regiment, 28th Infantry Division who held the castle from December 16 – 18, 1944.

Rue Schloff 4

9712 Clervaux

Inside Castle courtyard

Latitude: 50.054367

Longitude: 6.029938

IN HONOR
OF THE BRAVE MEN OF HQ COMPANY,
110th INFANTRY REGIMENT, 28th INFANTRY DIVISION,
WHO HELD THIS CASTLE
AGAINST SUPERIOR GERMAN FORCES
FROM 16 TO 18 DECEMBER 1944

CEBA

Consdorf

In October 1990, this 900 kg aerial bomb was found and defused in a field near Consdorf. The bomb was 1 of 2 dropped in the autumn of 1944 by a B17F "Flying Fortress" of the 8th Air Force to unballast the aircraft.

Route de Luxembourg 91

6210 Consdorf

Latitude: 49.7730

Longitude: 6.3414

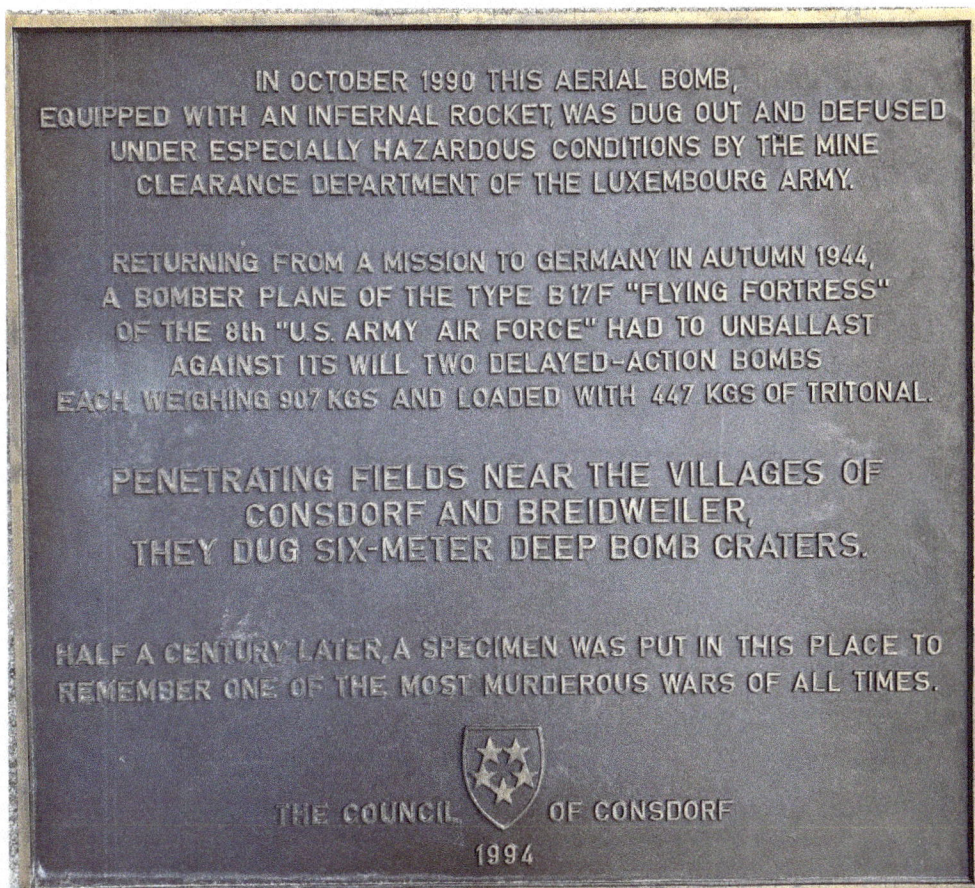

IN OCTOBER 1990 THIS AERIAL BOMB,
EQUIPPED WITH AN INFERNAL ROCKET, WAS DUG OUT AND DEFUSED
UNDER ESPECIALLY HAZARDOUS CONDITIONS BY THE MINE
CLEARANCE DEPARTMENT OF THE LUXEMBOURG ARMY.

RETURNING FROM A MISSION TO GERMANY IN AUTUMN 1944,
A BOMBER PLANE OF THE TYPE B17F "FLYING FORTRESS"
OF THE 8th "U.S. ARMY AIR FORCE" HAD TO UNBALLAST
AGAINST ITS WILL TWO DELAYED-ACTION BOMBS
EACH WEIGHING 907 KGS AND LOADED WITH 447 KGS OF TRITONAL.

PENETRATING FIELDS NEAR THE VILLAGES OF
CONSDORF AND BREIDWEILER,
THEY DUG SIX-METER DEEP BOMB CRATERS.

HALF A CENTURY LATER, A SPECIMEN WAS PUT IN THIS PLACE TO
REMEMBER ONE OF THE MOST MURDEROUS WARS OF ALL TIMES.

THE COUNCIL OF CONSDORF
1994

Consthum

In honor of Colonel Daniel B. Strickler and the 110th Regiment, 28th US Infantry Division.

Rue Knupp 3

Consthum

Latitude: 49.9739

Longitude: 6.0528

IN HONOR
TO COLONEL DANIEL B. STRICKLER AND
HIS SOLDIERS OF THE 110TH REGIMENT,
28TH US INFANTRY DIVISION,
WHO VALIANTLY DEFENDED OUR HOME TOWN
DURING THE BATTLE OF THE BULGE

1944

★ ★ ★

Information Panel Consthum

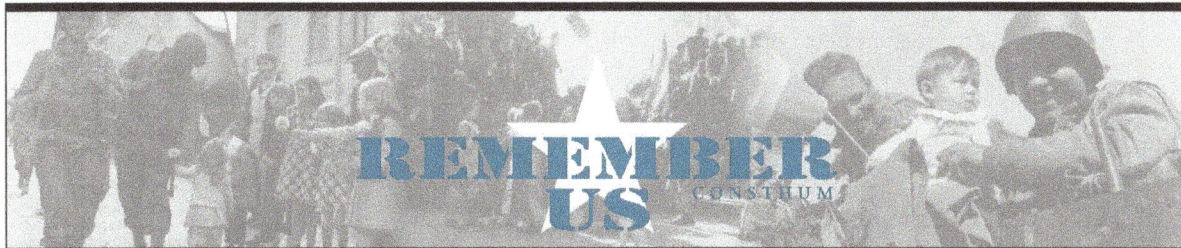

LIEUTENANT COLONEL DANIEL B. STRICKLER

On December 17, 1944, Colonel Hurley E. Fuller, Commander of the 110th Infantry Regiment, ordered Lieutenant Colonel Daniel B. Strickler, regimental executive officer, to Consthum, headquarters of the 3rd Battalion. Together with Major Harold F. Milton, 3rd Battalion's Commander, Lt. Colonel Strickler organized the defense of the town. He ordered to set out mines along the approaching ways to the town.

Lieutenant General Daniel b. STRICKLER, former Commander of 109th Infantry Regiment prior to December 12, 1944 and later Commanding General of the 28th Infantry Division.

Authorized for Release to the Public by PIO, PIO, Public Information Office

The garrison fought back all the German attacks on **December 16 and 17**. The troop was composed of the 80 men of 3rd Battalion Headquarters, the 20 men of Company M, the 20 survivors of Company L and the 25 men of Company I as well as the artillerymen of Battery C, 687th FA, the crews of the 447th AAA Battalion manning a quad .50 and a 40mm Bofors gun, three light armored cars M-8 of the 28th Reconnaissance troop and three effective tanks of the 707th Tank Battalion.

On **December 18**, about 01.00 p.m., fog wrapped up the village. Under this natural cover, German tanks of the Panzer Lehr Division's Reconnaissance detachment and infantrymen of the 39th Füsilier-Regiment, 26th Volksgrenadierdivision, penetrated into Consthum. The rests of the US 3rd Battalion retreated in good order to Nocher.

On **December 19**, Lt. Colonel Daniel B. Strickler retreated with the remaining 250 men to Wiltz. Major General Norman D. Cota, Commander of the 28th US Infantry Division, assigned Lt.-Col. Strickler as the acting commander of the 110th Regiment and charged him with the defense of Wiltz. Lt.-Col. Strickler assumed command under the name of "Task Force Strickler".

On **December 19, at 2:00 p.m.**, the Germans attacked Wiltz from three sides and succeeded in penetrating the centre of the town. The Americans defended their positions until they ran out of ammunition. In the evening, Lt. Col. Strickler gave the order to retreat in small groups and to try to reach Sibret and Bastogne, Belgium. As the town of Wiltz had meanwhile been encircled, many American soldiers were killed, hurt or taken prisoners by taking the road via Café Schumann to reach Belgium.

LIEUTENANT-COLONEL DANIEL B. STRICKLER

Le **17 décembre 1944**, Colonel Hurley E. Fuller, commandant du 110e Régiment, détacha Lieutenant-Colonel Daniel B. Strickler, officier exécutif du régiment, à Consthum, quartier-général du 3e Bataillon. Ensemble avec Major Harold B. Milton, commandant du 3e Bataillon, Lieutenant-Colonel Strickler organisa la défense de la localité. Il fit déposer des champs de mines sur les routes et les chemins d'approche de la localité.

La garnison repoussait toutes les attaques allemandes du **16 et 17 décembre**. Elle était composée des 80 hommes du quartier général du 3e Bataillon, des 20 hommes de la Compagnie M, des 20 survivants de la Compagnie L, des 25 soldats de la Compagnie I ainsi que des artilleurs de la Batterie C du 687e Bataillon de l'Artillerie de Campagne, de ceux du 447e Bataillon de l'Artillerie Anti-Aérienne ayant desservi une pièce à quatre canons .50 et un canon 40mm Bofors, de l'équipage de trois chars légers M-8 Greyhound de la Troupe de Reconnaissance de la 28e Division d'Infanterie et de trois chars opérationnels du 707e Bataillon de Chars.

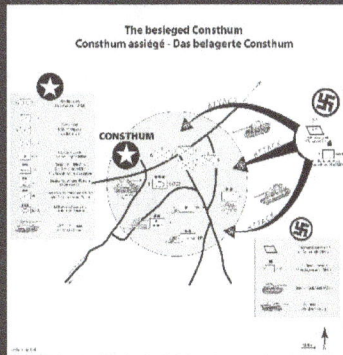

The besieged Consthum
Consthum assiégé - Das belagerte Consthum

The besieged Consthum - Consthum assiégé - das belagerte Consthum

Quand le **18 décembre** vers 13.00 heures, un épais brouillard enveloppa le village, les chars du détachement de reconnaissance de la Panzer-Lehr Division et l'infanterie du 39e Régiment, 26e Volksgrenadierdivision, pénétrèrent dans Consthum. Les restes du 3e Bataillon américain se retirèrent en bon ordre à Nocher.

Le **19 décembre**, Le Lieutenant-Colonel Strickler se retira avec ses 250 hommes restants à Wiltz. Général-Major Norman D. Cota, commandant de la 28e Division d'Infanterie américaine, assigna Lt.-Col. Strickler commandant du 110e Régiment et le chargea de la défense de Wiltz. Lt.-Col. Strickler assuma son commandement sous le nom de « Groupement de Combat Strickler ».

Le **19 décembre à 14.00 heures**, les Allemands attaquèrent Wiltz de trois côtés et réussirent à pénétrer dans le centre de la ville. Les Américains défendaient leurs positions jusqu'à épuisement de leurs munitions. Le soir, Lt.-Col. Strickler donna l'ordre de se retirer en de petits groupes et de rejoindre Sibret ou Bastogne en Belgique. La ville de Wiltz ayant entretemps été encerclée, beaucoup d'Américains furent tués, blessés ou faits prisonniers en voulant s'enfuir par la route via Café Schumann.

Sources - Quellen - To save Bastogne by Robert F Phillip, Borodino Books Barke, VA, USA. PIO PIO, Public Information Office.
- Die Ardennenschlacht 1944-1945 in Luxemburg, Jean Milmeister, Seite 256, 1. Edition Saint-Paul, 1994, ISBN 2-87963-216-3
- 26. Volksgrenadierdivision Ardennenoffensive 1944-1945 von Generalmajor Hans Kokott, 13.10.1949
Die Ardennen-Offensive, Pierre Fischer, nos cahiers, Kromm I, Orell, 2000/ 2/3

OBERSTLEUTNANT DANIEL B. STRICKLER

Am **17. Dezember 1944**, kommandierte Oberst Hurley E. Fuller, Kommandant des 110. Infanterieregimentes, Oberstleutnant Daniel B. Strickler, Stabsoffizier des Regimentes, nach Consthum ab, wo sich das Hauptquartier des 3. Bataillons befand. Zusammen mit Major Harold F. Milton, Kommandant des 3. Bataillons, organisierte Oberstleutnant Strickler die Verteidigung des Dorfes. Er ließ Minen auf die Zufahrtsstraßen und Wege zur Ortschaft legen.

Die Garnison schlug alle deutschen Angriffe am **16. und 17. Dezember** zurück. Sie bestand aus den 80 Mann des Hauptquartiers des 3. Bataillons, den 20 Mann der Kompanie M, den 20 Überlebenden der Kompanie L und den 25 Mann der Kompanie I sowie aus den Artilleristen der Batterie C, 687. Feldartilleriebataillon, den Mannschaften des 447. Fliegerabwehrbataillons, die ein Vierlings-Geschütz und eine 40mm Bofors-Kanone bedienten, drei drei leicht gepanzerten M-8 Schützenpanzern der 28. Aufklärungsgruppe und den drei einsatzbereiten Panzern des 707. Panzerbataillons.

Als am **18. Dezember** gegen 13.00 Uhr Nebel das Dorf überdeckte, drangen die Panzer der Aufklärungs-Abteilung der Panzer Lehr Division und die Infanterie des 39. Füsilier-Regimentes, 26. Volksgrenadierdivision, in Consthum ein. Die Reste des 3. US Bataillons zogen sich geordnet nach Nocher zurück.

Am **19. Dezember** zog sich Oberstleutnant Strickler mit den restlichen 250 Mann nach Wiltz zurück. Generalmajor Norman D. Cota, Kommandeur der 28. US Infanteriedivision, ernannte Oberstleutnant Strickler zum amtierenden Kommandeur des 110. Regimentes und beauftragte ihn mit der Verteidigung der Stadt Wiltz. Oberstleutnant Strickler übernahm das Kommando unter dem Namen „Kampfgruppe Strickler".

Am **19. Dezember um 14.00 Uhr**, griffen die Deutschen von drei Seiten Wiltz an und es gelang ihnen, bis in das Stadtzentrum vorzudringen. Die Amerikaner verteidigten ihre Stellungen so lange bis ihnen die Munition ausging. Am Abend gab Oberstleutnant Strickler den Befehl, sich in kleinen Gruppen nach Sibret oder Bastogne (Belgien) zurückzuziehen. Da die Stadt Wiltz inzwischen eingekesselt war, wurden viele Amerikaner getötet, verletzt oder gefangen genommen, indem sie versuchten, Belgien über die Straße, die am Café Schumann vorbei führte, zu erreichen.

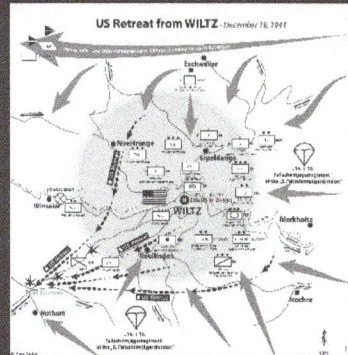

US Retreat from Wiltz - December 19, 1944

US retreat out of Wiltz and American retreat - Amerikanischer Angriff auf Wiltz und amerikanischer Rückzug

For more information
Pour en savoir plus
Für weitere Informationen

Information Panel Consthum

REMEMBER US
CONSTHUM

DANIEL B. STRICKLER

Daniel Bursk Strickler was born on May 17, 1897, in Columbia, Lancaster County, Pennsylvania. He attended the public schools of Columbia, graduated from Columbia High School and joined the Pennsylvania National Guard in 1916. The same year, he served as a sergeant in the Mexican Border Conflict.

In 1917, First Lieutenant Daniel B. Strickler commanded a company of the 109th Machine Gun Battalion of the 28th US Infantry Division. During World War I, his company was deployed to France. Daniel B. Strickler fought in five campaigns, including at Château-Thierry and the Meuse-Argonne Offensive. He was wounded, gassed and blinded for several days. He recovered quickly and returned to his unit in the frontline. After World War I, he was discharged from the Army with the rank of Captain.

After his return from World War I in 1918, Strickler studied law at Cornell University Law School in Ithaca, New York. He graduated to Bachelor of Laws in 1922. He worked with several law firms and, in 1930, established his own solicitor's office in Lancaster.

In 1930, Strickler, as a member of the Republican Party, won a seat in the Pennsylvania House of Representatives. In 1932, he was appointed Lancaster's police commissioner. From 1933 to 1942, he was Lancaster's city solicitor. At the same time, he graduated from the United States Army Command and General Staff College. He remained active in the reserves and held several commands in the United States Army Reserve. He was promoted to Major in 1922, to Lieutenant Colonel in 1928 and to Colonel in 1935.

When World War II broke out, he volunteered to command different battalions and regiments. Finally, he was in charge of the 110th Regiment, 28th Infantry Division. He was promoted to Brigadier General in 1946 and to Major General in 1947. After World War II, he returned to civil life and was elected as Lieutenant Governor of Pennsylvania in 1946.

During the Korean War from 1950 to 1953, he was the Commanding General of the 28th US Infantry Division in the U.S. and in Germany from 1950 to 1952. In 1960, his final commission to a Lieutenant General honored Strickler for his numerous services he had rendered to the nation. After the Korean War, he served as an advisor for the Army and a diplomat in Korea, Japan and Rome, Italy, until his retirement in 1957.

He was married to Caroline Bolton in 1924. The couple had a daughter, Nancy Cupper Strickler, and a son, Daniel B. Strickler Jr.

Strickler died in Lancaster on June 29, 1992. He was buried at Woodward Hill Cemetery in Lancaster.

Daniel B. Strickler was decorated with several medals, such as the Silver Star with oak leaf cluster, the Bronze Star for Valor, and the Legion of Merit.

Lt. Governor Strickler presiding over the Senate of state legislature of the Commonwealth of Pennsylvania, 1947-1950.

Lt. Governor Strickler présidant le Sénat du "Commonwealth" de Pennsylvanie, 1947-1950.

Lt. Governor Strickler Vorsitzende des Senats des "Commonwealth" von Pennsylvania, 1947-1950.

Photo: Daniel B. Strickler's archive at the Lancaster Historical Society.

Major General David Strickler, commanding the 28th Division, being greeted by NATO Commander General Dwight Eisenhower upon the Division's arrival in Germany, joining the NATO forces defending the country during the Cold War, November 21, 1951.

Le Général-Major David Strickler, Commandant de la 28e Division d'Infanterie accueilli estime, est salué par le Commandant de l'OTAN, le Général Dwight Eisenhower, lors de l'arrivée de la division en Allemagne le 21 novembre 1951, rejoignant les forces de l'OTAN.

Bei der Ankunft der 28. US Infanteriedivision in Deutschland am 21. November 1951, wird Generalmajor Daniel Strickler vom Oberbefehlshaber der NATO, General Dwight Eisenhower, begrüßt.

Photo: Daniel B. Strickler's archives.

Daniel Bursk Strickler est né le 17 mai 1897 à Columbia, Comté de Lancaster, Pennsylvanie. Il fréquentait les écoles publiques de Columbia et fut diplômé de la Columbia High School. Il s'engagea à la Garde Nationale de Pennsylvanie en 1916. La même année, il servait en tant que sergent dans le conflit frontalier avec le Mexique.

En 1917, Premier Lieutenant Daniel B. Strickler commandait une compagnie du 109e Bataillon de Mitrailleuses de la 28e US Division d'Infanterie. Pendant la Première Guerre Mondiale, sa compagnie fut déployée en France. Daniel B. Strickler combattait dans cinq campagnes, inclusivement les offensives de Château-Thierry et de Meuse-Argonne. Il fut blessé, gazéifié et perdit la vue pendant quelques jours. Il se remit rapidement et rejoignit son unité au front. Après la guerre, il fut démobilisé de l'armée au rang de capitaine. Après son retour de la Grande Guerre, il fit des études à l'École de Droit de l'Université Cornell à Ithaca, New York. Il obtint une licence en droit en 1922. Il a travaillé avec plusieurs cabinets juridiques et ouvrit son propre cabinet d'avocats à Lancaster en 1930.

En 1930, Strickler, comme membre du parti républicain, gagna un siège dans le Parlement de Pennsylvanie. En 1932, il fut nommé commissaire du département de la police de Lancaster. De 1933 à 1942, il était avocat du comté de Lancaster. En même temps, il promut au General Staff College du Commandement de l'Armée des États-Unis. Il restait actif dans les réserves et assumait plusieurs commandements dans la Réserve de l'Armée des États-Unis. Il fut promu major en 1922, lieutenant-colonel en 1928 et colonel en 1935.

Quand la Deuxième Guerre Mondiale éclata, Daniel B. Strickler se portait volontaire pour commander différents bataillons et régiments. Finalement il fut en charge du 110e Régiment de la 28e Division d'Infanterie. Il fut promu Brigadier Général en 1946 et Major Général en 1947. Après la Seconde Guerre Mondiale, il retourna dans la vie privée et fut élu Vice-Gouverneur de Pennsylvanie en 1946.

Dans la Guerre de Corée de 1950 à 1953, il était le Général Commandant de la 28e Division d'Infanterie Américaine aux États-Unis et en Allemagne de 1950 à 1952. En 1960, sa dernière promotion au grade de Lieutenant Général honorait Strickler pour ses nombreux services qu'il a rendus à la nation. Après la Guerre de Corée, il était conseiller militaire et diplomate en Corée, au Japon et à Rome, Italie, jusqu'à sa retraite.

Il était marié à Caroline Bolton en 1924. Le couple eut une fille, Nancy Cupper Strickler et un fils, Daniel Bursk Strickler Jr.

Strickler est décédé à Lancaster, le 29 juin 1992. Il fut enterré au Cimetière Woodward Hill à Lancaster.

Daniel B. Strickler fut décoré de plusieurs médailles, comme l'Étoile d'Argent avec feuilles de chêne, l'Étoile de Bronze pour bravoure et la Légion du Mérite.

DANIEL B. STRICKLER

Daniel Bursk Strickler wurde am 17. Mai 1897 in Columbia, Lancaster County, Pennsylvania, geboren. Er besuchte die öffentlichen Schulen von Columbia, absolvierte die Columbia High School und verpflichtete sich 1916 in der Nationalgarde von Pennsylvania. Im selben Jahr diente er als Feldwebel im mexikanischen Grenzkonflikt.

Im Jahre 1917 befehligte Leutnant Daniel B. Strickler eine Kompanie des 109. MG-Bataillons der 28. US Infanteriedivision. Während des Ersten Weltkrieges wurde seine Einheit in Frankreich eingesetzt. Daniel B. Strickler kämpfte in fünf Feldzügen, einschließlich der Château-Thierry und der Meuse-Argonne Offensiven. Er wurde verwundet, vergast und erblindete während Tagen. Er erholte sich schnell und kehrte zu seiner Einheit an die Front zurück. Er wurde im Rang eines Kapitäns ehrenvoll nach dem Ersten Weltkrieg entlassen.

Nach seiner Rückkehr aus dem Ersten Weltkrieg 1918, studierte Strickler Jura an der Cornell University Law School in Ithaca, New York. Er arbeitete als Rechtsanwalt mit verschiedenen Rechtskanzleien und eröffnete seine eigene Kanzlei 1930 in Lancaster.

1930, als Mitglied der republikanischen Partei, gewann er einen Sitz im Repräsentantenhaus von Pennsylvania. 1932 wurde er Polizeichef von Lancaster. Von 1933 bis 1942 war er Lancasters Stadtanwalt. Zugleich promovierte er am United States Army Command and General Staff College. Er blieb beim Militär aktiv und hielt mehrere Kommandoposten in der Reserve der Armee der Vereinigten Staaten. 1922 wurde er zum Major befördert, 1928 zum Oberstleutnant und 1935 zum Oberst.

Beim Ausbruch des Zweiten Weltkrieges, meldete er sich freiwillig, um das Kommando verschiedener Bataillone und Regimenter zu übernehmen. Schließlich befehligte er das 110. Regiment der 28. Infanteriedivision. 1946 wurde er zum Brigadegeneral und 1947 zum Generalmajor befördert.

Nach dem Zweiten Weltkrieg kehrte er zum zivilen Leben zurück und wurde 1946 zum Vizegouverneur des Staates Pennsylvania gewählt.

Während des Korea Krieges von 1950 – 1953, war er kommandierender General der 28. US Infanteriedivision in den Vereinigten Staaten und in Deutschland von 1950 bis 1952. Die letzte Beförderung zum Generalleutnant im Jahre 1960 ehrte Daniel B. Strickler für die vielen Dienste, die er dem Land geleistet hatte.

Nach dem Koreakrieg diente er als Berater der Armee und als Diplomat in Korea, Japan und in Rom, Italien, bis zu seinem Ruhestand im Jahre 1957.

Er war mit Caroline Bolton seit 1924 verheiratet. Das Ehepaar hatte eine Tochter, Nancy Cupper Strickler, und einen Sohn, Daniel B. Strickler Jr.

Strickler starb in Lancaster am 29. Juni 1992. Er wurde auf dem Friedhof Woodward Hill in Lancaster begraben.

Daniel B. Strickler war mit mehreren Medaillen ausgezeichnet, unter anderen mit dem Silbernen Stern mit Eichenlaub, dem bronzenen Stern für Tapferkeit und mit dem Verdienstorden.

Sources - Quellen:

- Lancaster History.org: Lt. General Daniel B. Strickler Collection 1916 - 1935
- Archives NY Times 1992/06/30 https://www.nytimes.com/.../lieut-gen-daniel-strickler-93-veteran...
- www.Dmva.pa.gov/ Department/ DMVA – Hall of Fame / Daniel B. Strickler pdf
- https://en.wikipedia.org/ Daniel B. Strickler)
- ©2008 American War Memorials Overseas, Inc. Website designed & development by AWPDC.com.

For more information
Pour en savoir plus
Für weitere Informationen

Cruchten

US 250 kg aerial bomb discovered in 2009 near Cruchten.

Rue de l'Alzette

7420 Nommern

Latitude: 49.8023

Longitude: 6.1299

Des amerikanisch Fliegerbomm aus dem 2. Weltkrich gouf den 17. März 2009 am Ausgang vu Cruchten op Schrondweiler, beim Planzen vun Stroossebeem, entdeckt.

Die 250 kg schweier Bomm war mat 135 kg TNT an Hexogen gefëllt.

Den ''service de déminage'' huet se op der Plaatz entschärft.

Inscription reads: This American aerial bomb from World War II was found on March 17, 2009 at the exit of Cruchten in direction of Schrondweiler when planting roadside trees. The 250 kg bomb was filled with 135 kg's of TNT and hexogen. The EOD (Explosive Ordnance Disposal) service defused the bomb on the site.

Dahl

In honor of Staff Sergeant Day G. Turner, Co. B, 319th Infantry Regiment, 80th Infantry Division who was posthumously awarded the Medal of Honor for action at this location on January 8, 1945.

Am Aastert 5

9644 Goesdorf

Latitude: 49.9358

Longitude: 5.9801

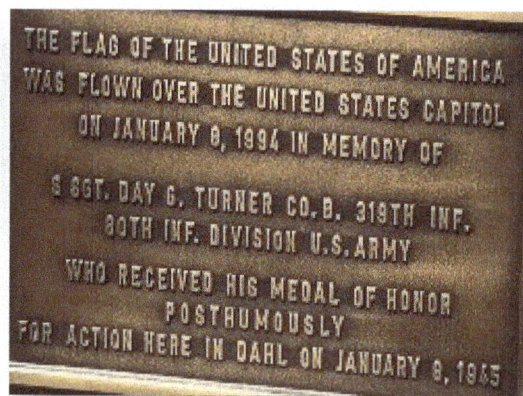

Plaque in French is located on the side of the "Aastert" farm house and its inscription reads: In memory of the Vaillant S/SGT Day G. Turner, Com B-319th Inf Regt, 80th Inf. Div. who for the heroic defense of this farm during the battle of the Ardennes was decorated with the Medal of Honor

Information Panel Dahl

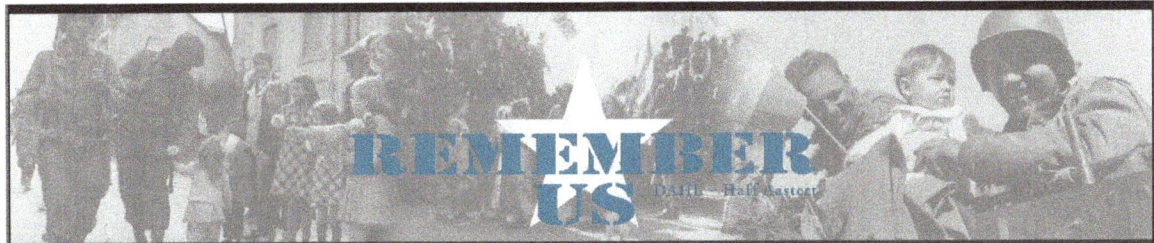

REMEMBER US
DAHL – Haff Aastert

On December 16, 1944, Hitler launched a surprise attack against the US soldiers holding the front from Echternach (L) to Monschau (Germ.). It was the Battle of the Bulge and it was intended to reach the port of Antwerp (B) within three days, to split the British and US troops and to force the Allies to an armistice.

General George S. Patton's Third Army counter-attacked the German south flank on December 22, 1944. The 4th Armored Division smashed the ring around the encircled town of Bastogne (B) on December 26, 1944. The 26th Infantry Division had Wiltz as an objective, which was liberated on January 21, 1945. The 80th Infantry Division liberated Ettelbruck already on December 25, 1944. The 5th Infantry Division operated in the area of Echternach, liberated the town on December 26, 1944 and pushed to the North from January 18, 1945.

On January 6, 1945, the US soldiers of the 3rd Battalion, 319th Infantry Regiment, 80th US Infantry Division, liberated the village of Dahl. The next days the Germans tried new assaults on the battered locality without success.

Staff Sergeant Day G. TURNER

CMH citation: "On January 8, 1945, Sgt. Day G. Turner commanded a 9-man squad, positioned in the open fields between Dahl and Nocher, with the mission of holding this critical flank position. When overwhelming numbers of the enemy attacked under cover of withering artillery, mortar and rocket fire, he withdrew his squad into the nearby Aastert farm house, determined to defend it to the last man. The enemy attacked again and again and was repulsed with heavy losses. Supported by direct tank fire, they gained entrance, but the intrepid sergeant refused to surrender although 5 of his men were wounded and one was killed. He boldly flung a can of flaming oil at the first wave of attackers, dispersing them, and fought doggedly from room to room, closing with the enemy in fierce hand-to-hand encounters. He hurled hand grenade for hand grenade, bayoneted two fanatical Germans who rushed to a doorway he was defending and fought on with the enemy's weapons when his own ammunition was expended. The savage fight raged for four hours, and finally, when only three men of the defending squad were left unwounded, the enemy surrendered. Twenty-five prisoners were taken, eleven enemy were dead and a great number of wounded were counted. Sgt. Turner's valiant stand will live on as a constant inspiration to his comrades. His heroic, inspiring leadership, his determination and courageous devotion to duty exemplify the highest tradition of the military service."

For his heroic action in Dahl, Sgt. Turner was awarded posthumously the highest decoration of the US Army, the Congressional Medal of Honor (CMH). On February 8, 1945, he was killed in action on German territory opposite Wallendorf, while attacking a bunker of the Westwall.

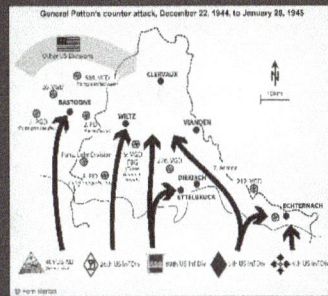

General Patton's counter attack, December 22, 1944, to January 28, 1945

Main drives of the US Third Army's counterattack
Axes principaux de la contre-attaque de la 3e Armée Américaine
Hauptstoßrichtungen des Gegenangriffs der Dritten US Armee

Sgt. Day G. Turner

Le 16 décembre 1944, Hitler lança une attaque surprise contre les soldats américains défendant le front à partir d'Echternach (L) jusque Monschau (Allem.). C'était la Bataille des Ardennes qui avait pour but de s'emparer du port d'Anvers (B) dans trois jours, de séparer les troupes britanniques et américaines et de forcer les Alliés à signer un armistice. La 3e Armée du général George S. Patton Jr. attaqua le flanc sud allemand le 22 décembre 1944 : La 4e Division Blindée rompit le siège de Bastogne (B) le 26 décembre 1944. La 26e Division d'Infanterie avait la ville de Wiltz comme objectif qu'elle libéra le 21 janvier 1945. La 80e Division d'Infanterie libéra Ettelbruck déjà le 25 décembre 1944. La 5e Division d'Infanterie opéra dans le secteur d'Echternach, libéra la ville le 26 décembre 1944 et poussa vers le nord à partir du 18 janvier 1945.

Le 6 janvier 1945, les soldats américains du 3ème Bataillon, 319e Régiment d'Infanterie, de la 80e Division, libérèrent le village de Dahl. Les jours suivants les Allemands attaquèrent la localité plusieurs fois sans succès.

Staff Sergeant Day G. TURNER

CMH citation : « Le 8 janvier 1945, le sergent Day G. Turner commandant une section de 9 hommes, était positionné dans la campagne entre Dahl et Nocher avec mission de tenir ce flanc critique à tout prix. Lorsque l'ennemi attaqua avec un nombre écrasant sous le couvert d'un feu d'artillerie, de mortier et de fusées dévastateur, il retira sa section dans la ferme Aastert toute proche, déterminé de défendre celle-ci jusqu'au dernier homme. L'ennemi attaqua sans cesse, mais fut repoussé avec de lourdes pertes. Supporté par le feu de char, il réussit à entrer dans la ferme. Mais l'intrépide sergent refusa de se rendre malgré que cinq de ses hommes fussent blessés et un fut mort. En jetant un bidon d'essence en feu sur la première vague des attaquants, il les dispersa. Il continuait obstinément à se battre corps à corps de chambre en chambre. Il jeta des grenades à main, tua deux Allemands qui voulaient forcer une porte qu'il défendait à la baïonnette et combattit avec les armes de ses ennemis quand il était à court de munitions. Le combat sauvage durait quatre heures ; finalement, quand seulement trois de ses hommes étaient encore indemnes, les ennemis se rendirent. Vingt-cinq soldats furent faits prisonniers, onze ennemis étaient morts et un grand nombre était blessé. La courageuse résistance du sergent Turner est un modèle à suivre pour ses camarades. Son héroïque leadership, sa détermination et son engagement courageux sont dignes des plus hauts faits militaires. »

Pour son action héroïque à Dahl, le sergent Turner fut récompensé posthume de la plus haute décoration de l'armée américaine, la Médaille d'Honneur du Congrès (CMH). Le sergent Turner fut tué en territoire allemand en face de Wallendorf.

Am 16. Dezember 1944, startete Hitler einen Überraschungsangriff gegen die amerikanischen Verteidiger, die von Echternach (L) bis Monschau (D) die Front hielten. Es war die Ardennen-Offensive, die zum Ziel hatte, den Hafen von Antwerpen (B) binnen drei Tagen zu erobern, die amerikanischen und britischen Truppen zu spalten und die Alliierten zu einem Waffenstillstand zu zwingen.

General George S. Patton's Dritte Armee griff am 22. Dezember 1944 die deutsche südliche Flanke an: Die 4. Panzerdivision sprengte den Ring um Bastnach/Bastogne (B) am 26. Dezember 1944. Die 26. Infanteriedivision hatte Wiltz als Objektiv, das sie am 21. Januar 1945 befreite. Die 80. Infanteriedivision befreite Ettelbrück schon am 25. Dezember 1944. Die 5. Infanteriedivision operierte im Raum Echternach und befreite die Abteistadt am 26. Dezember 1944, bevor sie ab dem 18. Januar 1945 die Dörfer nördlich von Diekirch befreite.

Am 6. Januar 1945, befreiten die Soldaten des 3. Bataillons, 319. Regiment, 80. Infanteriedivision, das Dorf Dahl. In den nächsten Tagen griffen die Deutschen das zerstörte Dorf immer wieder ohne Erfolg an.

Staff Sergeant Day G. TURNER

CMH-Zitat „Am 8. Januar 1945, befehligte Sgt. Turner eine 9-Mann-Abteilung, die zwischen Dahl und Nocher eine wichtige Flankenstellung halten sollte. Als eine übermäßige Anzahl Feinde unter Artillerie-, Mörser- und Werferfeuer vorrückte, zog er seine Abteilung in das nahe gelegene Aastert Gehöft zurück, entschlossen, es bis zum letzten Mann zu verteidigen. Der Feind griff immer wieder an und wurde unter schweren Verlusten zurückgeschlagen. Mit Panzerunterstützung konnten die Angreifer eindringen, doch der unerschrockene Sergeant weigerte sich, sich zu ergeben, obschon fünf seiner Männer verwundet waren und einer getötet worden war. Er warf einen brennenden Benzinkanister auf die erste Welle der Angreifer, um sie zu verteiben und kämpfte hartnäckig Mann gegen Mann von Raum zu Raum weiter. Er warf Handgranaten, erstach mit dem Bajonett zwei Deutsche, die zu einer Tür hereinstürmten und kämpfte mit den Waffen der Gegner weiter, als seine Munition zu Ende ging. Der wilde Kampf dauerte vier Stunden und schließlich, als nur noch drei seiner Männer unverletzt waren, ergab sich der Feind. Fünfundzwanzig Soldaten wurden gefangen genommen, elf waren tot und eine große Zahl der Angreifer war verwundet. Sgt. Turners Widerstand wird ein vorbildliches Beispiel für seine Kameraden bleiben. Seine heldenhafte Führung, seine Entschlossenheit und sein ausgesprochenes Pflichtbewusstsein zählen zu den höchsten militärischen Werten."

Sgt. Turner wurde postum für seine Heldentat mit der höchsten Auszeichnung der US-Armee, der Congressional Medal of Honor (CMH), ausgezeichnet. Er fiel am 8. Februar 1945 auf deutschem Boden, gegenüber Wallendorf, beim Ansturm auf einen Bunker des Westwalls.

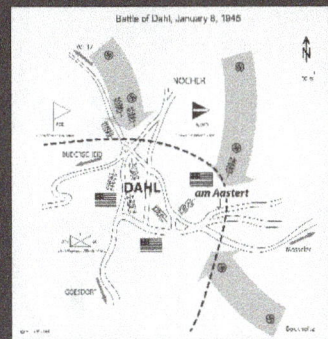

Battle of Dahl, January 8, 1945

Dahl

In honor of the 80th Infantry Division which fought in this area in January 1945.

Um Aale Wee

9644 Goesdorf

Latitude: 49.938936

Longitude: 5.972639

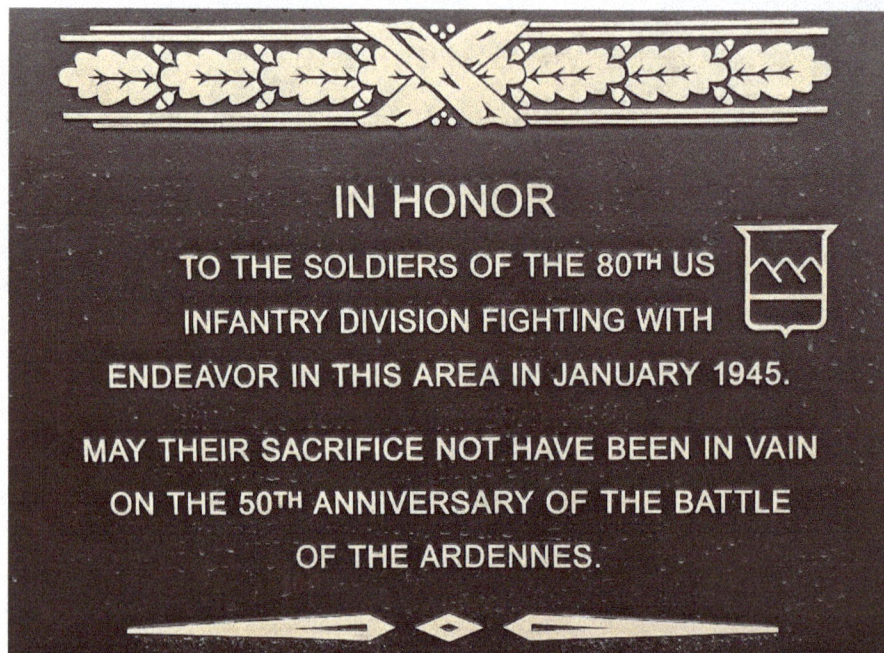

IN HONOR

TO THE SOLDIERS OF THE 80TH US
INFANTRY DIVISION FIGHTING WITH
ENDEAVOR IN THIS AREA IN JANUARY 1945.

MAY THEIR SACRIFICE NOT HAVE BEEN IN VAIN
ON THE 50TH ANNIVERSARY OF THE BATTLE
OF THE ARDENNES.

Dasbourg-Pont

Monument and information panel to the 249th Engineer Combat Battalion for construction of Bailey Bridge over the Our River, linking Luxembourg and Germany.

Monument is technically located in Germany at the end of the modern bridge but is included in this text since it is related to activities on the border between Luxembourg and Germany.

An der Brücke 2

54689 Dasburg (Germany)

Latitude: 50.04962

Longitude: 6.12689

★ ★ ★

We dedicate this Monument in recognition
of the valiant soldiers of the 249th U.S. Engineer Combat Battalion
General George S. Patton's 3rd Army.

Their sacrifices will never be forgotten.

Blacklions

U.S. Veterans, Friends, Luxembourg   Gemeinde Dasburg

Dasbourg-Pont Information Panel

At the end of January 1945, the Battle of the Bulge that started on December 16, 1944, came slowly to the end with the German forces being pushed back to the Germany-Luxembourg border. The Americans, especially General George S. Patton's 3rd Army, were preparing to enter Germany that was however heavily defended by bunkers and pillboxes along its border. This was the so-called "West Wall" or "Siegfried Line".

As all the bridges over the Our, Sauer and Moselle rivers were destroyed, the American engineer units had to built new ones. After crossing into Germany near Tintesmühle (L) on February 6, 1945, the US 6th Armored Division managed to make its way south and east and could enlarge its bridgehead into Germany. The town of Dasburg (D) was under American control by 5 p.m. on February 21, 1945. The same evening, the 25th Armored Engineer Battalion of the 6th Armored Division started building a footbridge over the Our River, which connected Luxembourg to Germany again. As this bridge could not hold any heavy loads, it was replaced by a Bailey Bridge on February 28, 1945 by C Company of the 249th Engineer Combat Battalion.

This unit had also been involved in the Battle of the Bulge, when its commander, Captain Arnold J. Cissna, was killed in action in Bilsdorf (L) on December 24, 1944. He was awarded the Distinguished Service Medal posthumously.
On March 22, 1945, the 249th ECB made a successful assault crossing over the Rhine River at Oppenheim (D),
After the war, they became a specialized unit for humanitarian services, such as restoring electrical power to disaster areas.

Vers la fin janvier 1945, la Bataille des Ardennes, qui avait commencé le 16 décembre 1944, se terminait avec les troupes allemandes repoussées à la frontière germano-luxembourgeoise. Les Américains, particulièrement la 3e Armée du Général George S. Patton, étaient en train de préparer leur attaque contre le «Westwall» ou la «Ligne Siegfried», une ceinture fortifiée le long des frontières allemandes.

Tous les ponts des rivières Our, Sûre et Moselle ayant été détruits, les pionniers américains devaient en construire des nouveaux. Après que la 6e Division Blindée US avait franchi l'Our près de Tintesmühle (L) le 6 février 1945, elle a pu élargir sa tête de pont vers le sud et l'est sur le territoire allemand et le 21 février 1945, vers 17.00 heures, Dasburg (D) était sous son contrôle. Le même soir, le 25e Bataillon Pionnier Blindé de la 6e Division Blindée US a commencé à construire un pont au-dessus de l'Our, connectant l'Allemagne de nouveau au Luxembourg. Ce pont ne supportant pas de charges lourdes, il était remplacé le 28 février 1945 par un pont Bailey, qui fut construit par la Compagnie C du 249e Bataillon Pionnier de Combat.

Cette unité avait déjà été engagée dans la Bataille des Ardennes. Leur commandant, le capitaine Arnold J. Cissna, fut tué à Bilsdorf (L) le 24 décembre 1944 et fut décoré du «Distinguished Service Medal» posthume. Le 22 mars 1945, le 249e Bataillon Pionnier de Combat a lancé avec succès un assaut à travers le Rhin, à Oppenheim (D),
Après la guerre, cette unité s'est spécialisée dans les missions humanitaires, entre autres pour restaurer les infrastructures d'électricité dans les régions sinistrées.

Ende Januar 1945 kam die Ardennenoffensive, die am 16. Dezember 1944 begonnen hatte, allmählich zu Ende, als die deutschen Truppen wieder an die deutsch-luxemburgische Grenze zurückgedrängt wurden. Die Amerikaner, vor allem General George S. Pattons 3. Armee, bereiteten sich darauf vor, die deutsche Grenze zu überschreiten, die allerdings durch eine Unzahl an Bunkern stark verteidigt war. Dies war der sogenannte „Westwall" oder „Siegfriedlinie".

Da alle Brücken über die Flüsse Our, Sauer und Mosel zerstört waren, mussten amerikanische Pioniere neue bauen. Nachdem die 6. amerikanische Panzerdivision am 6. Februar 1945 bei Tintesmühle (L) über die Our setzte, konnte sie ihren Brückenkopf nach Süden und Osten hin erweitern und am 21. Februar 1945, gegen 17.00 Uhr, war Dasburg (D) in amerikanischer Hand. Noch an demselben Abend begann das 25. Panzerpionierbataillon der 6. US Panzerdivision eine Brücke über die Our zu bauen, die Deutschland wieder mit Luxemburg verband. Da diese Brücke jedoch keine schweren Lasten aushielt, wurde sie am 28. Februar 1945 durch eine Bailey Brücke ersetzt, die von der C Kompanie des 249. Pionierkampfbataillons errichtet wurde.

Diese Einheit war ebenfalls in der Ardennenoffensive aktiv gewesen. Ihr Anführer, Hauptmann Arnold J. Cissna, kam in Bilsdorf (L) am 24. Dezember 1944 ums Leben. Für seinen Einsatz erhielt er postum die „Distinguished Service Medal".
Am 22. März 1945 gelang dem 249. Pionierkampfbataillon ein erfolgreicher Sturmangriff über den Rhein bei Oppenheim (D).

Nach dem Krieg spezialisierte diese Einheit sich in der humanitären Hilfe, um beispielsweise in Katastrophengebieten Elektrizitätsinfrastrukturen wiederherzustellen.

© Dan Roland

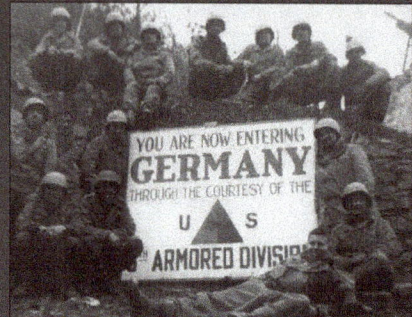
DASBURG: Soldiers of the 249th Engineer Combat Battalion pose around a sign erected by the 6th Armored Division on a destroyed pillbox on the German side of the Our River.
Source: Saive

DASBURG: Sign located on the Bailey Bridge built by C Company, 249th Engineer Combat Battalion, on February 28, 1945.
Source: D. Saive

DASBURG: Bridge built by the 25th Armored Engineer Battalion, 6th Armored Division on February 21 to 22, 1945.
Source: US Army

DASBURG: 120ft double-double Bailey Bridge built by C Company, 249th Engineer Combat Battalion.
Source: US Army

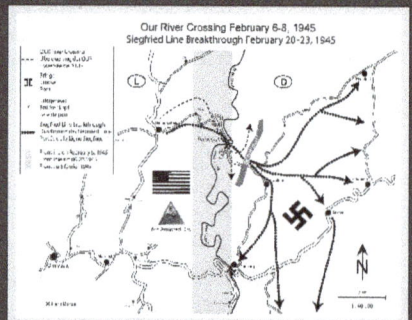
Our River Crossing February 6-8, 1945
Siegfried Line Breakthrough February 20-23, 1945

Diekirch

In gratitude to our American Liberators September 1944 – January 1945 from the citizens of Diekirch. There is also an information panel that outlines the evacuation of the civilian population of Diekirch on December 19-20, 1944. Captain Harry M. Kemp, 109th Infantry Regiment, 28th Infantry Division enabled the evacuation of +/-6000 civilians from Diekirch.

Avenue de la Gare 25

9233 Diekirch

Latitude: 49.8651

Longitude: 6.1586

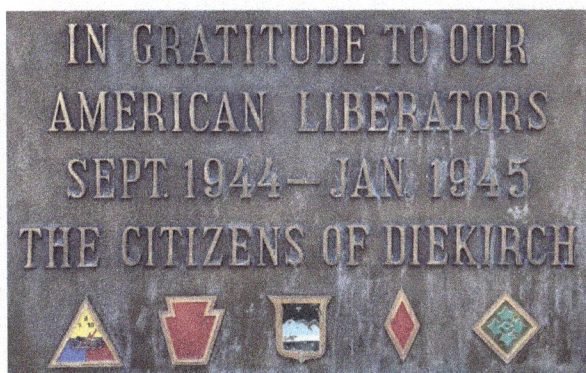

Information Panel Diekirch

THE EVACUATION OF THE CIVILIAN POPULATION OF DIEKIRCH DURING THE "BULGE" DEC 19/20, 1944

A tribute to Captain Harry M. Kemp and to the 109th US Infantry Regiment/28th US Infantry Division

On December 18, the initial "hold at all cost" order was lifted and the commander of the 109th Inf. Rgt., LTC Rudder, ordered the 3rd battalion from the "Our" sector towards the south banks of the "Sûre" to execute a delay fight thru Bettendorf and Gilsdorf towards Diekirch. 3rd Bn commander Major McCoy ordered the attached engineer platoon to destroy the bridges across the "Sûre" to "buy time". It was up to Captain Harry M. Kemp to transfer the battalion headquarters from Bettendorf back to Diekirch, into the centrally-located "Hôtel des Ardennes". On Dec. 19, 1944, Cpt. Kemp was directing the operations of the retreating front-line companies thru Diekirch, when the Mayor of Diekirch, Mr. Theis accompanied by Lt. Melchers, Luxembourg State police, came to draw his attention to the precarious fate of the +/- 6.000 local inhabitants and refugees.

Captain Harry M. Kemp, 1954
Photo : Archives Musée National
d'Histoire Militaire, Diekirch

Cpt. Kemp's orders from the regiment were to take the troops in a new position north of Diekirch they did not mention anything about civilians! Nonetheless, he took the initiative and received clearance from the division (General Norman "Dutch" Cota in Wiltz) to evacuate the town under his supervision! As he sent out Luxembourg police, resistance fighters and a few GIs to warn every household to gather before midnight in the courtyard of the Diekirch high-school, he quickly designed a plan that would not interfere with the military retreat: the road Diekirch-Ettelbruck was exclusively reserved for the troops. As the main Diekirch bridge had been blown-up in September 1944 by the retreating Germans, the only remaining way "out" would be the railroad bridge (today's "red" rope bridge), though it was already prepared for destruction.

Shortly after midnight, the civilians started getting across the bridge. The engineers held off on firing the demolition charges until all were safely on the other banks. Due to a miss fire, the bridge was only partially destroyed and later repaired by the Germans after capturing the town on December 20, 1944. But the civilian evacuation had worked – thanks to Cpt. Kemp's unselfish initiative, 97% of the population made it to safety and survived the war.

The Germans were able to keep the town until January 20, 1945, when Diekirch was re-liberated by units of the 5th US Infantry "Red Diamond" division.

Cpt. Harry Kemp was honored in 2004 by both the Luxembourg Government and by the City for saving many civilian lives. He became a honorary citizen of Diekirch. He passed away on Nov 19, 2006 in San Antonio, TX and is buried at Arlington cemetery.

Copyright: (RG/PV/SG-2018)

L'ÉVACUATION DE LA POPULATION CIVILE DE DIEKIRCH DURANT L'« OFFENSIVE DES ARDENNES » 19- 20 DÉCEMBRE 1944

En l'honneur du Capitaine Harry M. Kemp et du 109e régiment d'infanterie de la 28e division d'infanterie américaine

Le 18 décembre, l'ordre divisionnaire « Tenez à tout prix » fut levé et en vue de mettre sur pieds une nouvelle ligne de défense, le commandant du 109e régiment d'infanterie, le Lieutenant-colonel Rudder, ordonna au 3e bataillon de se retirer du secteur de l'Our en direction de Diekirch et de battre en retraite via la rive septentrionale de la Sûre et les villages de Bettendorf et Gilsdorf. Le commandant du 3e bataillon, le Major Jim McCoy, surveilla la destruction de tous les ponts sur la Sûre par ses propres pionniers, pour « gagner du temps ». Il incomba au Capitaine Harry M. Kemp, de retransférer le QG du bataillon de Bettendorf à Diekirch, dans l's Hôtel des Ardennes » situé en centre-ville. Le 19 décembre 1944, le Capitaine Kemp, fort occupé à diriger les opérations des compagnies du front à Diekirch, reçut la visite du maire de la ville, Monsieur Theis, accompagné du Lieutenant Melchers de la Gendarmerie luxembourgeoise, venus pour le rendre attentif au sort précaire des quelque 6.000 habitants et réfugiés de la ville.

Les ordres du Capitaine Kemp étaient de ramener le régiment dans une nouvelle position au nord de Diekirch ... la population civile n'était même pas mentionnée ! Peu importe, il prit lui-même l'initiative de pouvoir surveiller l'évacuation des civils, ce qui lui fut accordé par la division (dirigée par le Général Norman « Dutch » Cota établi à Wiltz). Il envoya des policiers et résistants luxembourgeois, flanqués de quelques GIs, enjoindre à chaque ménage de se réunir dans la cour du gymnase de Diekirch avant minuit. Lui-même conçut un plan génant en rien la retraite militaire : la route Diekirch-Ettelbruck restait réservée strictement aux troupes. Vu que le pont principal de Diekirch avait déjà été dynamité en septembre 1944 par la Wehrmacht, la seule issue fut le pont ferroviaire (de nos jours « le pont à ficelles rouge ») toujours debout, mais préparé à la destruction.

109e Infanterie
brass collar insignia

Label
"CIVES ARMA FERANT"

Peu après minuit, les habitants se ruèrent sur le pont. Les pionniers attendaient jusqu'à ce que tous aient atteint l'autre rive avant de le faire sauter. A cause d'un raté, le pont ne fut détruit qu'en partie et put être remis en état par les Allemands, après leur prise de la ville le 20 décembre. L'évacuation de la population civile fut couronnée de succès – grâce à l'initiative altruiste du Capitaine Kemp, 97% des citadins purent se mettre à l'abri et survivre à la guerre.

Les Allemands réussirent à tenir la ville jusqu'au 20 janvier 1945, avant que des unités de la 5e division d'infanterie américaine « Red Diamond » ne libèrent définitivement Diekirch.

En 2004, le Capitaine Harry Kemp fut honoré par le Gouvernement du Luxembourg et la Ville de Diekirch pour le sauvetage de nombreux civils. Nommé Citoyen d'Honneur de Diekirch, il décéda le 19 novembre 2006 à San Antonio, Texas. Il est enterré au cimetière militaire américain d'Arlington.

DIE EVAKUIERUNG DER ZIVILBE-VÖLKERUNG VON DIEKIRCH WÄH-REND DER „ARDENNENOFFENSIVE" 19./20. DEZEMBER 1944

Zu Ehren von Hauptmann Harry M. Kemp und des 109. Infanterieregiments der 28. US-Infanteriedivision

Am 18. Dezember wurde der Divisionsbefehl "Halten um jeden Preis" aufgehoben und der Kommandant des 109. U.S.-Infanterieregiments Rudder befahl dem 3. Bataillon, sich durch Rückzuggefechte aus dem Our-Sektor auf das südliche Ufer der Sauer, und dann durch Bettendorf und Gilsdorf bis nach Diekirch zurückzuziehen. Der Kommandant des 3. Bataillons, Major McCoy überwachte derweil die Zerstörung der Sauerbrücken durch die eigenen Pioniere, um "Zeit zu erkaufen". Es oblag Hauptmann Harry M. Kemp, das Bataillonshauptquartier von Bettendorf nach Diekirch, in das zentral gelegene „Hôtel des Ardennes" zurückzuversetzen. Am 19. Dezember 1944 war er dabei, die Operationen der Front-Kompanien durch Diekirch zu leiten, als der Bürgermeister der Stadt, Herr Theis, begleitet von Leutnant Melchers von der Luxemburger Gendarmerie, ihn aufsuchten, um seine Aufmerksamkeit auf das bronzlige Schicksal der rund 6.000 Einwohner und Flüchtlinge zu lenken.

„Bloody Bucket"
Insignie of the 28th US
Infantry Division

Die Befehle Hauptmann Kemps lauteten, das Regiment in eine neue Stellung nördlich von Diekirch zu bringen – die Zivilbevölkerung war hier nicht berücksichtigt! Nichts desto trotz übernahm er die Initiative und erhielt "grünes Licht" von der Division (unter General Norman „Dutch" Cota in Wiltz), die Stadt unter seiner Aufsicht zu evakuieren! So schickte er Luxemburger Polizisten, Widerstandskämpfer und einige GIs los, um jeden Haushalt zu veranlassen, sich um Mitternacht im Vorhof des Diekircher Gymnasiums einzufinden. Er selbst entwarf einen Plan, der den militärischen Rückzug nicht behindern würde – die Straße Diekirch-Ettelbruck blieb dabei ausschließlich den Truppen vorbehalten. Da die Hauptbrücke in Diekirch schon im September 1944 von den zurückflutenden Deutschen gesprengt worden war, war die schon seit der Sprengung vorbereitete Eisenbahnbrücke (die heutige „rote" Seilbrücke), der einzige Fluchtweg.

Kurz nach Mitternacht strömten die Einwohner zur Brücke. Die Pioniere warteten mit der Sprengung der Brücke bis Alle auf dem jenseitigen Ufer waren. Durch eine Fehlzündung konnte die Brücke nur teilweise zerstört und später von den Deutschen wieder repariert werden, nachdem die Stadt am 20. Dezember eingenommen worden war. Die Evakuierung der Zivilbevölkerung war ein Erfolg – dank Hauptmann Kemps selbstloser Initiative gelangten 97% der Einwohner in Sicherheit und überlebten den Krieg.

Die Deutschen konnten die Stadt bis zum 20. Januar 1945 halten, ehe Einheiten der 5. US "Red Diamond" Infanteriedivision Diekirch definitiv befreiten.

Hauptmann Harry Kemp wurde 2004 von der Luxemburger Regierung sowie der Stadt Diekirch für seine Rettung von zahlreichen Zivilisten geehrt. Er wurde Ehrenbürger der Stadt Diekirch und starb am 19. November 2006 in San Antonio Texas. Er liegt auf dem US-Friedhof in Arlington begraben.

For more information
Pour en savoir plus
Für weitere Informationen

Information Panel Diekirch

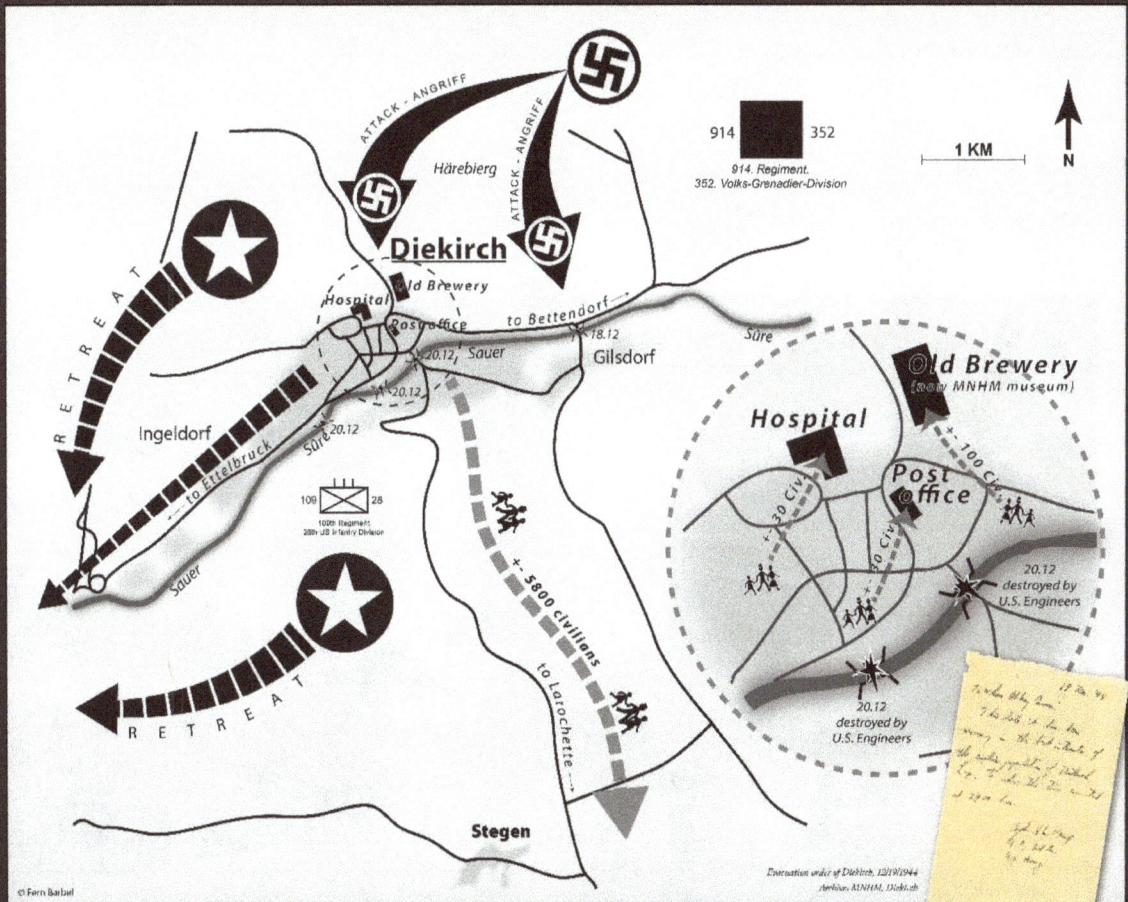

THE EVACUATION OF THE CIVILIAN POPULATION OF DIEKIRCH DURING THE "BULGE"

On December 16, 1944 by 05:30 a.m. the first German shells landed on the thinly-stretched American defense lines of the so-called "quiet sector", taking everybody by surprise. The 'Battle of the Bulge' had begun!

Between 16th–18th December U.S. soldiers tenaciously resisted the onslaught of overwhelming numbers of German attackers on the hills overlooking the Our. But it became clear that, eventually, they would be overrun.

Copyright: (RG/PVSG-2018)

L'ÉVACUATION DE LA POPULATION CIVILE DE DIEKIRCH DURANT L'« OFFENSIVE DES ARDENNES »

Le 16 décembre 1944 à 5h30 du matin, les premières grenades allemandes s'abattirent sur les lignes de défense clairsemées du « secteur calme », une surprise totale pour les Américains. L'offensive des Ardennes venait de commencer !

Entre le 16 et le 18 décembre, sur les hauts plateaux près de l'Our, les soldats américains résistèrent farouchement aux assauts des attaquants allemands en surnombre. Mais il était clair qu'ils allaient être débordés tôt ou tard.

DIE EVAKUIERUNG DER ZIVILBEVÖLKERUNG VON DIEKIRCH WÄHREND DER „ARDENNENOFFENSIVE"

Am 16. Dezember 1944 gegen 05:30 Uhr morgens explodierten im dünn besetzten „ruhigen Sektor" die ersten deutschen Granaten, zur totalen Überraschung der amerikanischen Verteidiger. Die Ardennenoffensive hatte begonnen!

Zwischen dem 16. und dem 18. Dezember wehrten die verbissen kämpfenden Amerikaner die Angriffe der zahlenmäßig haushoch überlegenen deutschen Angreifer auf den Anhöhen über der Our ab. Doch es war klar, dass sie ihre Stellungen nicht mehr lange halten würden.

For more information
Pour en savoir plus
Für weitere Informationen

Diekirch

Infopanel commemorating the 5th Infantry Division that recaptured of the greater Diekirch area after crossing the Sauer River on January 18, 1945. The intent of this offensive was to cut off the retreat of the German Army after Bastogne was liberated.

Route d'Ettelbruck

Diekirch

Located near foot bridge over Sauer

Latitude: 49.862011

Longitude: 6.150126

Information Panel at Sauer River Crossing

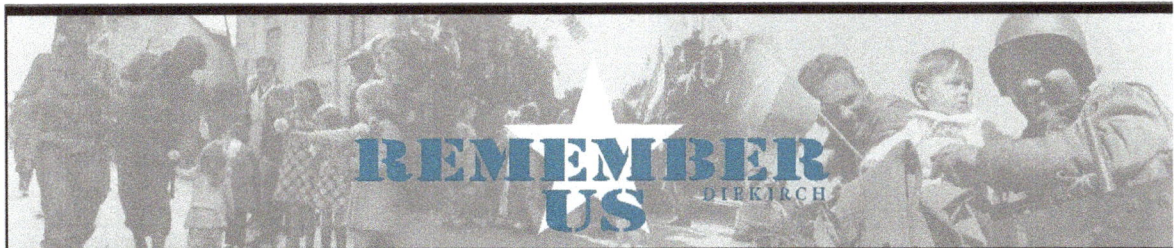

REMEMBER US DIEKIRCH

SAUER RIVER CROSSING
JANUARY 18, 1945
THE FINAL LIBERATION OF DIEKIRCH

After the U.S. retreat out of Diekirch, Dec. 20/21, 1944, the road lay open for the units of the 352nd German "Volksgrenadier division" in this sector. Around Christmas 1944, heavy snowfalls set in and the German advance came to a stop, while Lt. Gen. George S. Patton's Third Army counterattack from the south, in direction of Bastogne, gained momentum. By December 27, 1944 the remainder of the 352nd had retreated north of the "Sauer" river. This situation was to remain unchanged for almost 3 weeks.

The 5th U.S. Infantry Division recaptures the greater Diekirch area after crossing the "Sauer"

To deny another German attempt to cross the "Sauer" river, Gen. Patton shifted his own 5th U.S. Infantry "Red Diamond" division from the Echternach – Haller line to the Ingeldorf – Diekirch – Bettendorf sector, tying in with the 4th U.S. Infantry "Ivy Leaves" division to form a consistent line of defense. The front stabilized, as the temperatures dropped to –15 degrees Celsius daytime and often to –22 at nighttime.

After the complete liberation of Bastogne, the 5th U.S. Infantry division was tasked with crossing the "Sauer" to cut the enemy off from retreating to Germany.

Frontline troops developed an extensive night-time patrolling activity, clad in makeshift white camouflage suits supplied by the grateful local Luxembourg population. Intelligence revealed that Ingeldorf, Diekirch and Bettendorf were the key German strongholds.

*Hans Hennig,
8./KPGR/911, 352 VGD
(via QR-CODE)*

In January 1945, 18-year-old Hennig witnessed the fighting as a soldier in Diekirch.

LA TRAVERSÉE DE LA SÛRE
18 JANVIER 1945
LA LIBÉRATION FINALE DE DIEKIRCH

Du 20 au 21 décembre 1944, les troupes américaines furent contraintes de se retirer de Diekirch, laissant la voie libre aux unités allemandes de la 352e division « Volksgrenadier ». Mais autour de Noël, de fortes chutes de neige entravèrent l'avancée allemande, tandis que la contre-attaque lancée du sud par la troisième armée du Lieutenant–Général George S. Patton prenait de l'élan. Le 27 décembre 1944, les éléments restants de la 352e division se retirèrent au nord de la Sûre. Cette situation restait inchangée pendant trois semaines.

La 5e division d'infanterie américaine reconquiert la région de Diekirch après sa traversée de la Sûre

Pour éviter une nouvelle traversée allemande de la Sûre, le Général Patton décida de déplacer sa 5e division d'infanterie, connue sous le nom de « Red Diamond division », de la ligne Echternach – Haller vers le secteur d'Ingeldorf – Diekirch – Bettendorf, la liant ainsi à la 4e division d'infanterie « Ivy Leaves », en vue de former une robuste ligne de défense. La ligne de front commença à se figer tandis que les températures chutèrent davantage, atteignant les – 15° le jour et les – 22° la nuit. Après la libération de Bastogne, la 5e division d'infanterie fut chargée de traverser la Sûre pour contrecarrer toute retraite ennemie vers l'Allemagne.

Vêtus d'habits de camouflage blanc improvisés, fournis par la population locale reconnaissante, les troupes américaines de première ligne développèrent une intense activité de patrouilles nocturnes. Grâce au travail de reconnaissance, on identifia Ingeldorf, Diekirch et Bettendorf comme points névralgiques de la défense allemande.

DIE SAUER-ÜBERQUERUNG
18. JANUAR 1945
DIE ENDGÜLTIGE BEFREIUNG DIEKIRCHS

Am 20. und 21. Dezember 1944 mussten die amerikanischen Truppen den Rückzug aus Diekirch antreten und das Feld den deutschen Truppen der 352. Volksgrenadierdivision überlassen. Doch um Weihnachten behinderten starke Schneefälle den deutschen Vormarsch, während der aus südlicher Richtung erfolgende Gegenangriff der 3. U.S.-Armee unter Generalleutnant George S. Patton an Schwung gewann. Am 27. Dezember 1944 zogen sich die Reste der 352. Division auf die Nordseite der Sauer zurück. Diese Ausgangslage blieb während drei Wochen unverändert.

Die 5. U.S. Infanteriedivision erobert nach der Sauerüberquerung Diekirch und Umgebung zurück

Um ein weiteres deutsches Vordringen über die Sauer zu verhindern, entschied sich General Patton dazu, seine unter dem Namen „Red Diamond" bekannte 5. Infanteriedivision von der Linie Echternach-Haller in den Sektor Ingeldorf – Diekirch – Bettendorf zu verlegen und so die Verbindung zur 4. Infanteriedivision „Ivy Leaves" herzustellen und eine solide Verteidigungslage zu gewährleisten. Die Frontlinie begann sich zu stabilisieren und die Temperaturen fielen weiter unter den Gefrierpunkt, bis zu -15° tagsüber und -22° in der Nacht.

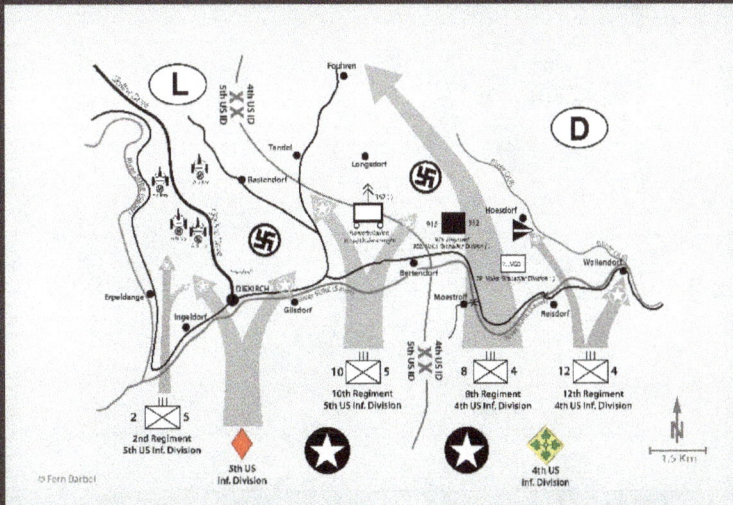

Nach der Befreiung der Stadt Bastogne wurde die 5. Infanteriedivision damit beauftragt, die Sauer zu überqueren und jeglichen deutschen Rückzug zur Reichsgrenze abzuschneiden.

In aus von einheimischer Bevölkerung gestiftetem Material improvisierten weißen Tarnkleidern unternahmen die amerikanischen Frontkompanien zahlreiche nächtliche Erkundungsgänge. Dadurch wurde in Erfahrung gebracht, dass Ingeldorf, Diekirch und Bettendorf die wesentlichen Bollwerke des deutschen Verteidigungsgürtels darstellten.

© Rol Gaul / Philippe Visot
Sources: NARA/Arch: MNHM – COLLECT: Rol Gaul

For more information
Pour en savoir plus
Für weitere Informationen

Information Panel at Sauer River Crossing

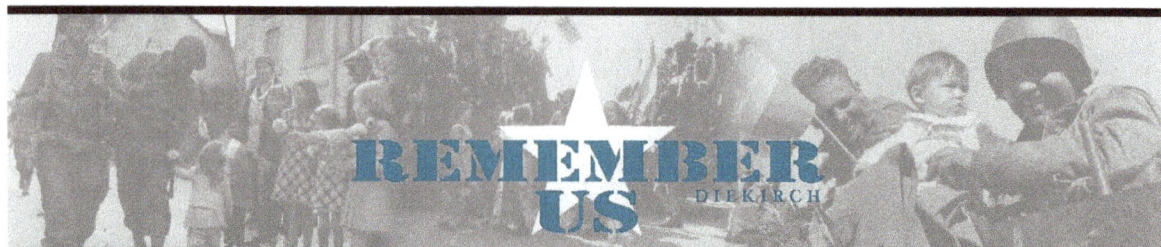

REMEMBER US
DIEKIRCH

SAUER RIVER CROSSING
JANUARY 18, 1945
THE FINAL LIBERATION OF DIEKIRCH

The vital crossing of the "Sauer" jumped off at 03:00 a.m. on January 18, 1945, without any artillery preparation. Two sectors were targeted. West: Ingeldorf – Diekirch and East: Diekirch – Bettendorf.

In the east, the frontline companies crossed by means of plywood assault boats and floating bridges without being detected. They were able to proceed directly to the high ground northwest of Bettendorf. At the same time, more to the west, reinforcements crossed the river near Ingeldorf, after two failed attempts to erect an assault bridge. Frontline companies were able to capture Erpeldange and advance towards "Friedhaff" despite knee-deep snow. Meanwhile, U.S. units in Diekirch were less lucky. Supported by U.S. 105 mm howitzers, 2 companies met strong resistance in the cemetery and station areas, while German artillery kept pounding the sector. Until later in the afternoon, supporting tanks had not managed to get across the Sauer. Also, German minefields in the area slowed down the U.S. advance.

Sauer River Crossing by US troops in the vicinity of Bettendorf over a floating bridge on January, 1945

Des troupes américaines traversent la rivière Sûre par un pont flottant aux alentours de Bettendorf en janvier 1945.

Amerikanische Truppen überqueren die Sauer über eine Ponton-Brücke in der Gegend von Bettendorf im Januar 1945

By nightfall of January 18, 1945, the 5th Division's bridgehead was about 1,5 KM deep, along a 6,5 KM front. In the end, the enemy had been caught by surprise. Especially in the Bettendorf sector, the Germans had been completely overrun, whole platoons being taken prisoner. Despite sporadic enemy fire and the handicapping elements of snow and mud, engineers were able to throw across 1 treadway bridge, 2 "class 40" bridges, 2 assault boat bridges and 2 footbridges that enabled some Sherman tanks and M-10 tank destroyers to cross the "Sauer".

NB: It is said that Gen. Patton swam thru the icy Sauer river at Bettendorf to inspire his men .... that it can be done! (see article and cartoon on QR-CODE)

© Rol Gaul / Philippe Victor
Sources: NARA/Arch. MNHM – COLLECT.: Rol Gaul

LA TRAVERSÉE DE LA SÛRE
18 JANVIER 1945
LA LIBÉRATION FINALE DE DIEKIRCH

La traversée cruciale de la Sûre fut lancée le 18 janvier 1945 à 3:00 heures du matin, sans aucune préparation d'artillerie. Deux axes furent visés, à savoir Ingeldorf – Diekirch à l'ouest et Diekirch – Bettendorf à l'est.

À l'est, les compagnies d'assaut réussirent à traverser la rivière à l'aide de barques d'assaut en bois et de ponts flottants, leur permettant de s'emparer rapidement des hauteurs au nord-est de Bettendorf, sans être détectées. Au même moment, plus à l'ouest, des renforts traversèrent le fleuve près d'Ingeldorf après avoir échoué à deux reprises d'ériger un pont d'assaut. Malgré une couche de neige à hauteur de genou, les compagnies du front parvinrent à prendre Erpeldange avant d'avancer en direction du « Friedhaff ».

Entretemps, les unités américaines engagées à Diekirch n'étaient pas si chanceuses. Épaulées par des canons d'artillerie de 105mm, deux compagnies américaines durent faire face à une résistance ennemie acharnée près du cimetière et de la gare, sous un feu nourri d'artillerie, de mortiers et de mitrailleuses allemandes. Jusqu'en fin d'après-midi, le support blindé n'arriva pas à franchir la Sûre et l'avancée fut compromise par des champs de mines allemands.

Le 18 janvier 1945, à la tombée de la nuit, la tête de pont de la 5e division d'infanterie américaine s'enfonça sur environ un kilomètre et demi le long d'une ligne de front de 6,5 km. En fin de compte, les Allemands furent pris par surprise, notamment dans le secteur de Bettendorf, où ils furent complètement submergés par les assaillants, à tel point que des pelotons entiers furent capturés. Malgré des sursauts de résistance ennemie et les aléas gênants de la neige et de la boue, le génie de combat réussit à établir un pont en acier, deux ponts « class 40 », deux pontons flottants et deux passerelles, permettant à plusieurs chars Sherman et chasseurs de char M10 de traverser la Sûre.

NB : On raconte que le Général Patton aurait traversé à la nage l'eau glacée de la Sûre près de Bettendorf pour motiver ses soldats et démontrer que c'était faisable ! (voir l'article et le croquis sur le QR-Code)

DIE SAUER-ÜBERQUERUNG
18. JANUAR 1945
DIE ENDGÜLTIGE BEFREIUNG DIEKIRCHS

Die entscheidende Überquerung der Sauer erfolgte ohne vorherige Artillerieunterstützung am 18. Januar 1945 um 3 Uhr morgens. Zwei Sektoren – Ingeldorf/Diekirch im Westen und Diekirch/Bettendorf im Osten - wurden angegriffen.

Die östlichen Sturmkompanien stießen mithilfe hölzerner Sturmboote und Pontonbrücken unentdeckt über den Fluss vor und bemannten rasch die Anhöhen nordwestlich Bettendorf. Gleichzeitig überquerte Verstärkung den Fluss im Westen bei Ingeldorf, nachdem zwei Versuche eine Sturmbrücke zu errichten, gescheitert waren. Trotz einer knietiefen Schneeschicht eroberten die Fronteinheiten Erpeldingen und marschierten weiter in Richtung „Friedhaff". Währenddessen hatten die in Diekirch kämpfenden amerikanischen Einheiten weniger Glück. Durch 105mm-Haubitzen unterstützt, trafen zwei amerikanische Kompanien in der Nähe des Friedhofs und des Bahnhofs auf verbitterten feindlichen Widerstand und waren dabei heftigem Artillerie-, Mörser- und Maschinengewehrfeuer ausgesetzt. Bis zum Nachmittag erreichte sie keine Panzerunterstützung und der Vormarsch kam durch deutsche Minenfelder ins Stocken.

Am 18. Januar 1945 bei Einbruch der Dämmerung, entsprach der Brückenkopf der 5. Infanteriedivision einer Tiefe von ungefähr 1,5 km entlang einer Frontlinie von 6,5 km. Die Deutschen waren überrascht worden, besonders im Raum Bettendorf, wo sie derart überrumpelt wurden, dass ganze Schützenzüge in Gefangenschaft gerieten. Trotz sporadisch aufflammendem feindlichen Widerstand und den erschwerenden Faktoren Schnee und Matsch, gelang es den U.S.-Pionieren, eine Stahlträgerbrücke, zwei „class 40" Brücken, zwei Pontonbrücken sowie zwei Stege zu errichten, welche mehreren Sherman-Panzern und M10 Jagdpanzern die Überquerung ermöglichten.

NB: Man erzählt, General Patton habe die eisige Sauer bei Bettendorf schwimmend durchquert, um seine Soldaten anzufeuern und zu zeigen, dass es „machbar" sei! (siehe Artikel und Cartoon im QR-Code)

For more Information
Pour en savoir plus
Für weitere Informationen

Diekirch – Caserne "Grand-Duc Jean" Härebierg (Luxembourg Army Barracks)

Dedicated to the Sauer River crossing on January 18, 1945 by the 5th Infantry Division. The panel overlooks the Sauer River and details the tactical situation in January 1945.

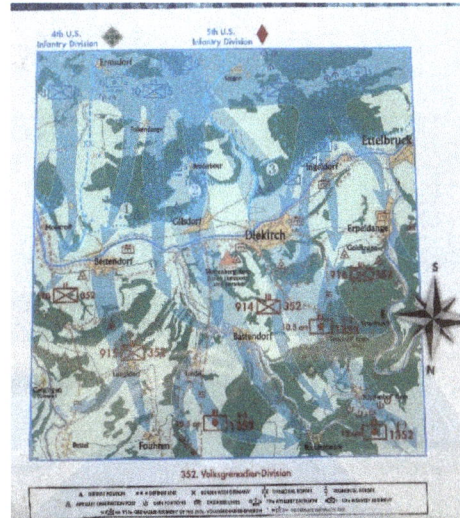

Caserne de l'Armée Herrenberg

9243 Diekirch

Latitude: 49.8733

Longitude: 6.1707

This information panel is located on a Luxembourg Army Barracks and is not open to the general public

Information Panel at Diekirch Caserne

"Sauer river Crossing" – January 18, 1945 and ensuing operations

The pleasant "Sauer" or "Sûre" river valley unfolding in front of you and visible from Erpeldange flat on the right to Bettendorf (left), was the site of a fierce battle at sub-zero temperatures in late January 1945.

The detailed map with tactical signs then present, reflects the situation of the January 18-21, 1945 "Sauer" river crossing operations in this sector by the 5th U.S. Infantry division, known as the "Red Diamond" division, commanded by Major-General Stafford LeRoy Irwin.

THE 5TH U.S. INFANTRY DIVISION RECAPTURES THE GREATER DIEKIRCH AREA AFTER CROSSING THE "SAUER"

NB: The exact location where you stand now, was back then an abandoned farmhouse with a German field observation post of a firing battery of four 15 cm howitzers and several 10.5 cm guns of the 1352nd artillery regiment of the 352–VGD, next to a battery of several 22cm "Nebelwerfer" (rocket projectors) around "Kippenhof" barracks, was manned day and night by artillery observers to scan the "Sauer" valley and its south banks and ridge.

After initial tactical successes and territorial gains during the December 16-22, 1944 time frame on the southern shoulder of the "Ardennes" offensive ("The Bulge"), the German units in this sector – primarily the 352– "Volksgrenadier Division" – were stopped in their advance and had to retreat to the northern banks of the "Sauer" river.

The pre-war "Herrenberg" farm

As Gen. Patton's Third Army was gaining momentum in its drive north to "rescue" Bastogne, part of his efforts focused on the southern shoulder to contain the enemy in this sector. To deny any further German attempt to cross the "Sauer" river, Gen. Patton after his entire Third Army had completed the move from the French-German border by December 22, 1944, shifted its own 5th U.S. Infantry "Red Diamond" division from the Echternach sector.

JANUARY 18, 1945

Early in the morning of 18 January, 1945, the 10th infantry regiment's 2nd and 3rd battalions jumped off, the 3rd being able to cross by means of plywood assault boats and floating infantry bridges without being detected by the Germans. Not having been detected it was able to proceed directly to high ground northwest of Bettendorf, and started patrolling to contact the 4th Division "by lanes" to its right.

The 2nd battalion crossed below Diekirch against mounting resistance, but cleared the strip of "Rue Clairefontaine" and moved on to occupy high ground. The 1st battalion crossed at mid-morning, moving out to relieve a company of the 2nd battalion at Gilsdorf, continuing mopping-up operations, tying the same time under sporadic German artillery, mortar, and rocket projector fire, the 2nd infantry regiment's own 2nd battalion crossed the river close by Diekirch after two columns to erect an assault bridge had been unsuccessful. Finally, a third attempt, in collaboration with the engineers of "B" company, 7th Engineer battalion was successful – as the 2nd battalion crossed over and immediately marched out on forward in northerly direction, captured Erpeldange. At this time, the 1st battalion was able to push hill before moving continuing to sustain and surrounded parts of the "Goldknapp"...

JANUARY 19, 1945

On the second day of the "crossing" – January 19, 1945 – the beachhead in the 10th infantry regiment's sector was extended to some 5 KMs in depth – very near the first objective, Bastendorf, including YOUR PRESENT STAND POINT on the "Herrenberg" farmhouse. German artillery observation near the Bastendorf reinforced well, continued its push while maintaining contact with the 4th Division on its right. The 2nd regiment took Bastendorf. On that same day, the 2nd regiment's 1st battalion got into vigorous exchange of fire at the "Friedbusch" on the "Erpeldinger Keppchen". "Two other" advance guard of the 2nd battalion had reported stiff enemy resistance around "Friedhof". After the 3rd battalion had cleaned out Diekirch, it was first called back to "Ingeldorf" as a reserve unit and some of its sub-units had to be detailed to U.Col Blabelfeld's 1st battalion. Around 3:00 p.m. the "Kippenhof" farm's "objective 12 – the "Kippenhof farm" in a coordinated attack German resistance became so strong that no control between the two units was possible.

JANUARY 20, 1945

January 20, 1945 the 1st battalion was the first to move in direction of "Kippenhof", where it ran into a local counterattack. In addition, numerous well dug-in German positions in the surrounding woods made progress without local support almost impossible. German artillery fire called for numerous losses in the 1st battalion rows. The 2nd battalion, although still not able to make contact...

OPERATION PREPARATIONS

To carefully plan this critical mission, the division's own frontline regiments, the 2nd, the 10th and the 11th, developed an extensive nighttime patrolling activity, clad in makeshift white camouflage suits consisting of bed sheets, nighttime, tabloth and so on, supplied by the grateful local Luxembourg population. Of those numerous I&R (intelligence and reconnaissance) patrols, one, the so-called "Bastendorf Raid", directed by the regiment's executive officer, Lt.Col. William M. Breckinridge, and carried out on January 10, 1945, whom on his NCO with valuable documents on him, should be highlighted. Intelligence gleaned from all those missions revealed that Ingeldorf, Diekirch and Bettendorf appeared to be the key locations of German defensive positions.

According to higher headquarters' planning, the vital crossing of the "Sauer" was to be a nighttime action, jumping off at 03:00 a.m. on January 18, 1945 without any artillery preparation (not to take away the element of surprise), and compelling the 2nd Dienbach (Rue Clairefontaine) – parts of Diekirch) and the 10th Infantry regiment (east of Diekirch held in reserve) to move on a 2-hour call. Initial support elements for this mission included the division's own 7 Engineer battalion (ferry the infantry cross the river), as well as the 50–, 46–, 19th Field artillery battalions (105 mm howitzer) the 21st 803rd Tank Destroyer battalion (3 inch gun M-10 tank destroyers), as well as company "C" of Third Army's own 91st chemical mortar (4.2" inch mortar) battalion.

By nightfall of January 18, 1945, the overall 5th Division's bridgehead was about 1.5 KM deep with a 6.5 KM front. Each regiment had crossed all its battalions, along with the 737th Tank battalion and tank destroyers from the 803rd Tank Destroyer battalion. The enemy had been caught by surprise, especially, the German 91 Stn regiment in the Bastendorf sector had been overrun, whole platoons being taken prisoner that claimed they never had a chance to fire a shot. Finally, the beachhead, but it seemed to have a lack of ammunition and the locomotion for their guns, so important to hitting moving targets. Despite the enemy fire and its handicapping elements of snow and mud, temporary bridges were able to throw across one treadway bridge, two class 40 bridges, two assault boat bridges and two foot bridges enabling some Sherman tanks and M-10 tank destroyers to cross the icy waters of the "Sauer river".

One of the many floating bridges across the Sauer river

On January 28, 1945, all assigned regimental objectives were achieved, and plans were formed by the division to establish a defensive line and outposts overlooking the Our river, as the 2nd Infantry regiment was ordered to relieve the exhausted elements of the 10th and 11th regiments. After a few days of much needed rest, the 5th Infantry Division's sector was gradually shifted at the beginning of February 1945 to take part in General Patton's Third army's strike across the "Sauer" river on an 8 KM front northeast of Echternach, thus marking the beginning of the "Rhineland campaign", on February 7, 1945.

Sauer: a prisoner of war was killed —

A dramatic recreation of the "Sauer river crossing" in form of a life-size diorama can be seen at the National Museum of Military History in Diekirch (www.nat-military-museum.lu)

In summary, the "Red Diamond" had thus crossed the "Sauer" twice – on January 18, 1945 in the Diekirch area and on February 7-10, 1945 operations resulted in a total of approximately 2,600 Battle and non-battle – due to extreme cold casualties!

"… How human beings could endure this continuous fighting at sub-zero temperatures is still beyond my comprehension."

(General George S. Patton, Jr. – ex- 'War as I knew it', 1945)

5th "Red Diamond" Infantry Division's staffing pattern:

C.O. 5th Infantry Division: Major General Stafford LeRoy IRWIN

C.O. 2nd Infantry Regiment: Col. Worrell ROFFE
  C.O. 1st Bn. Lt.Col. William BLAKEFIELD
  C.O. 2nd Bn. Lt.Col. Chester BALL
  C.O. 3rd Bn. Lt.Col. Robert CONNOR

C.O. 10th Infantry Regiment: Col. Robert P. BELL
  Exec. Off. Lt.Col William M. BRECKINRIDGE
  C.O. 1st Bn. Lt.Col. Frank LANGFITT, later Maj. Stanley HAYS
  C.O. 2nd Bn. Lt.Col. Harris C. WALKER
  C.O. 3rd Bn. Lt.Col. Alden P SHIPLEY, later Maj. Wilfred HAUGHEY

German 352nd Volksgrenadierdivision (VGD) staffing pattern:

C.O. Generalmajor Otto SCHMIDT (wounded Dec. 20, 1944); thereafter Generalmajor BAZING)

C.O. 914 G.R.: Major Theodor von LIECKEN
  C.O. 1st Bn. Hauptmann Siegfried KRAUSE
  C.O. 2nd Bn. Hauptmann Rudolf SCHNEIDER

C.O. 915 G.R.: Oberstleutnant Johannes DRAWE (wounded Dec. 17, 1944); thereafter Major HOFMEISTER
  C.O. 1st Bn. Hauptmann Behrend BERNARD
  C.O. 2nd Bn. Hauptmann Heinrich KOENIG

C.O. 916 G.R.: Major Von CIEGEREN (+ end of January, 1945)
  C.O. 1st Bn. Hauptmann Herbert KRUEGER
  C.O. 2nd Bn. Hauptmann der Reserve Hans SUCK (wounded Dec. 22, 1944); thereafter Maj. Friedrich SCHUBRING-GIESE

Differdange

Thanks from the people of Differdange to the Allied soldiers who liberated the town on September 10, 1944

Avenue Charlotte 40

4530 Differdange

Latitude: 49.5208489

Longitude: 5.8876368

Inscription reads: 10 September 1944. Our hearts breathe, our country is free! Differdange thinks back on this historic day and says thank you to the Allied soldiers who have given us back our freedom

Dillingen

In honor of the 80th Infantry Division which crossed the flooded Sauer river here on February 7, 1945 under intense enemy fire. This crossing marked the beginning of the Rhineland Campaign.

Route de Reisdorf 9

6350 Beaufort

Latitude: 49.8571

Longitude: 6.3159

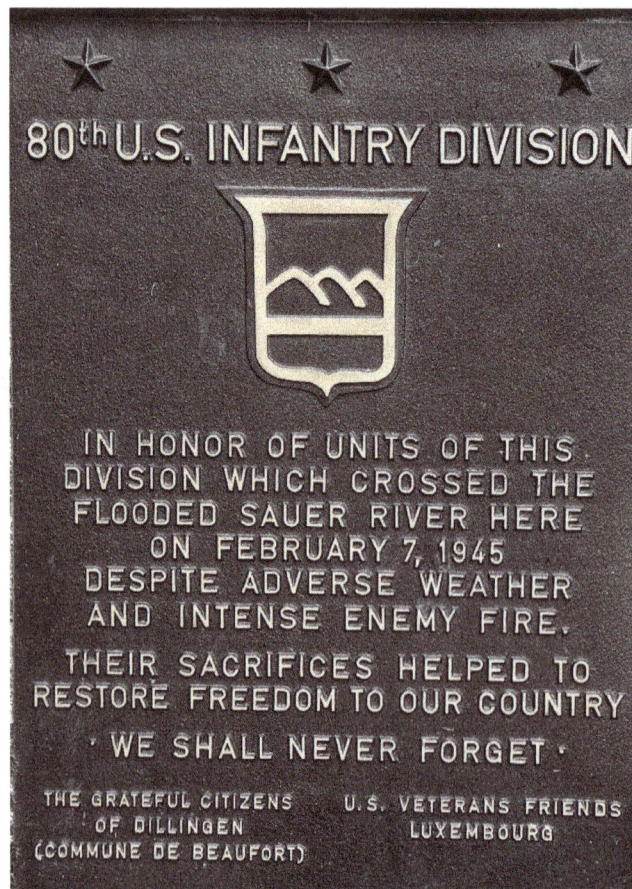

80th U.S. INFANTRY DIVISION

IN HONOR OF UNITS OF THIS
DIVISION WHICH CROSSED THE
FLOODED SAUER RIVER HERE
ON FEBRUARY 7, 1945
DESPITE ADVERSE WEATHER
AND INTENSE ENEMY FIRE.

THEIR SACRIFICES HELPED TO
RESTORE FREEDOM TO OUR COUNTRY

· WE SHALL NEVER FORGET ·

THE GRATEFUL CITIZENS   U.S. VETERANS FRIENDS
OF DILLINGEN   LUXEMBOURG
(COMMUNE DE BEAUFORT)

Dudelange

Liberation of Diddeléng (Dudelange) by the 3rd US Army

Route de Volmerange 145

3593 Dudelingen

Latitude: 49.4608

Longitude: 6.0810

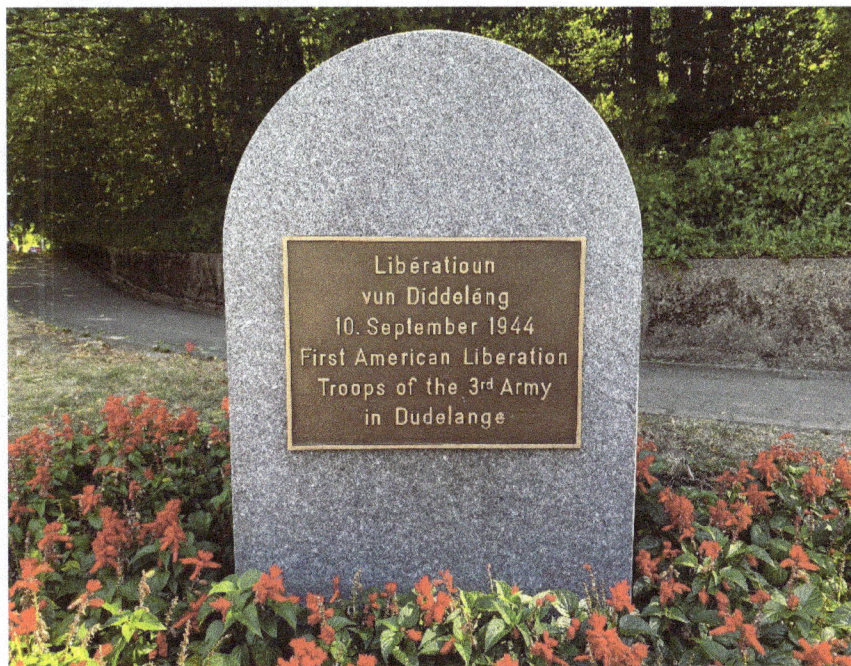

Echternach

Dedicated to the soldiers of the 83rd, 4th and 5th Infantry Division who liberated the city of Echternach October – December 1944 and to the 76th Infantry Division which crossed the river Sauer here on February 7, 1945, ending the Nazi oppression of our country.

By river, near large parking lot

6474 Echternach

Latitude: 49.8161

Longitude: 6.4167

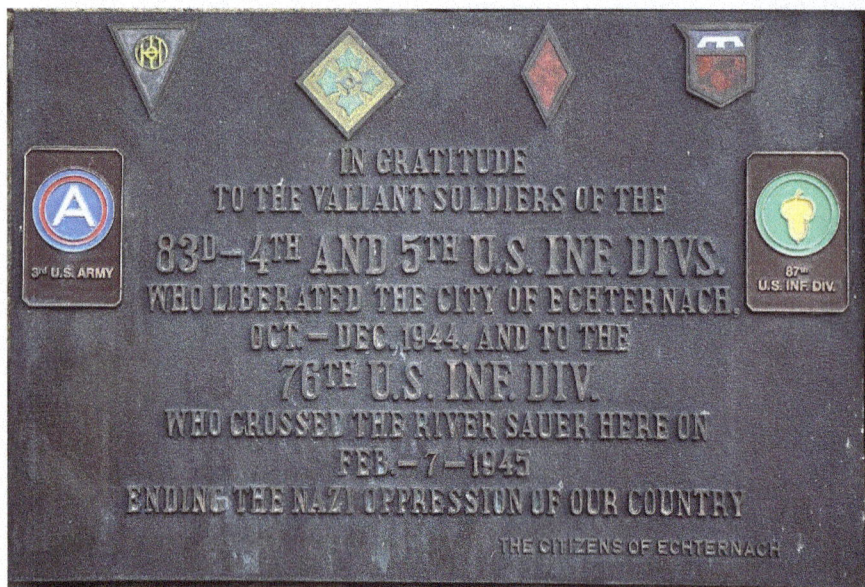

Echternach

During the Battle of the Bulge, in December 1944, this place was defended by soldiers of E Company, 12th Regiment, 4th Infantry Division.

Rue André Duchscher 17

6434 Echternach

Latitude: 49.8121

Longitude: 6.4174

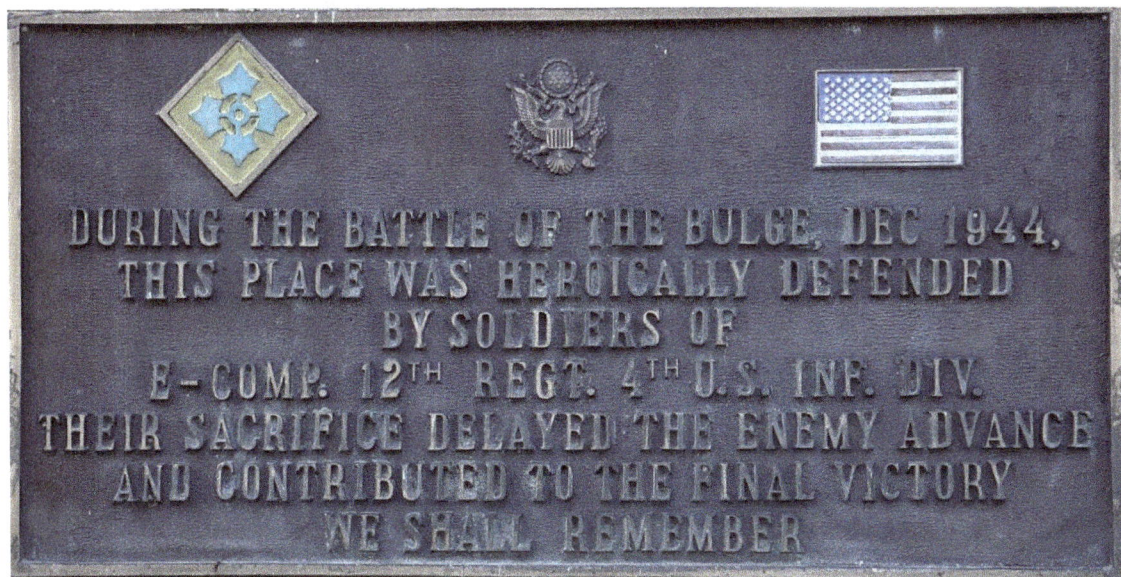

DURING THE BATTLE OF THE BULGE, DEC 1944,
THIS PLACE WAS HEROICALLY DEFENDED
BY SOLDIERS OF
E-COMP. 12TH REGT. 4TH U.S. INF. DIV.
THEIR SACRIFICE DELAYED THE ENEMY ADVANCE
AND CONTRIBUTED TO THE FINAL VICTORY
WE SHALL REMEMBER

Echternach – Lauterborn

Plaque to G company, 12th Infantry Regiment, 4th Infantry Division who defended the mill from 16 to 20 December 1944. Plaque is located on the side of the *Moulin Dieschbourg* mill building.

Moulin Dieschbourg

1, Lauterborn

6562 Echternach-Lauterborn

Latitude: 49.80211

Longitude 6.39749

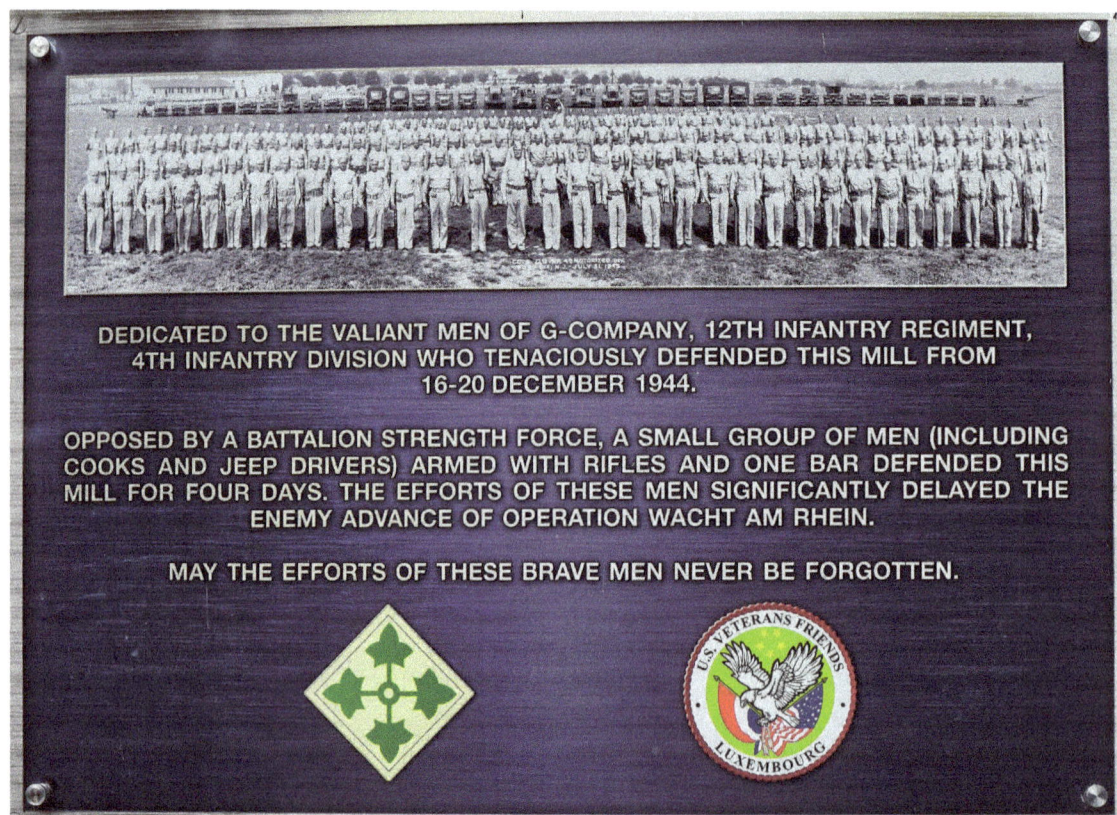

Plaque photo by Terry Farrelly

Echternach - Lauterborn

In remembrance of the valiant soldiers who defended this Hill – 313 in December 1944, thus delaying the enemy advance considerably and contributed to the final victory. Dedicated to A Company, 12th Regiment, 4th Infantry Division, and 159th Combat Engineer Bn, 10th Regiment 5th Infantry Division.

Park at picnic area on CR 118 near intersection of N11 and C118. Monument is located across C118 about 30 meters off road on the AP Echternach 1 trail.

CR 118

6562 Consdorf

Latitude: 49.7958

Longitude: 6.3839

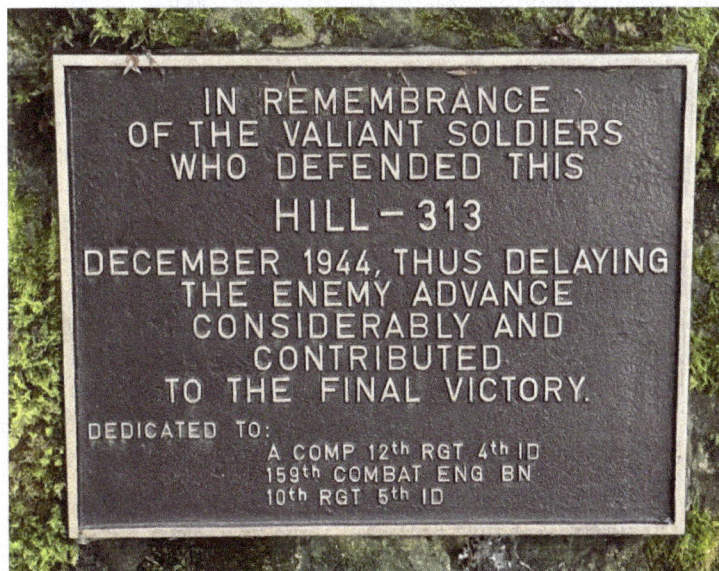

IN REMEMBRANCE
OF THE VALIANT SOLDIERS
WHO DEFENDED THIS
HILL – 313
DECEMBER 1944, THUS DELAYING
THE ENEMY ADVANCE
CONSIDERABLY AND
CONTRIBUTED
TO THE FINAL VICTORY.
DEDICATED TO:
A COMP 12th RGT 4th ID
159th COMBAT ENG BN
10th RGT 5th ID

Echternach - Lauterborn

Ermsdorf

Dedicated to Combat Command A (CCA), of the 9th US Armored Division who held the Ermsdorf-Savelborn-Waldbillig-Christnach line.

Suevelbuererstrooss 1

9366 Vallée de l'Ernz

Latitude: 49.8289

Longitude: 6.2231

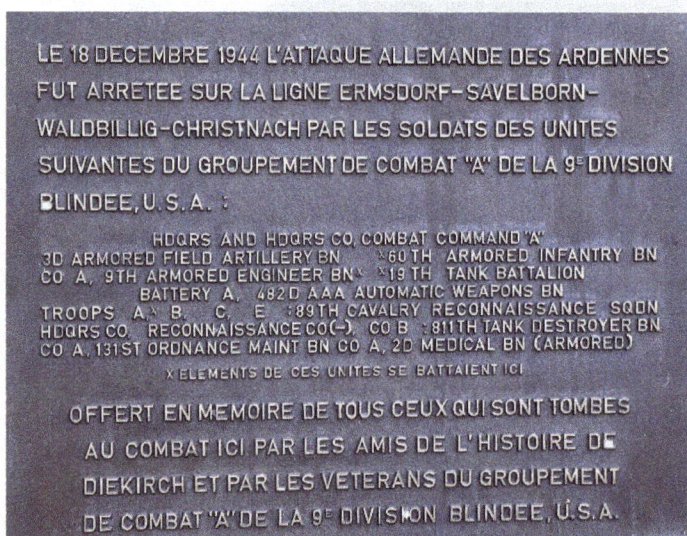

Inscription reads: On December 18, 1944 the German attack through the Ardennes was stopped on the Ermsdorf-Savelborn-Waldbillig-Christnach line by soldiers of the following units of Combat Command A of the 9th US Armored Division.

Headquarters and Headquarters Company, Combat Command A; 3rd Armored Field Artillery Battalion; 60th Armored Infantry Battalion; Company A, 9th Armored Engineer Battalion; 19th Tank Battalion; Battery A, 482nd Anti-Aircraft Artillery Automatic Weapons Battalion; Troops A[x], B[x], C, and E[x] 89th Cavalry Reconnaissance Squadron; Headquarters Company[x] Reconnaissance Company (-) Company B 811th Tank Destroyer Battalion; Company A 131st Ordnance Maintenance Battalion; Company A, 2nd Medical Battalion (Armored).

[x] Element of these units fought here

Offered in memory of all those who died in combat by the Diekirch Historical Society and the veterans of Combat Command A of the 9th US Armored Division.

Esch-sur-Alzette

In memory of the liberation of Esch by the US Army on September 10, 1944.

Intersection of Rue du Fossé and Rue de Luxembourg

4221 Esch-sur-Alzette

Latitude: 49.4987671

Longitude: 5.9861897

Esch-sur-Alzette

Eschdorf

To the 26th US Infantry Division.

An der Gaass 13

9150 Esch-sur-Sûre (Eschdorf)

Latitude: 49.8855

Longitude: 5.9349

Inscription reads: A tribute to the men of the 26th US Infantry Division who fought in this sector during the battle of the Ardennes 1944-1945

Eschette

In Memory of Lt. Robert H. Sovern, 9th Air Force, 362nd Fighter Group, 379th Fighter Squadron who crashed near Eschette on January 25, 1945.

Rue Principale 6

8817 Rambrouch

Latitude: 49.8229577

Longitude: 5.8867139

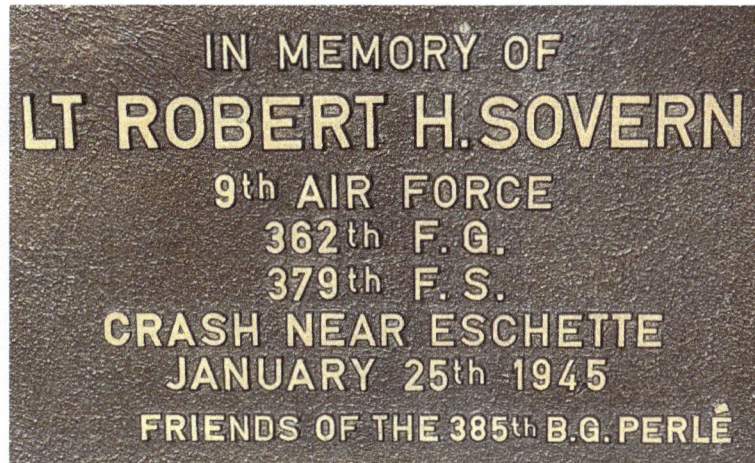

Eschette

Eschweiler

Information panel about the ambush at *Café Halt* and initial burial site of Private George O. Mergenthaler, killed in action on December 18, 1944. The site is located about 50 meters off the CR 328 about 1 km north of the intersection of the CR328 and CR 325.

Halt (CR328)

Wiltz

Latitude: 49.988433

Longitude: 5.959761

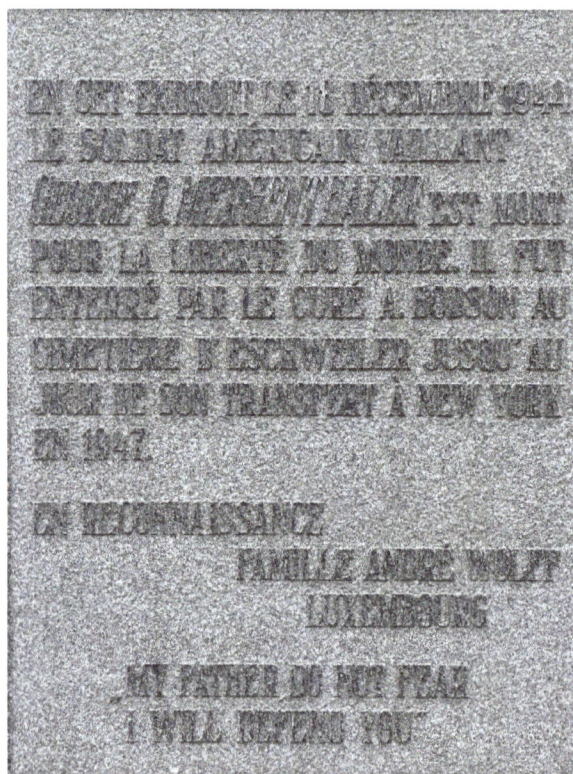

PVT George O. Mergenthaler

Inscription reads: In this place on December 18, 1944 the valiant American soldier George O. Mergenthaler died for the freedom of the world. He was initially buried in the Eschweiler cemetery until being transferred to New York in 1947.

Information Panel Eschweiler

AMBUSH AT - EMBUSCADE AU - HINTERHALT AM CAFÉ HALT DEC 18, 1944

On September 10, 1944, Luxembourg was liberated by the 1ˢᵗ US Army.

On December 16, 1944, Hitler launched a surprise attack against the US troops holding the front from Echternach (L) to Monschau (Germ.). This became known as the Battle of the Bulge.

In the morning of December 18, 1944, to prevent encirclement, the commanding officer in Eschweiler ordered to retreat to the 28ᵗʰ US Infantry Division's headquarters in Wiltz. A column of six vehicles took the shortest road to Wiltz by going south. In one of them was George Ottmar Mergenthaler, grandson of the inventor of the linotype machine. He was well known and beloved by the villagers of Eschweiler.

As the column passed the last bend before the "Café Halt", it ran into an ambush. The Germans fired from every available weapon: rifles, machine guns, mortars and tank guns. The column was blocked and the jeep ahead of Mergenthaler's was hit by a mortar shell. A German fired his submachine gun at the stopped vehicles. While Captain Lewis Meisenhelter, the company commander, Sergeant G. Raducovic, the driver, and Private Joe Vocasee, the mailman, took cover in the jeep, machine gunner Private George O. Mergenthaler fired at the attacker. Joe Vocasee hit a German's hand. At this moment Mergenthaler's machine gun jammed. George tried to fix the jam and ordered his companions to leave the vehicle. As the three soldiers ran for cover, George gave one more burst of gunfire, when his gun jammed again. The wounded German didn't give up and stroke George's neck. George slumped down.

After the thaw in March 1945, a villager found a soldier's grave. Reverend Antoine Bodson identified the body as his friend George O. Mergenthaler. In presence of all the villagers, he buried the brave soldier in the cemetery of Eschweiler. In October 1947, George's mortal remains were returned with 6.300 bodies to the U.S.A.

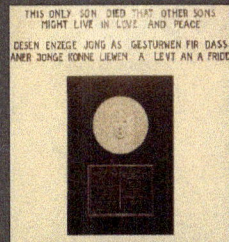

THIS ONLY SON DIED THAT OTHER SONS MIGHT LIVE IN LOVE AND PEACE

DESEN ENZEGE JONG AS GESTUERWEN FIR DASS ANER JONGE KONNE LIEWEN A LEVT AN A FRIDD

Memory panel in honor of George O. Mergenthaler in the entrance of the church of Eschweiler – Plaque en souvenir de George O. Mergenthaler à l'entrée de l'église d'Eschweiler – Erinnerungstafel zu Ehren von George O. Mergenthaler im Eingang der Kirche von Eschweiler
Photos: *Fernand Barbel /* ***Archives:*** *Gerry Klein*

Further information exists on the below QR-code on the left panel's side.

Le 10 septembre 1944, le Luxembourg fut libéré par la 1ʳᵉ Armée américaine.

Le 16 décembre 1944, Hitler lança une attaque surprise contre les soldats américains défendant le front d'Echternach (L) jusqu'à Montjoie (Allem.). C'était la Bataille des Ardennes.

Le matin du 18 décembre 1944, pour éviter un encerclement, l'officier commandant à Eschweiler ordonna le retrait vers le Quartier Général de la 28ᵉ Division d'Infanterie US à Wiltz. Une colonne de six véhicules prit la route la plus directe vers Wiltz par le sud. Dans l'une d'elle était George Ottmar Mergenthaler, petit-fils de l'inventeur de la machine linotype. Il était bien connu et aimé par les villageois d'Eschweiler.

Quand la colonne passa le dernier virage avant le « Café Halt », elle tomba dans une embuscade. Les Allemands tiraient avec toutes les armes dont ils disposaient : fusils, mitrailleuses, mortiers et canons de chars. La colonne fut bloquée et la jeep devant celle de Mergenthaler fut touchée par un obus de mortier. Un soldat allemand tira sa mitraillette sur les véhicules arrêtés. Tandis que le Capitaine Lewis Meisenhelter, le commandant de la compagnie, Sergent G. Raducovic, le chauffeur, et le soldat Joe Vocasee, le messager de la compagnie, se blottirent dans la jeep, le tireur de mitrailleuse George O. Mergenthaler fit feu sur les attaquants. Joe Vocasee toucha un Allemand à la main. A ce moment, la mitrailleuse de Mergenthaler eut des enrayages. George essaya de réparer le blocage et somma ses compagnons de quitter le véhicule. Quand les trois soldats prirent la fuite, George tira encore une salve. Mais son arme s'enraya de nouveau. Le soldat blessé allemand n'abandonna point et tira George dans la gorge. George s'écroula.

Après la fonte des neiges au mois de mars 1945, une villageoise trouva une tombe de soldat. Le curé d'Eschweiler, Révérend Antoine Bodson, identifia son ami George O. Mergenthaler. En présence des villageois, il enterra le courageux soldat au cimetière d'Eschweiler. En octobre 1947, les restes mortuaires de George retournèrent avec 6.300 corps aux USA.

Die 1. US Armee befreite Luxemburg am 10. September 1944.

Am 16. Dezember 1944, startete Hitler einen Überraschungsangriff gegen die amerikanischen Soldaten, die von Echternach (L) bis Monschau (D) die Front hielten. Es war die Ardennen-Offensive.

Um eine Einkesselung zu vermeiden, gab der kommandierende Offizier am Morgen des 18. Dezember 1944 in Eschweiler den Befehl, sich nach Wiltz, Sitz des Hauptquartiers der 28. US Infanteriedivision, zurückzuziehen. Eine aus sechs Wagen bestehende Kolonne nahm den kürzesten Weg südwärts nach Wiltz. In einem saß George Ottmar Mergenthaler, Enkel des Erfinders der Linotypesetzmaschine. Er war im Dorf bekannt und er wurde von den Eschweiler Einwohnern geliebt.

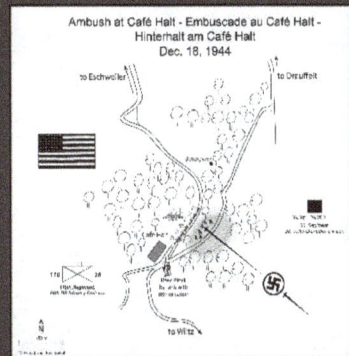

Battle of the Bulge - Bataille des Ardennes - Ardennenoffensive
16.12.1944 28.01.1945

Nach der letzten Kurve vor dem „Café Halt", geriet die Kolonne in einen Hinterhalt. Die Deutschen schlugen sofort mit allen ihnen zur Verfügung stehenden Mitteln zu: mit Gewehren, Maschinengewehren, Granatwerfern und Panzerkanonen. Die Kolonne wurde gestoppt und ein Granatwerfergeschoss traf einen vor Mergenthaler fahrenden Jeep. Ein Deutscher feuerte seine Maschinenpistole auf die anhaltende Fahrzeugkolonne. Während der Kompaniekommandant Kapitän Lewis Meisenhelter, der Fahrer Sergeant G. Raducovic, und der Postbote der Truppe Joe Vocasee im Jeep die Köpfe einzogen, feuerte MG-Schütze Mergenthaler auf die Angreifer. Joe Vocasee traf einen Deutschen an der Hand. Plötzlich hatte Mergenthalers MG eine Ladehemmung. George versuchte, den Schaden zu beheben und forderte seine Kampfgefährten auf, das Gefährt zu verlassen. Während die drei Soldaten um ihr Leben liefen, gab George noch einen Feuerstoß ab, als seine Waffe wieder klemmte. Der verwundete Deutsche gab nicht auf und schoss George in den Hals. Der Amerikaner sank zu Boden.

Nach der Schneeschmelze im März 1945, fand eine Dorfbewohnerin ein Soldatengrab am Waldesrand. Pfarrer Antoine Bodson identifizierte die Leiche als die seines Freundes George O. Mergenthaler. In Anwesenheit aller Einwohner von Eschweiler, bestattete er den Soldaten auf dem Dorffriedhof. Im Oktober 1947, brachte man Georges sterbliche Überreste mit denjenigen von 6.300 amerikanischen Soldaten in die Vereinigten Staaten zurück.

Weitere Informationen sind mittels QR-Code links unten am Rahmen der Tafel abrufbar.

Sources/Quelle: Pierre Eicher, Zum tragischen Tode von George Mergenthaler, in „Schumanns Eck. 1944-1945 Liberation Memorial", Imprimerie rapidpress, Luxembourg, 1994.

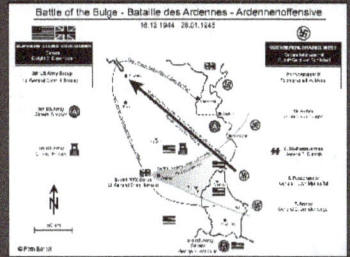

Ambush at Café Halt - Embuscade au Café Halt - Hinterhalt am Café Halt
Dec. 18, 1944

to Eschweiler
to Drauffelt
to Wiltz

Des informations supplémentaires sont disponibles au QR-code en bas à gauche du cadre du tableau.

Fedef

Information Panel Eschweiler

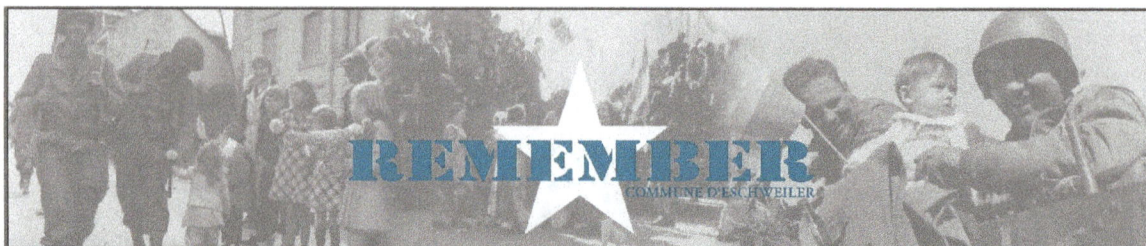

SIGHTS OF INTEREST - POINTS D'INTÉRÊT - SEHENSWÜRDIGKEITEN

1. The parish church of Saint-Maurice of Eschweiler
G.I. George Ottmar Mergenthaler killed in action on December 18, 1944, is eternalized in a memory panel located at the entrance of the church of Eschweiler. Behind the high altar, on the wall paintings representing the multiplication of the loaves, he embodies an apostle. His parents, who contributed to the restoration of the church after World War II, are also honored in the church.

2. Ambush at Café Halt
A monument located off-road in the woods bordering the road Eschweiler – Wiltz indicates George O. Mergenthaler's temporary burial place. A panel explains the military situation on December 18, 1944 and the circumstances of George's death.

3. Ambush at the crossroads Wiltz - Wilwerwiltz - Clervaux
On September 10, 1944, 1st Lieutenant Jesse P. Swear, Infantry & Reconnaissance Team, 109th Regiment, 28th US Infantry Division, and his driver Corporal Leonard A. Deluga, were killed in action as they ran into a German ambush with their jeep.

The villagers of Erpeldange, Eschweiler, Knaphoscheid and Selscheid erected a three-fold monument at the roadside Wilwerwiltz – Clervaux where the two US soldiers lost their lives for the liberation of Luxembourg.

Please, pay attention while visiting this site: The traffic is dense and there is no parking for your car!

4. The "Bunker"
A "Bunker" (underground hiding shelter) is located in the woods between Knaphoscheid and Selscheid, aside from the so-called "Massepad" (worship walk), where four young boys from Knaphoscheid, Selscheid and Luxembourg, who refused to join the "Wehrmacht" (German Army), hid themselves from April to September 1944. The deserters' families restored the dugout.

5. The "Massepad" (L) – Worship walk
It is a popular saying that the cross-country trail Knaphoscheid – Selscheid is known as the "Massepad", (worship walk) as the villagers of Selscheid followed this track to attend church services on Sundays in the parish-church of Knaphoscheid.

1. L'église paroissiale Saint-Maurice d'Eschweiler
Le soldat américain George Ottmar Mergenthaler tué au combat le 18 décembre 1944, est immortalisé dans un médaillon en bas-relief à l'entrée de l'église d'Eschweiler. Derrière le maître-autel, il incorpore un apôtre dans la peinture murale représentant la multiplication des pains. Ses parents qui ont contribué à la restauration de l'église après la Deuxième Guerre Mondiale, sont également honorés en cette église.

2. Embuscade au Café Halt
Un monument situé à une cinquantaine de mètres dans la forêt bordant la route Eschweiler – Wiltz indique la tombe initiale de George O. Mergenthaler. Un panneau explique la situation militaire au 18 décembre 1944 ainsi que les circonstances de la mort de George.

3. Embuscade au croisement des routes Wiltz - Wilwerwiltz – Clervaux
Le 10 septembre 1944, 1er Lieutenant Jesse P. Swear, de l'unité de Reconnaissance du 109e régiment, 28e US Division d'Infanterie, et son chauffeur, Caporal Leonard A. Deluga, perdirent leur vie en tombant avec leur jeep dans une embuscade allemande.

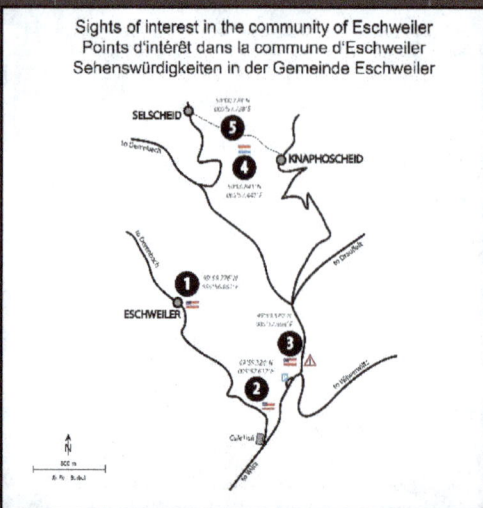

Sights of interest in the community of Eschweiler
Points d'intérêt dans la commune d'Eschweiler
Sehenswürdigkeiten in der Gemeinde Eschweiler

Les villageois d'Erpeldange, d'Eschweiler, de Knaphoscheid et de Selscheid érigèrent un monument composé de 3 parties au bord de la route Wilwerwiltz – Clervaux en l'honneur des deux soldats américains tués au combat pour la liberté du Grand-Duché de Luxembourg.

Veuillez veiller à votre sécurité en visitant ce site : La circulation y est intense et il n'y a pas de parking pour votre voiture.

4. Le "Bunker"
Un abri est caché dans la forêt entre Knaphoscheid et Selscheid à l'écart du sentier « Massepad » (sentier des messes), où quatre jeunes gens de Knaphoscheid, Selscheid et de Luxembourg qui refusaient de joindre la « Wehrmacht » (l'armée allemande) se cachèrent pendant les mois d'avril à septembre 1944. Les familles des déserteurs restaurèrent la cachette.

5. Le « Massepad » - le sentier des messes
Le sentier qui relie Knaphoscheid et Selscheid à travers les champs est usuellement appelé « Massepad » (sentier des messes), parce que les villageois de Selscheid le suivaient les dimanches et les jours de fêtes pour assister aux offices religieux qui se tenaient en l'église paroissiale de Knaphoscheid.

1. Die Sankt-Mauritius-Pfarrkirche von Eschweiler
Der im Kampf am 18. Dezember gefallene amerikanische Soldat George Ottmar Mergenthaler ist in einer Erinnerungsplatte am Eingang der Kirche von Eschweiler verewigt. Im Raumbild hinter dem Hauptaltar, das die Brotvermehrung darstellt, nimmt George den Platz eines Apostels ein. Seine Eltern, die nach dem Zweiten Weltkrieg zur Renovierung der Kirche beitrugen, werden ebenfalls in der Kirche geehrt.

2. Hinterhalt am Café Halt
Ein Gedenkstein, im Wald an der Strecke Eschweiler – Wiltz gelegen, zeigt G. O. Mergenthalers Grabstelle an. Eine Tafel erklärt die militärische Lage am 18. Dezember 1944, und die Umstände, die zu Mergenthalers Tod geführt haben.

3. Hinterhalt an der Straßenkreuzung Wiltz - Wilwerwiltz - Clervaux
Am 10. September 1944, verloren 1. Leutnant Jesse A. Swear der Aufklärungseinheit des 109. Regiments der 28. US Infanteriedivision und sein Fahrer, Korporal Leonard A. Deluga, ihr Leben, als sie mit ihrem Jeep in einen deutschen Hinterhalt gerieten.

Die Dorfbewohner von Erpeldingen, Eschweiler, Knaphoscheid und Selscheid richteten am Ort des Geschehens den zwei im Kampf für die Freiheit Luxemburgs gefallenen Amerikanern ein dreiteiliges Monument am Rand der Straße Wilwerwiltz – Clervaux auf.

Achtung! Größte Vorsicht ist beim Besuch dieser Stätte anzuraten: Der Verkehr ist sehr dicht und es sind keine Parkplätze vorhanden!

4. Der „Bunker"
Im Wald zwischen Knaphoscheid und Selscheid, abseits des im Volksmund genannten „Messen-Pfades", liegt ein „Bunker", wo vier Jungen aus Knaphoscheid, Selscheid und Luxemburg-Stadt, die sich weigerten, dem Stellungsbefehl der deutschen Wehrmacht Folge zu leisten, sich von April bis September 1944 versteckt hielten. Die Familien der Fahnenflüchtigen stellten das Versteck in seiner alten Form wieder her.

5. Der „Massepad"
Im Volksmund wurde die Querfeldein-Strecke Knaphoscheid – Selscheid „Massepad" (Messen-Pfad) genannt, weil früher die Selscheider Dorfbewohner diesen Weg einschlugen, um an den Sonn- und Feiertagen den Gottesdiensten in der Pfarrkirche von Knaphoscheid beizuwohnen.

Eschweiler (Church)

In memory of Private George Ottmar Mergenthaler who fell in action at Eschweiler, Luxembourg on December 18, 1944

Kiirchewee 1

9651 Wilz

Latitude: 49.9963

Longitude: 5.9480

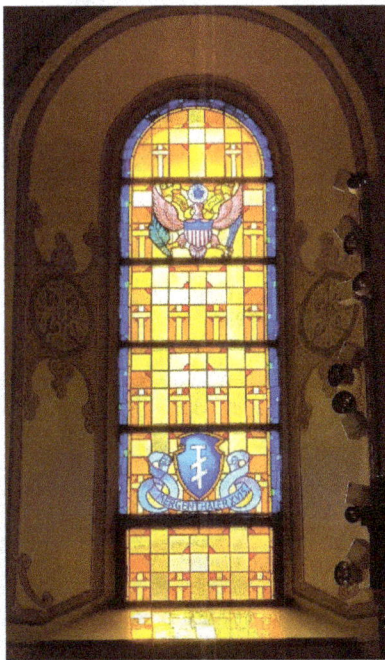

IN MEMORY OF

GEORGE OTTMAR MERGENTHALER
AUGUST 5,1920 - DECEMBER 18,1944

AMERICAN SOLDIER

GRANDSON OF OTTMAR MERGENTHALER
INVENTOR OF THE LINOTYPE

WHO FELL IN ACTION AT
ESCHWEILER, LUXEMBOURG
DECEMBER 18, 1944

MAY THE RESTORATION OF THIS CHAPEL
BE OF SOLACE TO ALL
AS IT GAVE COMFORT TO OUR SON

ALICE AND HERMAN MERGENTHALER
RYE, NEW YORK

MCMXLVI

REV. ANTHONY BODSON, CURE

Information Panel Eschweiler Church

REMEMBER ESCHWEILER

CHURCH - ÉGLISE - KIRCHE SAINT-MAURICE

The parish of Eschweiler probably dates back to the 10th century. It is the only church in Luxembourg that is under the protection of Saint Maurice (280/300), a Roman officer of the Theban Legion who died a martyrs' death with his Christian soldiers.

Prosper Biwer, the architect of the district of Diekirch, drew the sketches of the present-day church that was entirely built in the neo-Gothic style in 1870. It was consecrated on May 26, 1877. The baroque furniture of high quality which dates back to 1730 was taken over from the anterior church. It mainly comes from the workshop of the Luxembourg artist Jean-Georges Scholtus (1680 – 1754) of Bastogne.

The choir with the high altar - Choeur avec le maître-autel - Chor mit Hauptaltar - Photo: Fernand Barbel

In 1924, the decorator Jean Neumanns († 1973) of Beaufort, Luxembourg, painted the mural freskos, two of which we can still admire; they are the multiplication of bread (Jn 6, 5-13) and the announcement of the Kingdom of God by Jesus Christ (Lc 12,31).

After the Battle of the Bulge, as the church had been restored, the apostle holding the basket with some bread on the painting showing the multiplication of the loaves received the face of G.I. George Otmar Mergenthaler who was killed in Eschweiler on December 18, 1944. In the entrance of the church a panel commemorates the Mergenthaler couple's unique son, dead at the age of 24 years.

The artistic glass paintings created after World War II by the Luxembourg artists Julien and Nina Lefèvre are due to the generosity of the Mergenthaler family in memory of their only son George Otmar. Opposite the saints Antonio of Padua, Isidor and George are represented saint Alice whose face is the one of the fallen soldier's mother, Saint Maurice and the blessed Hermann Joseph. The coats of arms of the Grand Duchy of Luxembourg, of the Pforzheim family, of the United States of America and of the Mergenthaler family are visible in the glass paintings of the choir.

In 2004 a new organ was consecrated that had been built by the Eisenbarth manufactory of Passau, Bavaria. A register named in honor of George Mergenthaler and made in an American design, was added to the new organ.

Further information exists on the below QR-code on the left panel's side.

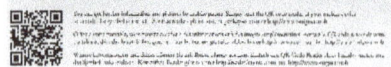

La paroisse d'Eschweiler remonte probablement jusqu'au 10e siècle. Elle est l'unique église du Luxembourg qui est sous la protection de saint Maurice (280/300), un officier romain de la légion thébaïde qui subit le martyr avec ses soldats chrétiens.

L'architecte Prosper Biwer de Diekirch dessina les plans de l'actuelle église qui fut construite entièrement en style néo-gothique en 1870. Elle fut consacrée le 26.05.1877. Le mobilier baroque de qualité remarquable date de 1730 et fut repris de l'église antérieure. Il provient en grande partie de l'atelier de l'artiste luxembourgeois Jean-Georges Scholtus de Bastogne (1680 – 1754).

En 1924, le peintre-décorateur Jean Neumanns († 1973) de Beaufort (Luxembourg) réalisa des peintures murales dont on peut encore admirer les deux peintures du choeur, à savoir la multiplication des pains (Jn 6, 5-13) et la prédication du Royaume de Dieu par le Christ (Lc 12,31).

Lors de la restauration de l'église après l'Offensive des Ardennes, l'apôtre tenant le panier avec les pains sur la peinture de la multiplication des pains reçut le visage du soldat américain George Otmar Mergenthaler, tombé à Eschweiler le 18.12.1944. À l'entrée de l'église, une plaquette commémore l'unique fils des époux Mergenthaler mort à l'âge de 24 ans.

Les vitraux d'art, créés par les artistes luxembourgeois Julien et Nina Lefèvre après la Deuxième Guerre Mondiale, sont dus à la générosité de la famille Mergenthaler des États-Unis et ont été réalisés en souvenir de leur fils unique George Otmar. Vis-à-vis des saints Antoine de Padou, Isidore et Georges, sont représentés la sainte Alice, dont le visage est celui de la mère du soldat tombé, saint Maurice et le bienheureux Hermann Joseph. Les armoiries du Grand-Duché de Luxembourg, celles de la famille de Pforzheim et des États-Unis ainsi que celles de la famille Mergenthaler sont visibles dans les vitraux du choeur.

En 2004 fut consacré le nouvel orgue construit par la manufacture Eisenbarth de Passau en Bavière. Un registre en l'honneur de George Mergenthaler fut réalisé en style américain et fut inclus dans le nouvel orgue.

The coat of arms of the Mergenthaler family - les armoiries de la famille Mergenthaler - das Mergenthaler Familienwappen - Photo: Fernand Barbel

Des informations supplémentaires sont disponibles au QR-code en bas à gauche du cadre du tableau.

Sources/Quelle:
Die Kirchen & Kapellen im Pfarrverband Kiischpelt von Prof. Michel Schmitt, Verlag Schnell & Steiner, GmbH Regensburg, 2003 - Amerikanische Kriegsdenkmäler in Luxemburg zur Erinnerung 1940 bis 1945 von Prof. Norbert Thill, Verlag Heimat und Mission, Imprimerie Saint Paul, Luxemburg, 1995 - Internet: Gemeng Eschweiler, Die St. Mauritius-Pfarrkirche in Eschweiler und ihre Orgeln von Francis Erasmy

Die Pfarrei Eschweiler reicht vermutlich ins 10. Jahrhundert zurück. Als einzige Pfarrkirche Luxemburgs steht sie unter dem Schutz des heiligen Mauritius († 280/300), eines römischen Offiziers, der mit seinen christlichen Soldaten der Thebäischen Legion, den Märtyrertod erlitt.

Distriktsarchitekt von Diekirch Prosper Biwer zeichnete die Pläne der heutigen Pfarrkirche, die 1870 im Stil der Neugotik erbaut wurde. Sie wurde am 26.05.1877 konsekriert. Das bemerkenswerte Barockmobiliar stammt größtenteils aus der Werkstatt des bekannten altluxemburgischen Bildhauers Jean-Georges Scholtus (1680 – 1754) aus Bastnach. Es wurde aus dem Vorgängerbau übernommen und entstand spätestens im Jahre 1730.

Im Jahre 1924 führte Jean Neumanns († 1973) aus Beforт, Luxemburg, großflächige Wandmalereien aus, von denen nur die biblische Brotvermehrung (Joh 6,5 – 13) und die Verkündigung des Gottesreiches durch Jesus Christus (Lk 12,31) am Chorabschluss noch sichtbar sind.

The coat of arms of the USA – les armoiries des USA – das Staatswappen der Vereinigten Staaten von Amerika - Photo: Fernand Barbel

Nach Ende der Ardennen-Offensive wurde die Kirche renoviert. Bei der Gelegenheit wurde am Bild der Brotvermehrung dem Brot tragenden Apostel die Gesichtszüge des in Eschweiler am 18.12.1944 gefallenen US Soldaten George Otmar Mergenthaler verliehen. In der Vorhalle der Kirche erinnert eine Tafel an den toten GI, der als 24-jähriger einziger Sohn der Eheleute Mergenthaler sein Leben verlor. Die monumental wirkenden Glasmalereien der luxemburgischen Künstler Julien und Nina Lefèvre entstanden nach Kriegsende und sind der Großzügigkeit der Familie Mergenthaler aus den Vereinigten Staaten von Amerika zu verdanken. Gegenüber der Heiligen Antonius von Padua, Isidor und Georg, sind die selige Alexia, die die Gesichtszüge der Mutter des Gefallenen trägt, der heilige Mauritius und der selige Hermann Joseph dargestellt. In den Chorfenstern erkennt man das Luxemburger Staatswappen, dasjenige der Familie Pforzheim, das der Vereinigten Staaten sowie das Familienwappen Mergenthaler.

Im Jahre 2004 wurde die in der Orgelmanufaktur Eisenbarth in Passau, Bayern, entstandene neue Orgel eingeweiht. Als Hommage an George Mergenthaler wurde ein Register nach ihm benannt und in amerikanischer Bauart in die neue Orgel eingebaut.

Weitere Informationen sind mittels QR-Code links unten am Rahmen der Tafel abrufbar.

Eschweiler

Dedicated to 1st Lt Jesse P. Sweat, and CPL Leonard A. Deluga, killed in Action on September 10, 1944. Both were part of an intelligence and reconnaissance (I & R) team, 109th Regiment, 28th Infantry Division.

CR 325

4885 Wilz

Latitude: 49.992328

Longitude: 5.965110

Note: this monument is on the side of the road near the intersection of CR 325 and CR 324. There is no parking so you must park on the street and be very careful when walking to the monument!

Ettelbruck

Dedicated to Lieutenant-Colonel James Rudder, Commanding Officer 109th Infantry Regiment, 28th Infantry Division (formerly commander of 2nd Ranger Infantry Battalion, during Normandy landing). From late November until December 29, 1944 LTC Rudder established his headquarters in the 'St. Ann' Girls school in Ettelbruck.

Grand-Rue 102

9051 Ettelbruck

Latitude: 49.8474

Longitude: 6.1024

Lieutenant Colonel James Earl RUDDER, Commanding Officer of the 109th Regiment, 28th US Infantry Division

Inaugurated on December 15, 2019

Information Panel Ettelbruck

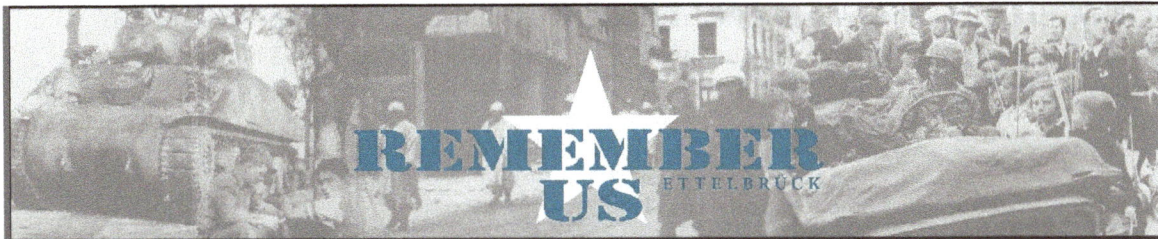

REMEMBER US ETTELBRÜCK

LIEUTENANT-COLONEL JAMES RUDDER, COMMANDING OFFICER OF THE 109TH REG./28TH US INF. DIV.

From late November until December 20, 1944, Lt.Col. James Rudder, Commanding Officer of the 109th regiment, 28th US Infantry Division, had its headquarters established at the "St. Ann" girl's school in Ettelbruck.

"St. Ann" girl's school in Ettelbruck. Photo: Archiv GREG

Lt. Col. Rudder became famous, as he had climbed together with his 2nd ranger battalion the cliffs of the "Pointe du Hoc" on D-Day (June 6, 1944) to eliminate the German coastal artillery batteries.

Lieutenant Colonel James Earl RUDDER, Commanding Officer of the 109th Regiment, 28th US Infantry Division. Photo Credit: NARA/US Signal Corps-coll. MNHM (R.GAUL)

He later took over the 109th regiment of the 28th US Infantry Division, one of the two divisions that had liberated Luxembourg on September 10, 1944.

In October 1944, the regiment was committed as part of the 28th Infantry Division to "Huertgen Forest", where it suffered so many casualties during the bitter fighting that it was pulled back to Luxembourg to "refresh".

When the "Battle of the Bulge" began on December 16, 1944, two battalions of the 109th regiment held the frontline between Stolzembourg and Wallendorf. Thru their tough and stubborn resistance as well as the skilled leadership by Lt.Col. Rudder, the men of the 109th regiment were able to disrupt the German time table. By doing so, they contributed that the defense of the frontline sectors of Vianden-Wallendorf and later of the high-ground Ettelbruck-Mertzig-Grosbous and made sure that the German 352nd Volksgrenadier-division ("people's grenadier division) did not reach the objectives fixed by their leaders.

The 109th US infantry regiment, just as the entire division that defended the line from the northernmost tip of Luxembourg down to Wallendorf, had to bear the bulk of the German attack during the first days. The heroic stand by the division was decisive for the disruption of the German time schedule, so that the main road hub at Bastogne, which was critical to the German supply train, could not be captured by them!

LT. COLONEL JAMES RUDDER, COMMANDANT DU 109E RÉGIMENT DE LA 28E DIVISION D'INFANTERIE AMÉRICAINE

Du début de la Bataille des Ardennes au 20 décembre 1944, le Lt. Colonel James Rudder, commandant du 109e Régiment de la 28e Division d'Infanterie américaine, eut son quartier général dans le pensionnat Sainte-Anne à Ettelbruck.

Le Lt. Colonel Rudder s'était taillé une réputation par son assaut, le 6 juin 1944, des falaises de la « Pointe du Hoc », lancé avec le 2e Bataillon des Rangers en vue d'y éliminer les positions d'artillerie allemandes.

C'est par la suite qu'il prit le commandement du 109e Régiment de la 28e Division d'Infanterie, une des deux divisions libérant le Grand-Duché de Luxembourg le 10 septembre 1944.

En octobre 1944, ce régiment intervint dans la Bataille de la Forêt de Huertgen, où la 28e Division subir, au terme de combats très meurtriers, des pertes tellement lourdes qu'elle dut être transférée au Luxembourg pour y reprendre ses forces.

Le 16 décembre 1944, le jour même du lancement de l'Offensive des Ardennes, deux bataillons du 109e Régiment tenaient le front entre Stolzembourg et Wallendorf. Se battant avec acharnement et dirigés avec talent par le Lt. Colonel Rudder, ces unités du 109e Régiment réussirent à obstruer l'avance allemande : à la suite de leur défense de la ligne de front entre Vianden et Stolzembourg, puis de la ligne Ettelbrück-Mertzig-Grosbous, la 352e Division Volksgrenadier allemande échoua à atteindre les objectifs fixés par son haut commandement. Ce fut le 109e Régiment tout comme le reste de la division défendant le secteur s'étendant de l'extrémité nord du pays jusqu'à Wallendorf qui durent faire face à la charge principale de l'assaut allemand. La résistance héroïque de la division fut décisive quant au retardement général des troupes allemandes, de manière à ce que Bastogne, plaque tournante essentielle pour l'approvisionnement allemand, put être défendu avec succès malgré les assauts répétés de l'ennemi.

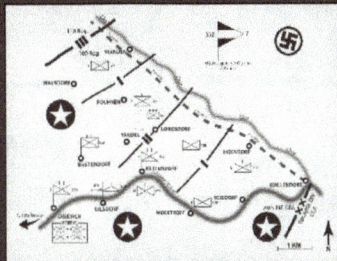

Situation of the 109th Regiment, 28th Infantry Division on Dec. 16, 1944 © Roland Gaul / Fern Bärbel

Sources, Quellen: 1944 - 1945 Schicksale zwischen Sauer und Our, Soldaten und Zivilpersonen erzählen von Roland Gaul, Band II, Sankt-Paulus-Druckerei AG Luxemburg 1987

LT. COLONEL JAMES RUDDER, KOMMANDEUR DES 109. REGIMENTS DER 28. US-INFANTERIE-DIVISION

Vom Beginn der Ardennenoffensive an bis zum 20. Dezember 1944 hatte Lt. Colonel James Rudder, Kommandeur des 109. Regiments der 28. US-Infanteriedivision, sein Hauptquartier im Pensionat St. Anne in Ettelbrück.

Lt. Colonel Rudder hatte sich einen Namen gemacht, als er am 6. Juni 1944 mit dem 2. Rangerbataillon die Klippen zur „Pointe du Hoc" in der Normandie erklomm, um die deutschen Artilleriestellungen auszuschalten. Später übernahm er das 109. Regiment der 28. US-Infanterie-Division, eine der beiden Divisionen, die das Großherzogtum am 10. September 1944 befreiten. Im Oktober 1944 wurde das Regiment als Teil der 28. Infanterie-Division im Hürtgenwald eingesetzt, wo sie bei den dort ausgetragenen erbitterten Kämpfen so schwere Verluste erlitt, dass sie zur Auffrischung nach Luxemburg verlegt wurde.

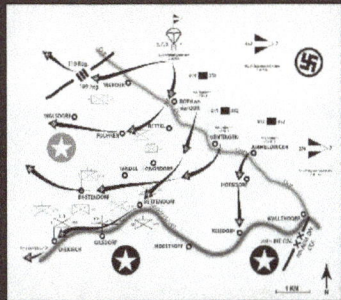

Germans Attack and U.S. Retreat on December 16 - 19, 1944 © Roland Gaul / Fern Bärbel

Als die Ardennenoffensive am 16. Dezember 1944 begann, hielten zwei Bataillone des 109. Regiments die Front zwischen Stolzemburg und Wallendorf. Durch ihren zähen und verbissenen Abwehrkampf und die geschickte Führung von Lt. Colonel Rudder gelang es den Männern des 109. Regiments, den deutschen Zeitplan zu durchkreuzen, so dass die Verteidigung der Frontlinie Vianden-Stolzemburg und später die Verteidigung des Höhenzuges Ettelbrück-Mertzig-Grosbous entscheidend dazu beitrugen, dass es der 352. deutschen-Volksgrenadier-Division nicht gelang, die ihnen von der Führung gesetzten Ziele zu erreichen. Das 109. US-Infanterie-Regiment sowie die gesamte Division, die den Sektor von der nördlichsten Spitze des Landes bis nach Wallendorf verteidigten, trugen in den ersten Tagen die Hauptlast des deutschen Angriffs. Der heldenmütige Widerstand der Division war entscheidend für den zeitlichen Verzug der deutschen Truppen, so dass der für den deutschen Nachschub äußerst wichtige Verkehrsknotenpunkt Bastogne, trotz wochenlangem Ansturm des Gegners, nicht eingenommen werden konnte.

Ettelbréck VILLE D'ETTELBRÜCK   norTIC

Ettelbruck

Colonel Lansing McVickar Square, Commander 318th Regiment, 80th Infantry Division who liberated the City of Ettelbruck on Christmas 1944. Col McVickar was Killed in Action on 14 January 1945.

Boulevard Grande-Duchesse Charlotte 10-24

9024 Ettelbruck

Latitude: 49.8444501

Longitude: 6.0987672

Ettelbruck, General Patton Memorial Park

Dedicated to General George S. Patton, Jr.

6162 Ettelbruck

Latitude: 49.85128

Longitude: 6.11027

Ettelbruck, General Patton Memorial Park

IN MEMORY
OF
GEORGE S. PATTON JR.
COMMANDING GENERAL
3RD U. S. ARMY

WHOSE FORCES
LIBERATED ETTELBRUCK
ON 27 DECEMBER 1944
DURING THE ARDENNES
COUNTEROFFENSIVE

PRESENTED BY HIS SON
GEORGE S. PATTON
CAPTAIN
UNITED STATES ARMY

ON THE 50th ANNIVERSARY OF
THE BATTLE OF THE BULGE
WE RENEW OUR GRATITUDE TO
GENERAL GEORGE S. PATTON JR
AND THE BRAVE MEN OF
THE THIRD UNITED STATES ARMY
WHO BROUGHT US BACK FREEDOM
IN 1944 – 1945

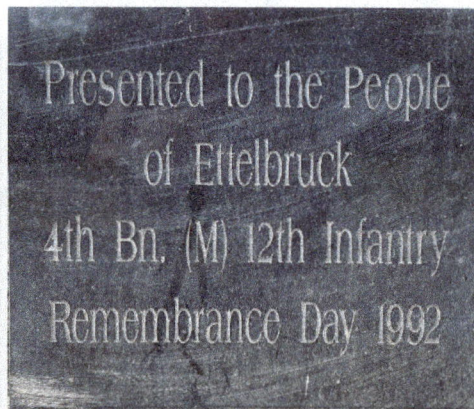

Presented to the People
of Ettelbruck
4th Bn. (M) 12th Infantry
Remembrance Day 1992

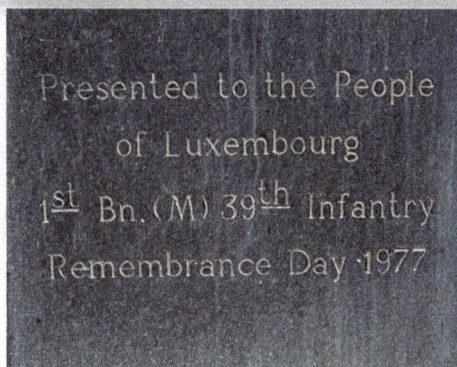

Presented to the People
of Luxembourg
1st Bn. (M) 39th Infantry
Remembrance Day 1977

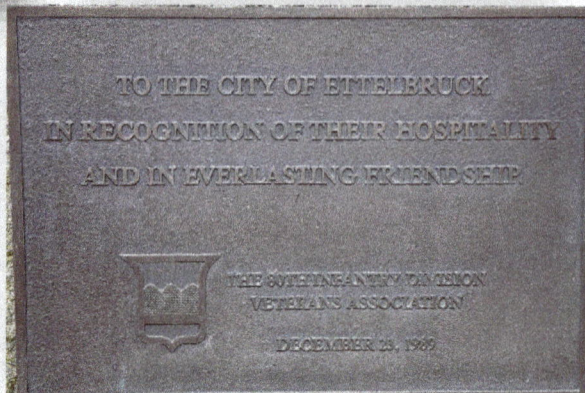

TO THE CITY OF ETTELBRUCK
IN RECOGNITION OF THEIR HOSPITALITY
AND IN EVERLASTING FRIENDSHIP

THE 80TH INFANTRY DIVISION
VETERANS ASSOCIATION

DECEMBER 23, 1989

Information Panel Ettelbruck Patton Memorial Park

REMEMBER US ETTELBRÜCK

THE TWO LIBERATIONS OF LUXEMBOURG

On September 9, 10 and 11, 1944, the 5th US Armored Division (AD) and the 28th US Infantry Division (ID) liberated the Grand Duchy of Luxembourg from the German occupation that had begun on May 10, 1940.

STOLZEMBOURG
September 11, 1944

First crossing into the Third Reich, breakthrough at Wallendorf
Première entrée dans le Troisième Reich, percée à Wallendorf
Erste Überquerung der deutschen Grenze, Durchbruch bei Wallendorf

On Sept. 11, a unit of the 5th AD crossed the border of the Third Reich for the first time at Stolzembourg (L).

On Sept. 14, Combat Command R, 5th AD and the 112th Regiment of the 28th ID, broke through the German West Wall near Wallendorf (D). After a successful push in the beginning, the US troops had to retreat on Sept 22 under heavy losses due to the lack of flanking protection and additional support by further infantry units.

The front stabilized at the German border and along the road Diekirch to Wemperhardt ("Skyline Drive" as called by the GIs). It was relatively calm up to December 16, 1944, when Hitler launched a surprise attack with three armies, totalling 240,000 men, against the 83,000 US soldiers holding the front from Monschau (D) to Echternach (L). It was the beginning of the Ardennes Offensive, the Battle of the Bulge.

Thanks to the stubborn resistance by the US garrisons, especially by the 28th US ID, holding their positions at all costs, reinforcements arrived in Bastogne (B) and St. Vith (B). On Dec 22, General George S. Patton Jr., Commander of 3rd US Army, counter attacked on the German south flank with three divisions: The 4th AD pushed towards Bastogne, the 26th ID had Wiltz as a goal, and the 80th ID liberated Ettelbruck on December 25, 1944. The 5th ID operated around Echternach. In January 45, it liberated the towns north of Diekirch. After a six-week battle, Luxembourg was free again.

Sources, Quellen: Die Ardennenschlacht 1944 – 19454 in Luxemburg von Jean Milmeister, Editions St. Paul, Luxembourg, 1994
Lëtzebuerg am Zweete Weltkrich, Léonie Thill, Marcel Scheidweiler, Fernand Barbel, 2014, ISBN 978-99959-0-147-9

LES DEUX LIBÉRATIONS DE LUXEMBOURG

Le 9, 10 et 11 septembre 1944, la 5e Division Blindée américaine (DB) et la 28e Division d'Infanterie (DI) libérèrent le Grand-Duché de Luxembourg de l'occupation allemande qui avait commencé le 10 mai 1940.

Le 11.09.1944, une unité de la 5e Division Blindée américaine traversa la frontière du Troisième Reich pour la première fois à Stolzembourg.

Le 14.09, le Groupement de Combat R (Brigade) de la 5e DB et le 112e Régiment de la 28e Division d'Infanterie pénétrèrent dans la Muraille de l'Ouest (Ligne Siegfried) près de Wallendorf (D). Au début, les troupes américaines avaient beaucoup de succès, mais elles devaient se retirer sous de lourdes pertes le 22 sept. faute de protection suffisante des flancs et manque de support par des unités d'infanterie supplémentaires.

Le front fut stabilisé à la frontière allemande et le long de la route Diekirch–Wemperhardt (appelée « Skyline Drive » par les GIs). La situation militaire resta relativement calme sur ce secteur jusqu'au 16 décembre 1944. A cette date, Hitler lança une attaque surprise avec trois armées, totalisant 240.000 hommes, contre les 83.000 soldats américains qui tenaient le front de Monschau (D) jusqu'à Echternach (L). C'était le commencement de la Bataille des Ardennes.

Battle of the Bulge - Bataille des Ardennes - Ardennenoffensive
18.12.1944 - 28.01.1945

German plan (←) and German actual advance (in grey)
Plan allemand (←) et avance réelle (en gris)
Deutscher Plan der Offensive (←) und wirklicher Vorstoß (in grau)

Grâce à la résistance farouche des garnisons américaines, spécialement de celle de la 28e Division d'Infanterie, de tenir les positions à tout prix, des renforts arrivèrent à Bastogne (B) et à St. Vith (B). Le 22 décembre, le Général George S. Patton Jr., Commandant de la 3e Armée américaine, contre-attaqua le flanc sud allemand avec trois divisions : La 4e Division Blindée avança vers Bastogne (B), la 26e Division d'Infanterie se dirigea vers Wiltz, la 80e Division d'Infanterie libéra la ville d'Ettelbruck le 25 décembre 1944, la 5e Division d'Infanterie opéra autour d'Echternach. En janvier 45, elle libéra les villages au nord d. Diekirch. Après une bataille qui dura six semaines, le Luxembourg fut de nouveau libre.

DIE BEIDEN BEFREIUNGEN VON LUXEMBURG

Am 9., 10. und 11. September 1944 befreiten die 5. US Panzerdivision (PD) und die 28. US Infanteriedivision (ID) das Großherzogtum Luxemburg von der deutschen Besatzung, die am 10. Mai 1940 begonnen hatte.

Am 11. September überquerte eine Einheit der 5. US Panzerdivision zum ersten Mal die Grenze zum Deutschen Reich bei Stolzemburg (L).

Am 14. September 44 durchbrachen die 5. US PD und die 28. US ID den deutschen Westwall (Siegfried Linie) bei Wallendorf (D). Nach einem erfolgreichen Beginn zogen sich die US Truppen unter schweren Verlusten am 22. September zurück, da sie unzureichenden Flankenschutz hatten und keine zusätzliche Unterstützung von weiteren Infanterieeinheiten erhielten.

Die Front stabilisierte sich an der deutschen Grenze und entlang der Verbindungsstraße Diekirch-Wemperhardt („Skyline Drive"). An diesem Frontabschnitt war es bis zum 16. Dezember 44 ziemlich ruhig, bis drei deutsche Armeen mit 240.000 Mann die 83.000 US-Soldaten, die die Front von Monschau (D) bis Echternach (L) hielten, angriffen. Es war der Beginn der Ardennenoffensive.

Dank des zähen Widerstandes der amerikanischen Garnisonen, besonders der 28. ID, im Bestreben, ihre Stellungen um jeden Preis zu halten, kamen Verstärkungen nach Bastogne (B) und St. Vith (B). Am 22. Dezember 44 griff General George S. Patton Jr. die deutsche Südflanke mit drei Divisionen an: Die 4. US Panzerdivision stieß nach Bastogne (B) vor, die 26. US Infanteriedivision hatte Wiltz als Ziel und die 80. US Infanteriedivision befreite Ettelbrück am 25. Dezember 44. Die 5. US Infanteriedivision operierte rund um Echternach. Im Januar befreite sie die Dörfer nördlich von Diekirch. Nach einem sechswöchigen Kampf war Luxemburg wieder frei.

Général George S. PATTON Jr.
Photo: Archives Musée Patton, Ettelbruck

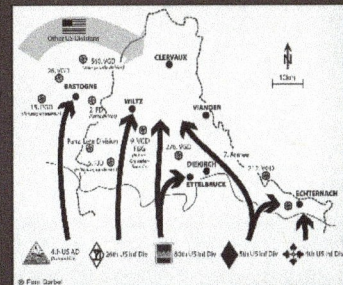

General Patton's counter-attack - la contre-attaque du Général Patton
General Pattons Gegenangriff

Ettelbréck VILLE D'ETTELBRUCK · norTIC

Information Panel Ettelbruck Patton Memorial Park

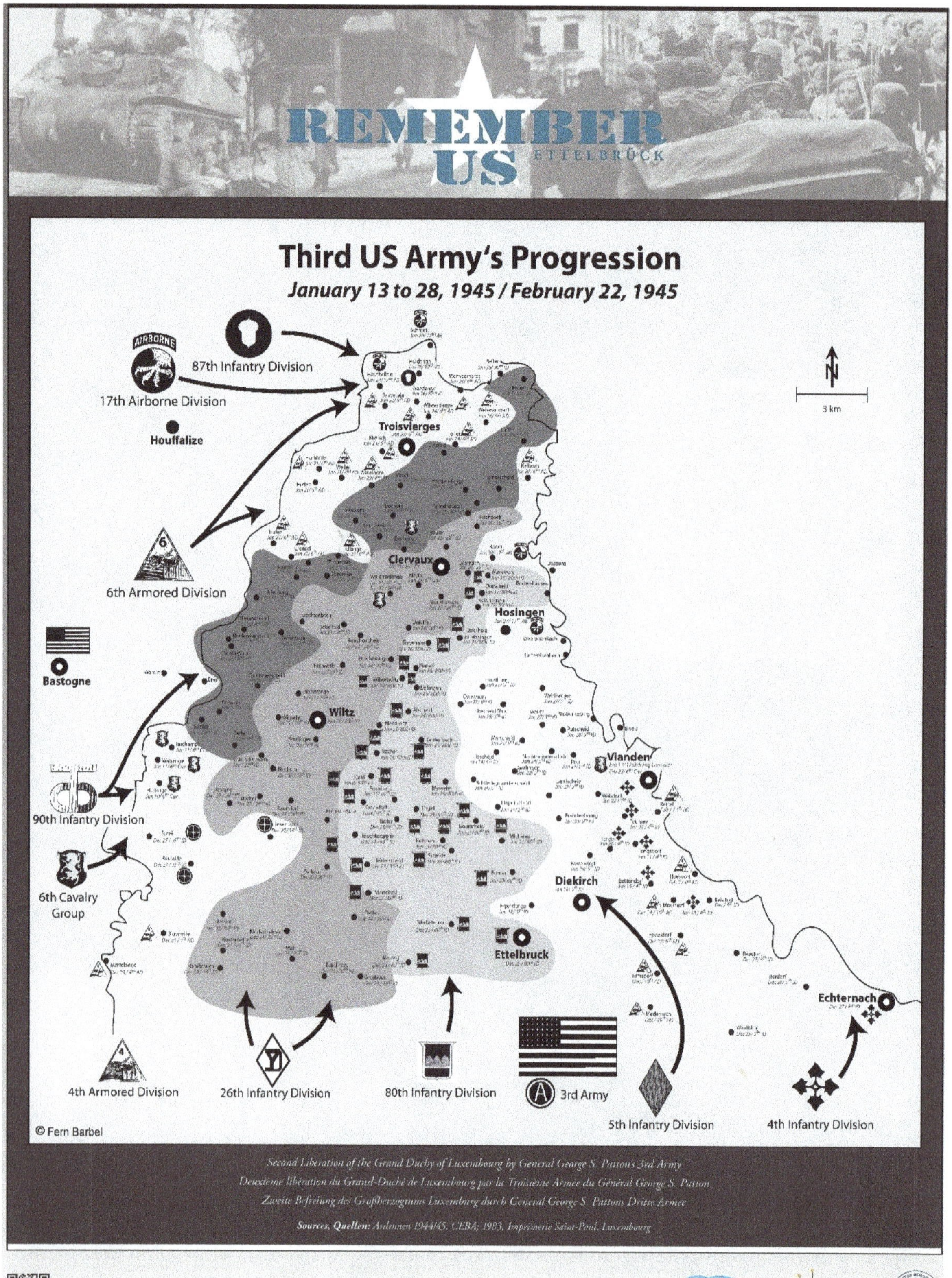

REMEMBER US ETTELBRÜCK

Third US Army's Progression
January 13 to 28, 1945 / February 22, 1945

87th Infantry Division

17th Airborne Division

Houffalize

6th Armored Division

Troisvierges

Clervaux

Hosingen

Bastogne

Wiltz

Vianden

90th Infantry Division

6th Cavalry Group

Diekirch

Ettelbruck

4th Armored Division

26th Infantry Division

80th Infantry Division

3rd Army

5th Infantry Division

Echternach

4th Infantry Division

© Fern Barbel

3 km

Second Liberation of the Grand Duchy of Luxembourg by General George S. Patton's 3rd Army
Deuxième libération du Grand-Duché de Luxembourg par la Troisième Armée du Général George S. Patton
Zweite Befreiung des Großherzogtums Luxemburg durch General George S. Patton Dritte Armee

Sources, Quellen: Ardennen 1944/45, CEBA, 1983, Imprimerie Saint-Paul, Luxembourg

Ettelbréck VILLE D'ETTELBRÜCK   noTIC centre de services

Information Panel Ettelbruck Patton Memorial Park

SHERMAN TANK M4 A1

The Sherman tank was the most important tank of the Allies. It played a decisive part in the military operations in World War II. Sherman tanks of the M4 series were involved in both liberations of Luxembourg. They are a symbol of the efforts the American liberators made for our freedom.

Since June 28, 1970, a Sherman M4A1 is displayed at the Patton Square. The armored units under Patton's command used the same type of tank.

Depuis le 28 juin 1970 un char Sherman, modèle M4A1, est exposé au Patton Square. Les divisions blindées sous le commandement du Général Patton utilisaient le même type de char.

Seit dem 28. Juni 1970 ist ein Sherman, Panzer, Modell M4A1, am Patton Square aufgestellt. Die Panzerdivisionen unter General George S. Pattons Befehl verwendeten denselben Panzertyp.

Photo: Archives Musée Patton, Ettelbruck

During the war, 49,234 Shermans of all types were built in the United States of America. The medium tank M4 (75) was equipped with a 75 mm gun, two calibre .30 machineguns (7.62 mm) and one calibre .50 machinegun (12,7 mm). It weighed 66,500 lb. (30,5 tons). Its engine had 350 HP, and its speed limit was 40km/h. It was operated by a 5-man-crew. The medium Sherman tank was inferior to the German "Panther" and "Tiger" tanks. It only became a match when upgraded with a 76 mm cannon that could defy its adversaries.

A US Armored Division numbered 10,937 men, 501 halftracks and 263 tanks:

- 186 medium and 77 light tanks organized into 3 Tank Battalions (2,187 men)
- 3 Armored Infantry Battalions (3,003 men)
- 3 Armored Field Artillery Battalions (1,502 men)
- 1 Cavalry Reconnaissance Squadron (935 men)
- 1 Engineer Combat Battalion (693 men)
- 1 Armored Ordnance Battalion (762 men)
- 1 Armored Medical Battalion (417)
- 1 Signal Company (226 men)
- Civil Affairs, Band, chaplain, MP-Platoon, Division Head Quarters, Combat Command A, B, R Head Quarters (695 men)
- Tank Destroyer- and Antiaircraft Artillery Battalions were normally attached to the division.

Sources, Quellen: U.S. Army (ETO 1944 - 45) Marquages et organisation de E. Becker et de J. Milmeister, Imp. Willems, Dudelange, 1988 US Army Handbook 1939-1945 by George Forty, Alan Sutton Publishing Limited, 1996, ISBN 0-7509-1078-X Panzer 1939-1945 von Jim Winchester, Gondrom Verlag, GmbH 2000, ISBN 3-8112-1797-6 The American Arsenal, The World War II Official Standard Ordnance Catalogue, ISBN 978-1-84832-726-9, Frontline Books

CHAR SHERMAN M4 A1

Le char Sherman était le char le plus important des Forces Alliées. Il jouait un rôle décisif dans les opérations militaires de la Deuxième Guerre Mondiale. Des chars des séries M4 étaient impliqués dans les deux libérations du Luxembourg. Ils sont un symbole des efforts que les libérateurs américains ont faits pour notre liberté.

Pendant la guerre, les Américains en construisirent 49.234 exemplaires en différentes versions. Le char médium M4 (75) était équipé d'un canon de 75 mm, de 2 mitrailleuses calibre .30 (7.62 mm) et d'une mitrailleuse calibre .50 (12,7 mm). Il pesait 30,5 tonnes; son moteur avait une puissance de 350 CV, sa vitesse maximale était de 40 km/h. Son équipage comptait 5 hommes. Le char Sherman était inférieur aux chars allemands « Panther » et « Tiger ». Ce n'est qu'une fois équipé d'un canon de 76 mm qu'il était à même de se mesurer à ses adversaires.

Une division blindée américaine comptait 10.937 hommes, 501 véhicules semi-chenillés et 263 chars

- 186 chars médiums et 77 chars légers organisés en 3 bataillons de chars (2.187 hommes)
- 3 bataillons d'Infanterie Blindée (3.003 hommes)
- 3 bataillons d'Artillerie de Campagne Blindée (1.502 hommes)
- 1 Escadron de Cavalerie de Reconnaissance (935 hommes)
- 1 bataillon de Combat de Génie (693 hommes)
- 1 bataillon d'Ordonnance Blindé (762 hommes)
- 1 bataillon Médical Blindé (417 hommes)
- 1 compagnie de Transmissions (226 hommes)
- Affaires Civiles, Musique Militaire, Aumônier, Police Militaire, État-major de la Division, États-majors des Groupements de Combat A, B, R (695 hommes)
- Des bataillons de Chasseurs de Chars et d'Artillerie Anti-Aérienne furent normalement attachés à la division

SHERMAN PANZER M4 A1

Der Sherman-Panzer war der wichtigste Panzer der Alliierten. Er spielte eine entscheidende Rolle in den militärischen Operationen im Zweiten Weltkrieg. Sherman Panzer des M4 Typs waren an beiden Befreiungen Luxemburgs beteiligt. Sie sind ein Symbol für die Anstrengungen, die die Amerikaner für unsere Freiheit machten.

1954 to 2004, every year a "Remembrance Day" was celebrated in honor to the US soldiers of WWII. Sherman tank in the parade 1960.

Une journée de commémoration, appelée « Remembrance Day », fut célébrée annuellement, de 1954 à 2004 à Ettelbruck.

1954 bis 2004 wurde jedes Jahr ein "Remembrance Day" zu Ehren der US Soldaten des Zweiten Weltkrieges in Ettelbrück gefeiert.

Photo Archives Musée Patton, Ettelbruck

Während des Krieges stellten die Amerikaner 49.234 Sherman Panzer in verschiedenen Versionen her. Der „Medium Tank M4" (75) war mit einer 75-mm-Kanone, 2 MGs .30 (7,62 mm) und einem MG .50 (12,7 mm) ausgerüstet. Er wog 30,5 Tonnen, sein Motor hatte eine Stärke von 350 PS, seine Höchstgeschwindigkeit betrug 40 km/h. Seine Besatzung belief sich auf 5 Mann. Der Sherman-Panzer war den deutschen „Panther"- und „Tiger"-Panzern unterlegen. Erst als er mit einer 76 mm Kanone ausgerüstet wurde, konnte er es mit seinen Gegnern aufnehmen.

Eine US Panzerdivision zählte 10.937 Mann, 501 Halbkettenfahrzeuge und 263 Panzer:

- 186 mittlere und 77 leichte Panzer waren in 3 Bataillone eingeteilt (2.187 Mann)
- 3 Panzergrenadierbataillone (3.003 Mann)
- 3 Panzer-Feldartilleriebataillone (1.502 Mann)
- 1 motorisiertes Kavalleriesschwadron (935 Mann)
- 1 Pionierkampfbataillon (693 Mann)
- 1 Panzer-Instandhaltungsbataillon (762 Mann)
- 1 Panzer-Sanitätsbataillon (417 Mann)
- 1 Fernmeldekompanie (226 Mann)
- Zivile Angelegenheiten, Divisionsorchester, Geistlicher, Zug der Militärpolizei, Divisionshauptquartier, Hauptquartiere der Kampfgruppen A, B und R (695 Mann)
- Panzerjäger- und Flak-Artilleriebataillone wurden der Division normalerweise angegliedert

Information Panel Ettelbruck Patton Memorial Park

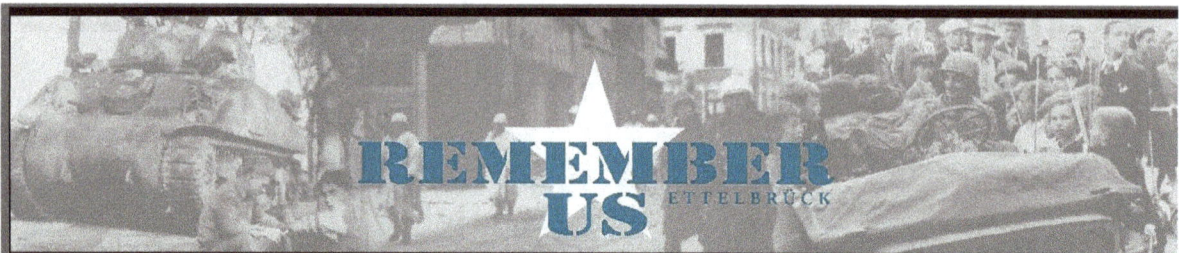

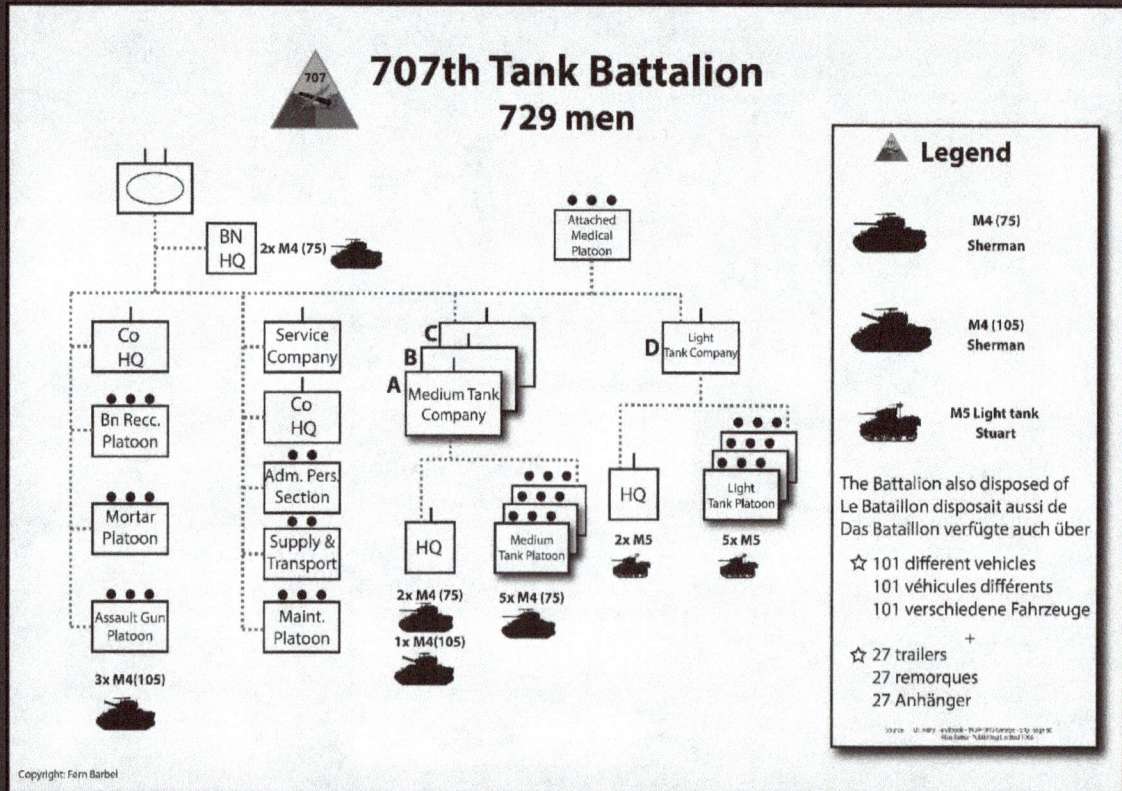

707th Tank Battalion
729 men

Legend

M4 (75) Sherman

M4 (105) Sherman

M5 Light tank Stuart

The Battalion also disposed of
Le Bataillon disposait aussi de
Das Bataillon verfügte auch über

☆ 101 different vehicles
101 véhicules différents
101 verschiedene Fahrzeuge
+
☆ 27 trailers
27 remorques
27 Anhänger

BN HQ — 2x M4 (75)

Attached Medical Platoon

Co HQ

Service Company

B Medium Tank Company
A
C

D Light Tank Company

Bn Recc. Platoon

Co HQ

Adm. Pers. Section

HQ

Light Tank Platoon

Mortar Platoon

Supply & Transport

HQ

Medium Tank Platoon
2x M5

5x M5

Assault Gun Platoon

Maint. Platoon

2x M4 (75)

5x M4 (75)

1x M4(105)

3x M4(105)

Copyright: Fern Barbel

THE 707TH TANK BATTALION IN SUPPORT OF THE 28TH US INFANTRY DIVISION

LE 707e BATAILLON DE CHARS EN SUPPORT DE LA 28e DIVISION D'INFANTERIE AMÉRICAINE

DAS 707. PANZERBATAILLON IN UNTERSTÜTZUNG DER 28. US INFANTERIEDIVISION

The 707th Tank Battalion was divided into four companies and numbered 729 men. It was equipped with 53 medium Sherman M4 (75 mm gun), 6 Sherman M4 (105 mm Howitzer) and 17 light Stuart tanks.

Companies A and B were assigned to the 110th Regiment (Clervaux). Company C came under the orders of the 109th Regiment (Diekirch), Company D (light tanks) was attached to the 112th Regiment (Weiswampach).

Le 707e Bataillon de Chars était divisé en quatre compagnies et comptait 729 hommes. Il était équipé de 53 chars médium Sherman M4 (canon de 75 mm), 6 chars Sherman M4 (105 mm obusier) et de 17 chars légers Stuart.

Les Compagnies A et B furent attribuées au 110e Régiment (Clervaux), la Compagnie C passa sous les ordres du 109e Régiment (Diekirch), la Compagnie D (chars légers) rejoignit le 112e Régiment (Weiswampach).

Das 707. Panzerbataillon war in vier Kompanien aufgeteilt und zählte 729 Mann. Es war mit 53 mittelschweren Sherman M4 (75 mm Kanone), 6 Sherman M4 (105 mm Haubitze) und 17 leichten Stuart Panzern ausgerüstet.

Kompanie A und B wurden dem 110. Regiment (Clerf) zugeordnet. Kompanie C wurde dem 109. Regiment (Diekirch) zugeteilt. Kompanie D (leichte Panzer) wurde dem 112. Regiment (Weiswampach) unterstellt.

Sources, Quellen: U.S. Army (ETO 1944 – 45) Marquages et organisation de Emile Becker et de Jean Milmeister, Imprimerie G. Willems, Dudelange, 1988. US Army Handbook 1939-1945 by George Forty, Alan Sutton Publishing Limited, 1996, ISBN 0.-7509-1078-X

Féitsch Crossroads (Wincrange)

Monuments to the events at *Antoniushaff* and at the *Féitsch* crossroads on December 17 and 18, 1944. The 9th Armored Division attempted to stop the 2nd German Panzer Division on its way to Bastogne.

Monument is located about a 15-minute walk off of the CR 333 and overlooks the Féitsch crossroads. It is best to use the Latitude/Longitude listed below to find the site.

Naturpark Our

9752 Wincrange

Latitude: 50.04239

Longitude: 5.88617

The events at Antoniushaff and at the Féitsch crossroads in December 1944

On December 17th, two units of the 9th Armoured Division CCR were deployed to Antoniushaff and the important crossroads at Féitsch to stop the 2nd German Panzer division on its way to Bastogne. Two Task Forces were set up, Task Force Rose took position at Antoniushaff and Task Force Harper took position at Féitsch crossroads, where an important road leads in a direct way to Bastogne.

In the morning of December 18th, German recon units of the 2nd Panzer division reached the american outposts of the Antoniushaff roadblock. They attacked the American positions twice but they were pushed back each time by the support of the American Artillery.

At noon, the German tanks spread around the fields of Antoniushaff with a higher number of tanks than the Task Force, before attacking from three sides. Rose tried to hold its positions, but when darkness fell he had to withdraw its five remaining tanks in direction of Houffalize.

Now the German tanks moved to the south in direction of the Féitsch crossroads, to reach the road to Bastogne. Unfortunately a German recon patrol found out that an important road block was set up at Féitsch. About thirty tanks with armoured infantry were waiting around the crossroads to ambush the Germans.

At Heisdorf, the Germans didn't use the main road, but they took a small trail on the left, which leads behind a ridge just a half mile opposite the American positions of Task Force Harper. They only left their cover to fire, then they drew back immediately. After a short time the first American tank was burning which illuminated the whole roadblock and turned the American tanks into good targets for the Germans. Under these circumstances Harper lost most of his tanks in a very short time. Those tanks which hadn't been hit by the Germans had to be blown up by their own crew because they couldn't move back through the dense wooden area behind. Like many of his men, LT Col Ralph Harper didn't survive that day.

The two roadblocks, the one at Antoniushaff ant the other at Féitsch crossroads, held the German advance long enough to allow the 10th Armoured Division to set up three new roadblocks closer to Bastogne so that the 101th Airborne Division had enough time to gain this town with its most important crossroads in this area of the offensive. The fact that the Germans never took Bastogne meaned the failure of the whole offensive.

Findel - Airport

Monument to the 4th US Infantry Division, who halted the left shoulder of the German thrust into the American lines and saved the city of Luxembourg.

Aéroport de Luxembourg

2633 Niederanven

Latitude: 49.6363

Longitude: 6.2174

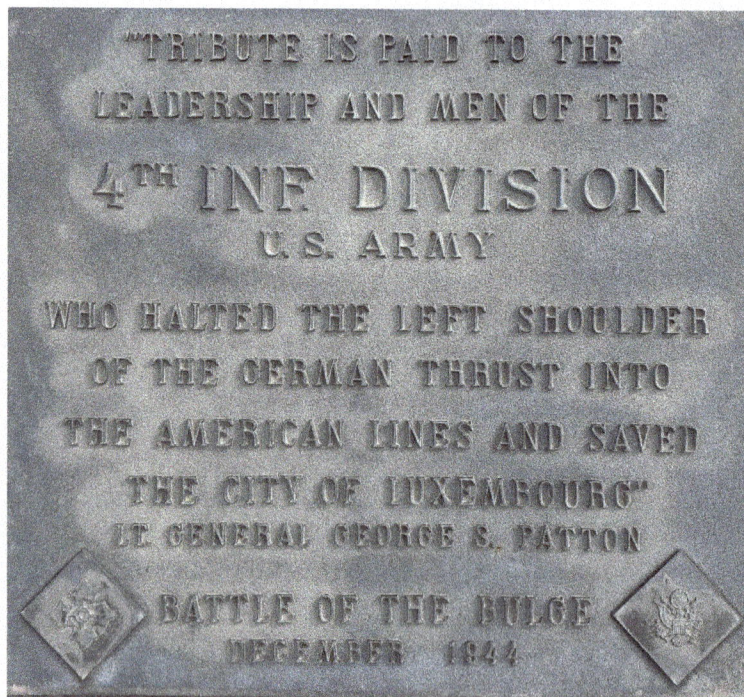

Fouhren

Dedicated to the 5th Armored Division and the 8th, 28th, 4th and 5th Infantry Divisions.

Jean-Baptiste Zewenstrooss 7

9454 Tandel

Latitude: 49.91148

Longitude: 6.19558

OUR DEAD 1940-1945: And to the memory of the brave American Soldiers of the 5th Armored Division and the 8th, 28th, 4th and 5th Infantry Divisions who suffered, bled, and died liberating our villages Fouhren, Bettel, Longsdorf, and Walsdorf.

A plaque inaugurated in 2023 in Fouhren located uphill of the chapel (not church) remembers several civilians who were killed by German artillery shells in Nov 1944 from the "Westwall" during field works.

Frisange Freedom Square

Munnerëferstrooss 1

5750 Frisange

Latitude: 49.5160361

Longitude: 6.1881156

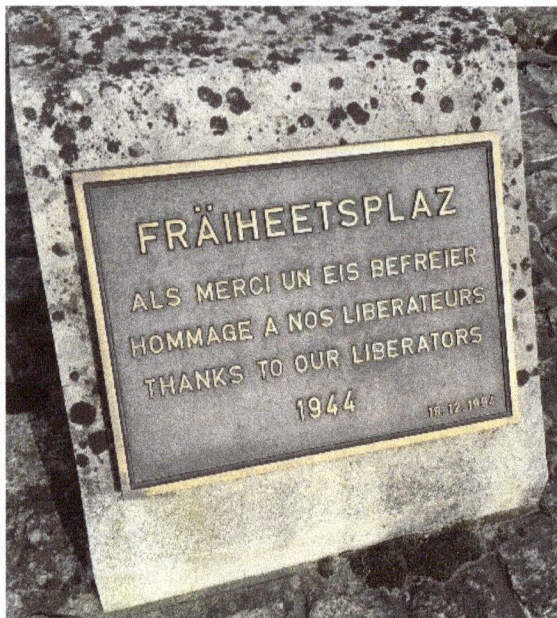

Inscription
Fräiheetsplaz
Als Merci un eis Befreier
hommage à nos liberateurs
Thanks to our Liberators 1944

Freedom Square
As a thank you to our liberators
1944

The monument is at the intersection of route N3 and N13. There is no parking directly at the monument, but street parking is available nearby.

Gilsdorf

Dedicated to "Task Force Rudder" consisting of elements of the 28th Infantry Division and the 10th Armored Division who liberated Gilsdorf on December 24, 1944.

Route de Broderbour 2A

9373 Bettendorf

Latitude: 49.8659

Longitude: 6.1814

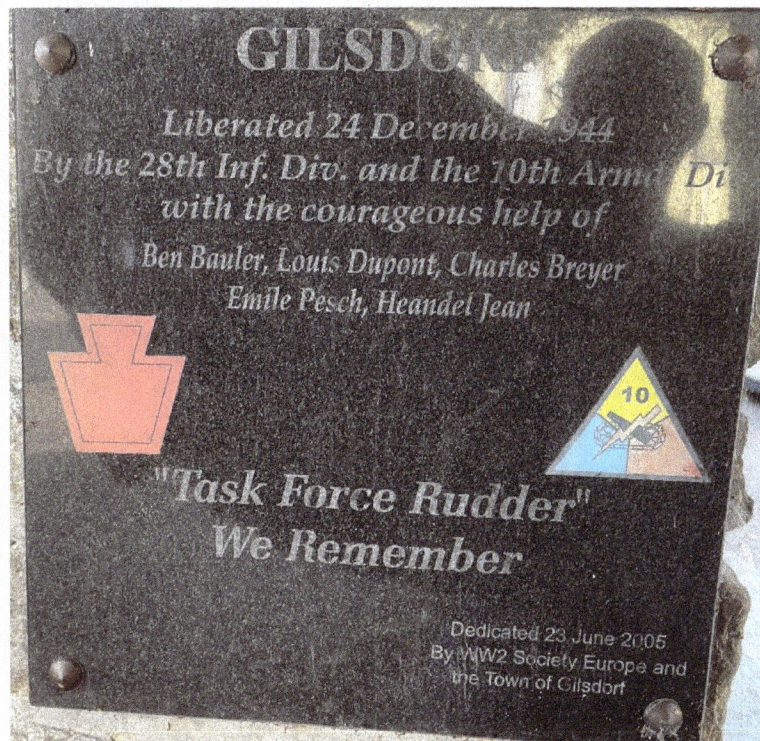

Grevels (Wahl)

Monument located in front of *Centre Nei Brasilien* near intersection of Rue Principale (308) and Rue Kinigshaff in Grevels. Monument in Luxembourgish dedicated to the 26th Infantry Division.

Rue Principale,

8838 Grevels (Wahl)

Latitude: 49.848040

Longitude 5.907818

Mir hunn eis amerikanesch Frënn
vun der 26ter Infanterie-Divisioun net vergiess,

déi am Dezember 1944 d'Gemeng Wahl vun der däitscher
Besatzung befreit hunn

Éier hinnen a Merci fir hiren Asaz

D'Gemeng Wahl an
d'Friends of Patton's 26th Infantry-Division

Night Vigil 2016

We have not forgotten our American friends of the 26th Infantry Division, which in December 1944 liberated the commune of Wahl from the German occupiers.

We honor and thank them.

The commune of Wahl and the Friends of Patton's 26th Infantry Division

Grevenmacher

Dedicated to the members of the 331st Infantry Combat Team, 83rd US Infantry Division.

Route du Vin 22

6794 Grevenmacher

Latitude: 49.6756

Longitude: 6.4416

83rd INFANTRY DIVISION WW II
331st INFANTRY COMBAT TEAM

★ Normandy   ★ Ardennes
★ Brittany   ★ Rhineland
        ★ Central Europe

THIS PLAQUE IS DEDICATED TO THE MEMORY OF
THE MEMBERS OF THE 331st INFANTRY COMBAT TEAM-
83rd INFANTRY DIVISION, U.S. ARMY, WORLD WAR II,
WHOSE SACRIFICES AND SUFFERING DURING THE
CONFLICT, WILL BE REMEMBERED FOREVER.

DEDICATED THIS 26th DAY OF SEPTEMBER
by SAMUEL KLIPPA
1981

Goesdorf

To Private Alfred
Etcheverry, 80th US
Infantry Division

Moettelsergaass 6

9653 Goesdorf

Latitude 49.920864

Longitude: 5.966592

Luxembourgish inscription reads: Jean Thilges 1926-1945 killed by shell shrapnel while speaking to Pvt. Alfred Etcheverry

Gralingen

Dedicated to the 5th US Infantry Division who liberated the villages of the community of Putscheid in January 1945. Site includes information panel about combat operations in the area.

Rue Principale 30

9375 Putscheid (Gralingen)

Latitude: 49.93615

Longitude: 6.100027

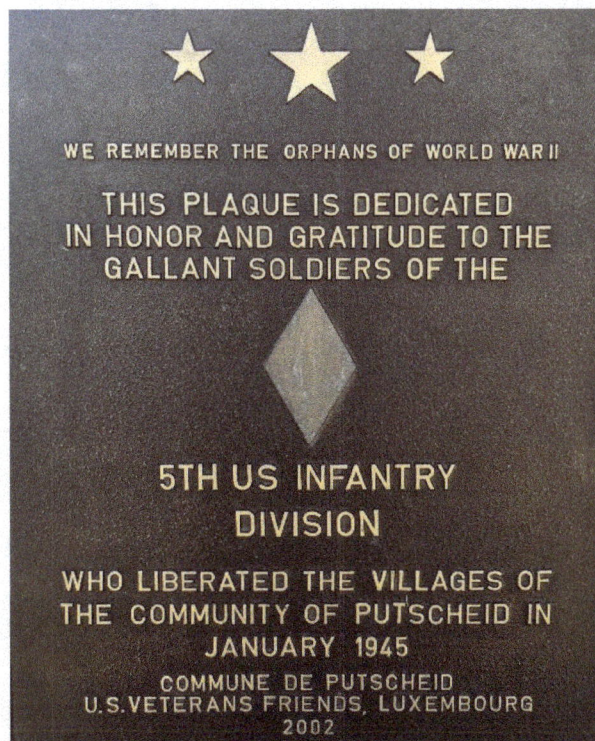

WE REMEMBER THE ORPHANS OF WORLD WAR II

THIS PLAQUE IS DEDICATED IN HONOR AND GRATITUDE TO THE GALLANT SOLDIERS OF THE

5TH US INFANTRY DIVISION

WHO LIBERATED THE VILLAGES OF THE COMMUNITY OF PUTSCHEID IN JANUARY 1945

COMMUNE DE PUTSCHEID
U.S. VETERANS FRIENDS, LUXEMBOURG
2002

Information Panel Gralingen (Putscheid)

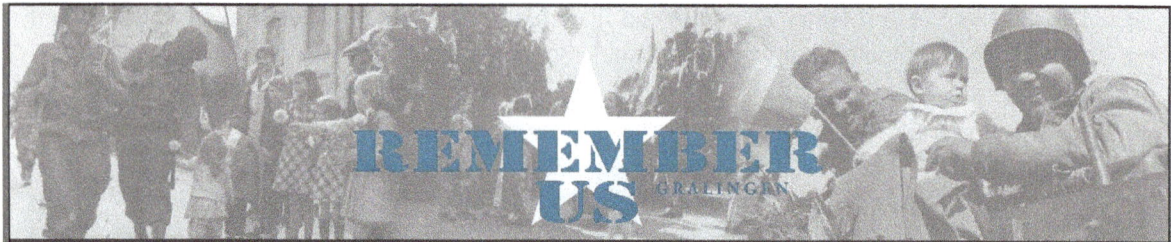

REMEMBER US GRALINGEN

The German surprise attack on December 16, 1944, was directed to capturing the port of Antwerp (Belgium) within three days. General George S. PATTON Jr. ordered his 3rd Army, prepared to attack in the Saar region at that time, to turn 90° north. Some 130.000 vehicles drove within 48 hours about 100 km to the north towards Luxembourg. Four of his key divisions attacked on December 22, 1944:

- The 4th Armored Division smashed the ring around the encircled town of Bastogne (B) on December 26, 1944.
- The 26th Infantry Division had Wiltz as an objective, which was liberated on January 21, 1945.
- The 80th Infantry Division liberated Ettelbruck already on December 25, 1944.

- The 5th Infantry Division operated in the area of Echternach and liberated the town on December 26, 1944.

In January 1945, the 5th Infantry Division liberated the town of Diekirch as well as the villages north of Diekirch up to Hoscheid, Wahlhausen, Gralingen, Merscheid, Weiler, Nachtmanderscheid, and Putscheid.

US military historians state the end of the Battle of the Bulge on January 25 or 28, 1945. In fact, the village of Leithum was liberated on February 1, 1945, by units of the 90th US Infantry Division. The upper town of Vianden was only liberated on February 12 and the lower part on February 22, 1945. It was the last town on the Luxembourg territory to be liberated.

Après l'attaque surprise allemande du 16 décembre 1944, qui avait pour but de conquérir le port d'Anvers (B) endéans les trois premiers jours, le général George S. PATTON Jr qui s'apprêtait à une offensive du côté de la Sarre, tourna sa 3e Armée de 90°, ordonna à 130.000 véhicules de se diriger dans les 48 heures à 100 km vers le Luxembourg et attaqua le 22 décembre 1944 avec quatre divisions de la manière suivante:

- La 4e Division Blindée rompit le siège de Bastogne (B) le 26 décembre 1944.
- La 26e Division d'Infanterie avait la ville de Wiltz comme objectif qu'elle libéra le 21 janvier 1945.
- La 80e Division d'Infanterie libéra Ettelbruck le 25 décembre 1944.
- La 5e Division d'Infanterie opéra dans le secteur

d'Echternach et libéra la ville le 26 décembre 1944.

Au mois de janvier 1945, la 5e Division d'Infanterie libéra la ville de Diekirch et les villages au nord de celle-ci, à savoir Hoscheid, Wahlhausen, Gralingen, Merscheid, Weiler, Nachtmanderscheid et Putscheid.

Les historiens militaires américains considèrent le 25 ou le 28 janvier 1945 comme étant la fin de la Bataille des Ardennes. Toutefois le village de Leithum à l'extrémité nord du pays a été libéré le 1er février 1945 par des unités de la 90e Division d'Infanterie. La partie supérieure de la ville de Vianden fut libérée le 12 février et la partie inférieure le 22 février 1945. C'était la dernière ville qui fut libérée sur le territoire luxembourgeois.

Nach dem deutschen Überraschungsangriff am 16. Dezember 1944, der zum Ziel hatte, den Hafen von Antwerpen (B) in den drei ersten Tagen einzunehmen, drehte General George S. PATTON Jr. seine 3. Armee, die zur Offensive an der Saar bereitstand, um 90°, verlegte 130.000 Fahrzeuge 100 km nördlich nach Luxemburg und griff am 22. Dezember 1944 mit vier Divisionen an:

- Die 4. Panzerdivision sprengte den Ring um Bastnach/Bastogne (B) am 26. Dezember 1944.
- Die 26. Infanteriedivision hatte Wiltz als Objektiv, das sie am 21. Januar 1945 befreite.
- Die 80. Infanteriedivision befreite Ettelbrück am 25. Dezember 1944.
- Die 5. Infanteriedivision operierte im Raum Echternach und befreite die Abteistadt am 26. Dezember 1944.

Im Januar 1945 befreite die 5. Infanteriedivision Diekirch sowie die Dörfer nördlich dieser Stadt, Hoscheid, Wahlhausen, Gralingen, Merscheid, Weiler, Nachtmanderscheid und Pütscheid.

Amerikanische Militärhistoriker erwägen das Ende der Ardennenoffensive am 25. oder 28. Januar 1945. Das Dorf Leithum, an der Nordspitze des Landes gelegen, wurde jedoch erst am 1. Februar 1945 durch Einheiten der 90. Infanteriedivision eingenommen. Die Oberstadt von Vianden wurde am 12. und die Unterstadt am 22. Februar 1945 befreit. Es war dies die letzte Stadt, die auf luxemburgischem Territorium befreit wurde.

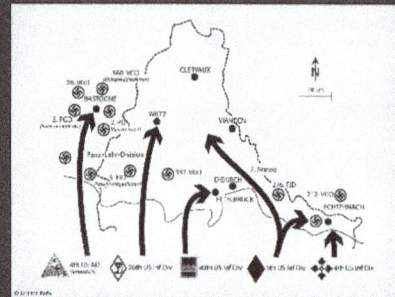

General Patton's counter-attack, December 22, 1944, to January 28, 1945

Hoscheid, 24.01.1945   Merscheid, 1946   Nachtmanderscheid, May 1945

Putscheid, Oct. 1945. "Panther" tank near Imm. Nosbusch

Liberation of the area north of Diekirch, January 18 - February 22, 1945

Haller

In honor of the brave soldiers of the 5th Infantry Division fighting for our freedom around Christmas 1944 and to Private Louis F. Schwall, killed in action in Haller on December 25, 1944.

Rue des Romains 1

6370 Waldbillig

Latitude: 49.819797

Longitude: 6.281472

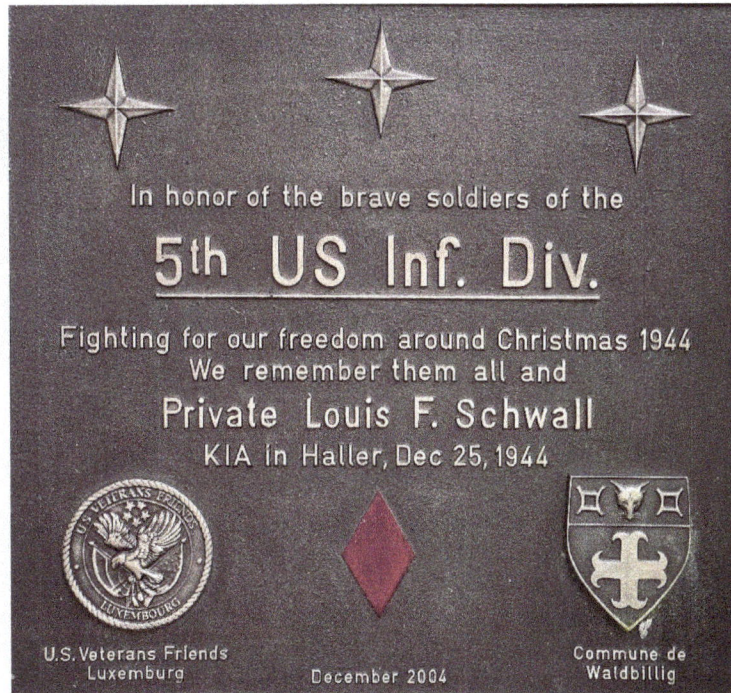

In honor of the brave soldiers of the
5th US Inf. Div.
Fighting for our freedom around Christmas 1944
We remember them all and
Private Louis F. Schwall
KIA in Haller, Dec 25, 1944

U.S. Veterans Friends
Luxemburg

December 2004

Commune de
Waldbillig

Harlange

Memorial park located on the outskirts of Harlange.

CR 309

Harlange 9655

Located on 309 Southeast of Harlange

Latitude 49.92646

Longitude: 5.8018

Heiderscheid

In memory of the 80th US Infantry Division.

Haaptstrooss 37

9158 Esch-sur-Sûre

Latitude 49.88839

Longitude: 5.965601

Inscription in French reads: In memory of the valiant soldiers of the 80th US Infantry Division who fought on these hills during the Battle of the Ardennes.

Heinerscheid

Monument to the 6th US Armored Division. A US 105 mm Howitzer M2A1 is located at the site along with a German 8.8cm PAK 43/41 anti-tank gun.

Route N7

Heinerscheid

Latitude: 50.0937

Longitude: 6.0867

Extreme weather damage to the large white concrete plaque which shows the advance of the 6th US Armored Division through Luxembourg in January 1945.

Heinerscheid (Bunker)

Monument and two information panels about the Heinerscheid *bunker* where five young Luxembourgers hid to avoid forced conscription into the Germany military. On 13 April 1944 two US Army Air Corps aviators who had been shot down earlier in the day were hid in the bunker by the Luxembourg resistance until they could be evacuated (See Hupperdange monument page).

Monument and information panels are located about 3km to the southeast of Heinerscheid on a farm road, an extension of *Kierchestrooss*. The bunker is located about 200 meters from the information panels.

Auto-Pédestre Heinerscheid

Heinerscheid

Latitude: 50.07669

Longitude: 6.11220

Heinerscheid (Bunker) Information Panel

REMEMBER US
HEINERSCHEID

THE HEINERSCHEID "BUNKER" DISASTER (1)

On August 2, 1940, the occupied Grand-Duchy of Luxembourg came from a military administration under a German civilian, party-led, administration. Under the leadership of the Gauleiter Gustav Simon, head of the neighbouring German district, the Gau Koblenz-Trier. Under the name of Zivilverwaltung the German administration implemented a policy of strict Germanisation and Gleichschaltung, the subjugation of every area of public life in Luxembourg, under the rules of the National Socialist Party.

A little more than two years later, on August 30, 1942, Gauleiter Simon declared the introduction of compulsory military service in Luxembourg. At that time the classes of 1920 to 1926, all in all 10,211 men, were supposed to join the ranks of the hated Wehrmacht. The announcement of the Zwangsrekrutierung, forced conscription, was answered by a series of impromptu strikes, as well in Luxembourg as in the French regions of Alsace and Lorraine where the forced service in the German army had also been implemented. The strikes were harshly quelled by the NAZIs, many of the leaders were executed or deported. While many young Luxembourgers accepted their fate, 3,501 choose to desert and either tried to join the allied forces or hid in secret caches.

In Heinerscheid five young Luxembourgers, Pierre Jungels, Alphonse Jungels, Willy Kremer, Alphonse Kremer and Josy Frères tried to avoid forced conscription by going underground. When in August 1943, the number of raids and searches by the German security services in the area of Heinerscheid increased the Jungels brothers and the Kremer twins were forced to leave their hideout inside the village. They moved to a stand of firs in the Am Pesch forest, deep in the Our river valley, where they started building a large dugout, a so-called Bunker. There they were joined by Josy Frères, another refractaire, deserter from the German army. To erase all traces of their building activity they filled the soil from the diggings into buckets and carried it to a bog, about 200 yards away. Some villagers provided food, and other necessities like clothes, so the young men could subsist.

Picture of the inside of a "Bunker" built by Luxembourgish internees.
Photo de l'intérieur d'un "Bunker" construit par des réfractaires.
Bild vom Innern eines "Bunkers" so den sich Refraktäre versteckt hatten.

On April 13, 1944 they were rejoined in their hideaway by two US Army Air Corps aviators, Staff Sergeant Joe Kerpan and Second Lieutenant Robert Korth, who were lead into security by members of the Luxembourgish resistance. Earlier that day their B-24 Liberator bomber, nicknamed Tenovus, had started, on a bombing mission to Munich, from a base in England. After having accomplished their mission Tenovus came under attack of German fighter planes on the flight back home. All tough the 10 men crew succeed in repulsing the attacks of the Luftwaffe fighter planes, their B-24 was so badly damaged that they had to abandon their aircraft, which crashed in-between the villages of Huppergange and Grindhausen. Today a monument in honour of the US pilots stands at the crash site. Staff Sergeant Joe Kerpan and Second Lieutenant Robert Korth spend a few days with the réfractaires in their Bunker, before being picked up by members of the Resistenz, who put them on a rat line to rejoin the United Kingdom by the way of Belgium and France. But after a few days they ran out of luck and were caught by the Germans when they were stopped at a checkpoint.

LE DÉSASTRE DU BUNKER DE HEINERSCHEID (1)

Dès le 2 août 1940, le Luxembourg passa d'une administration militaire à une administration civile (Zivilverwaltung) sous le chef de district (Gauleiter) de la région Trèves-Coblence, à savoir Gustav Simon. Ce chef de l'administration civile (Chef der Zivilverwaltung) entama une véritable politique de germanisation et de mise au pas dans tous les domaines de la vie publique de la population luxembourgeoise.

Le 30 août 1942, Gustav Simon décida d'instaurer le service militaire obligatoire, qui concerna à l'époque 10,211 hommes des classes 1920 à 1926. Cette disposition entraîna des grèves au Luxembourg, mais aussi en Alsace et en Lorraine, qui durent également se plier au service militaire obligatoire dans l'armée allemande (Wehrmacht). Ces grèves furent néanmoins sévèrement réprimées, notamment par des déportations et des exécutions. Tandis qu'une grande partie des jeunes luxembourgeois intégrèrent l'armée allemande, 3,501 réfractaires essayèrent d'échapper à l'enrôlement de force, en fuyant ou en se cachant.

À Heinerscheid, cinq jeunes luxembourgeois, à savoir Pierre Jungels, Alphonse Jungels, Willy Kremer, Alphonse Kremer et Josy Frères rentèrent eux-aussi d'échapper au service militaire en se cachant. Lorsque les perquisitions par la police allemande se multiplièrent dans la région de Heinerscheid dès août 1943, les frères Jungels et les jumeaux Kremer furent obligés de quitter leur planque au village. Ils se rendirent dans une forêt de sapins à proximité, au lieu-dit « Am Pesch », où ils décidèrent de creuser une cachette. Ils furent rejoints par un autre réfractaire en fuite, à savoir Josy Frères. Pour effacer toute trace de leur construction, ils transportèrent la terre enlevée dans un seau vers une mare, se trouvant à environ 180 mètres de leur cachette. Certains habitants leur apportèrent régulièrement des provisions et d'autres biens de première nécessité, comme par exemple des vêtements, permettant ainsi de couvrir leurs besoins.

Le 13 avril 1944, ils furent rejoints par deux aviateurs américains, le Staff Sergeant Joe Kerpan et le Second Lieutenant Robert Korth, qui y furent guidés par des membres de la résistance. Plus tôt ce même jour, leur bombardier, un B-24 Liberator, auquel ils arrivèrent le surnom Tenovus, décolla d'Angleterre avec un équipage de dix hommes ayant pour mission de bombarder certaines positions à Munich et aux alentours. Bien que leur but fût accompli, ils furent attaqués par des chasseurs allemands sur le retour, qu'ils réussirent néanmoins à repousser. Cependant, leur avion avait subi de tels dégâts qu'ils furent obligés de l'abandonner. Leur B-24 s'écrasa donc entre Huppardange et Grindhausen, où se trouve aujourd'hui un monument dédié à l'équipage. Le Staff Sergeant Joe Kerpan et le Second Lieutenant Robert Korth restèrent quelques jours auprès des réfractaires, avant de les quitter pour la France, d'où ils tentèrent de rejoindre la Grande-Bretagne. Malheureusement ils furent capturés quelques jours plus tard lors d'un contrôle d'identité.

DIE TRAGÖDIE IM BUNKER VON HEINERSCHEID (1)

Ab dem 2. August 1940 wurde die deutsche Militärverwaltung durch eine Zivilverwaltung, an deren Spitze Gustav Simon, Gauleiter der Region Trier-Koblenz (D) stand, abgelöst. Simon verfolgte eine regelrechte Germanisierungspolitik und ordnete zugleich die Gleichschaltung des öffentlichen Lebens an.

Am 30. August 1942 entschied Gustav Simon, den obligatorischen Militärdienst einzuführen, wovon damals 10.211 Männer der Jahrgänge 1920 bis 1926 betroffen waren. Diese Anweisung zog mehrere Streiks im Land nach sich, genauso wie auch in der Region Elsass-Lothringen (F), welche sich ebenfalls dem obligatorischen Militärdienst beugen musste. Diese Streiks wurden mit aller Strenge niedergeschlagen, besonders durch Deportationen und Erschießungen. Während ein Großteil der jungen Luxemburger sich in ihr Schicksal ergab, haben 3501 Refraktäre versucht, dieser Zwangsrekrutierung zu umgehen, indem sie davor flüchteten oder sich versteckten.

In Heinerscheid versuchten fünf junge Luxemburger, namentlich Pierre Jungels, Alphonse Jungels, Willy Kremer, Alphonse Kremer und Josy Frères der Zwangsrekrutierung zu entgehen, indem sie sich versteckten. Als die Hausdurchsuchungen der deutschen Polizei in der Region von Heinerscheid im August 1943 zunahmen, sahen sich die Gebrüder Jungels und die Zwillingsbrüder Kremer gezwungen ihr Versteck im Dorf zu verlassen. Sie begaben sich in einen, sich in der Nähe befindlichen Tannenwald, in der Örtlichkeit „Am Pesch". Hier entschieden die vier Refraktäre ein Versteck in den Boden zu graben. Kurz darauf gesellte sich auch Josy Frères zu ihnen. Um jede Spur dabei zu verwischen, trugen sie den Aushub mit einem Eimer zu einem Weiher, der sich zirka 180 Meter vom ihrem Versteck entfernt befand. Einwohner aus der Umgegend versorgten sie regelmäßig mit Lebensmitteln und anderen dringend benötigten Gegenständen, wie beispielsweise Kleider, damit es ihnen an nichts mangelte.

Am 13. April 1944 führten Mitglieder der lokalen Resistenzorganisation die beiden abgestürzten Flieger Staff Sergeant Joe Kerpan und Second Lieutenant Robert Korth zu dem Bunker. Ihr Flugzeug, ein B-24 Liberator Bomberflugzeug mit dem Namen Tenovus, war am selben Tag in England mit einer Besatzung von 10 Mann gestartet, um einige Positionen in München (D) und Umgegend zu bombardieren. Nachdem sie ihre Bomben ins Ziel gebracht hatten, wurden sie auf dem Rückflug von deutschen Abfangjägern angegriffen. Es gelang der Besatzung zwar, die feindlichen Angriffe abzuwehren, allerdings wurde ihr Flugzeug derart beschädigt, dass sie es aufgeben mussten. Die B-24 Liberator zerschellte anschließend zwischen Hupperdingen und Grindhausen. Heute befindet sich dort ein Monument zu Ehren der Besatzung. Staff Sergeant Joe Kerpan und Second Lieutenant Robert Korth blieben einige Tage im Versteck der Refraktäre, ehe sie, mit Hilfe der Resistenz, die gefährliche Flucht durch Belgien und Frankreich, quer durch feindliches Territorium, antraten. Es sollte den US Piloten jedoch nicht gelingen Großbritannien zu erreichen, bei einer Ausweiskontrolle im besetzten Frankreich wurden sie von den Deutschen verhaftet.

The five refractaires who were killed in the "Bunker".
Les cinq réfractaires tués au "Bunker".
Die fünf Refraktäre die im Bunker ihr Leben ließen.

The crew of the B-24 bomber "Tenovus".
L'équipage du bombardier US B-24 "Tenovus".
Die Besatzung des B-24 Bombers "Tenovus".

For more information
Pour en savoir plus
Für weitere Informationen

Heinerscheid (Bunker) Information Panel

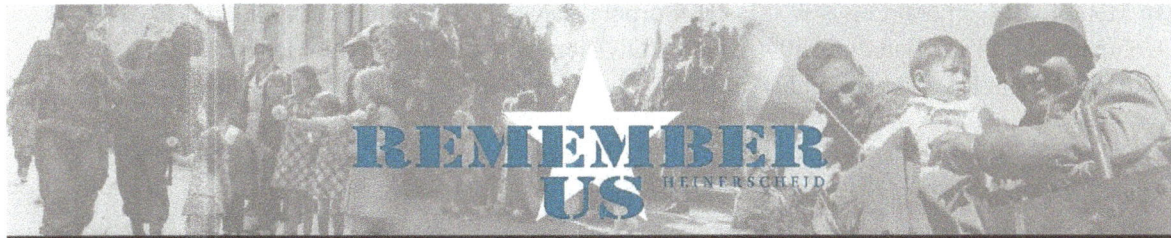

REMEMBER US HEINERSCHEID

THE HEINERSCHEID "BUNKER" DISASTER (2)

For the réfractaires their life in hiding seemed to go on without any difficulties, but they were careless and committed a string of errors that were to cost them dearly. The young men started lighting fires in order to cook their meals and warm up. On several evenings they went back into the village to visit friends and families.

One day German soldiers, who were on guard in one of the Siegfriedline pillboxes, on the opposite slopes of the river valley, noticed smoke rising form the stand of fir trees where the young Luxembourgers were hiding. They informed the Ortsgruppenleiter Egidius Wolter, the local party hack. The quisling and his son Franz rallied a few German Policemen and went into the Our river valley to investigate. On the evening of April 23, 1943 they discovered the deserters and opened fire on the Luxembourgers who answered the shots before fleeing to Heinerscheid where they warmed to re-supply on food and clothing, before trying to organise a new hiding place. But in the night of April 24, to April 25, the deserters were surrounded in their dugout in the Posch forest by a large force of German Policemen. It came to another shootout and the Luxembourgers killed a traitor named Weber, an auxiliary Policeman. The réfractaires took a submachine gun from the killed Auxiliary Police Officer and managed to wound several other assailants who were then forced to retreat. Giving up was no option for the Germans though and they came back in force. at 14:30, on April 25, 1944, the German security forces launched another attack on the Bunker! The Luxembourgish boys hit back and managed to wound even more Policemen, but after a while the Germans succeeded in placing a breaching charge onto the top of the dugout. The force of the explosion that ensued was so strong that the hideout was completely destroyed and the neighbouring trees were uprooted. None of the five young men was to survive this final onslaught.

LE DÉSASTRE DU BUNKER DE HEINERSCHEID (2)

Pour les réfractaires, la vie en cachette semblait continuer sans difficultés, mais ils commirent certaines imprudences, qui leur coûtèrent cher. En effet, ils allumèrent parfois un feu pour cuisiner ou se réchauffer, et ils montrèrent à plusieurs reprises le soir dans le village.

Un jour, des soldats allemands stationnés de l'autre côté de l'Our, donc de la rivière frontière entre l'Allemagne et le Luxembourg, dans les fortifications du Westwall aperçurent de la fumée au-dessus de la forêt de sapins. Ceux-ci en informèrent l'Ortsgruppenleiter de Heinerscheid, Egidius Wolter, qui se rendit avec son fils Franz et plusieurs gendarmes allemands dans le secteur indiqué. Le 23 avril 1944 au soir, Wolter et son fils aperçurent les réfractaires et ouvrirent le feu sur les jeunes hommes, qui ripostèrent avant de se rendre à Heinerscheid pour se ravitailler et changer de cachette. Pourtant, dans la nuit du 24 au 25 avril 1944, ils se voyaient encerclés par d'importants contingents de gendarmerie allemande. Les réfractaires et les gendarmes allemands se livrèrent donc à un échange de tirs, lors duquel les réfractaires réussirent à abattre un traître luxembourgeois du nom de Weber, qui servait en tant qu'agent de police auxiliaire. Ils s'emparèrent de sa mitraillette et réussirent à blesser d'autres gendarmes, qui se voyaient obligés de battre en retraite. Néanmoins, les Allemands ne se montèrent pas vaincu et revinrent avec un grand nombre d'hommes armés, qui ouvrirent le feu vers 14.30 heures le 25 avril 1944.

Les jeunes réfractaires se défendirent et blessèrent quelques Allemands, avant que ceux-ci ne réussirent à placer une charge explosive sur leur abris. La détonation fut si violente que la terre fut projetée en l'air et que des arbres furent déracinés. Aucun des cinq Luxembourgeois ne survécut.

DIE TRAGÖDIE IM BUNKER VON HEINERSCHEID (2)

Für die Refraktäre sah es eine Zeit lang so aus, als ob sie ihr Leben im Versteck ohne Schwierigkeiten weiterführen könnten. Doch einige Unachtsamkeiten sollten sie teuer zu stehen kommen. Tatsächlich machten sie Feuer zum Kochen oder um sich zu wärmen, oder sie zeigten sich mehrfach abends im Dorf. Eines Tages bemerkten deutsche Soldaten auf der gegenüberliegenden Seite des Grenzflusses Our, im sogenannten Westwall, eine Rauchsäule über dem Tannenwald, in welchem sich das Versteck der fünf Refraktäre befand. Daraufhin informierten sie den Ortsgruppenleiter aus Heinerscheid, Egidius Wolter, welcher sich zusammen mit seinem Sohn Franz und mehreren deutschen Gendarmen an die vorgegebene Stelle begab. Am Abend des 23. April 1944 entdeckten Wolter und sein Sohn die Refraktäre und eröffneten das Feuer auf die jungen Männer, die sich zur Wehr setzten und sich später nach Heinerscheid absetzten, um sich mit Lebensmitteln zu versorgen, sowie das Versteck zu wechseln. Trotzdem konnten sie die Umzingelung durch ein bedeutendes Kontingent von deutschen Polizisten in der Nacht vom 24. auf den 25. April 1944 nicht verhindern. Es kam zu einem Schusswechsel zwischen den Refraktären und der deutschen Polizei, dabei wurde ein luxemburgischer Verräter namens Weber, der als Hilfsgendarm diente, getötet. Sie bemächtigen sich dessen Maschinenpistole und weitere deutsche Polizeibeamte erlitten Schussverletzungen. Letztere sahen sich schließlich gezwungen, den Rückzug anzutreten. Nichtsdestotrotz waren die Deutschen damit nicht besiegt und sie kamen mit einer höheren Anzahl bewaffneter Männer zurück, die am 25. April 1944 gegen 14.30 Uhr erneut das Feuer eröffneten. Die jungen Refraktäre verteidigten sich weiter und verletzten wiederum einige der Angreifer, ehe es diesen gelang eine Sprengladung auf dem Dach des Bunkers anzubringen. Die Explosion war derart heftig, dass die Walderde in die Luft gewirbelt wurde und Bäume entwurzelt wurden. Hierbei kamen alle fünf Refraktäre ums Leben.

Frères Joiy

Jungels Alphonse

Jungels Pierre

Kreiner Alphonse

Kreiner Willy

The mortal remains of the réfractaires were taken to Germany where they were secretly buried. In 1946 the inhabitants of Heinerscheid came to know about the resting place of their fellow countrymen and had them repatriated to their native village where they were put to rest with full military honors.

Leurs dépouilles furent transportées et enterrées clandestinement en Allemagne. Mais en 1946, après la fin de la Seconde Guerre mondiale, les habitants de Heinerscheid et des alentours apprirent la localisation des dépouilles de leur compatriotes et les firent rapidement rapatrier au village, où ils eurent droit à des funérailles avec honneurs militaires.

Ihre sterblichen Überreste wurden heimlich nach Deutschland transportiert und dort begraben. Erst im Jahre 1946 nach dem Ende des Zweiten Weltkrieges erfuhren die Einwohner aus Heinerscheid und Umgegend wo ihre Landsleute begraben lagen. Sehr schnell wurden ihre sterblichen Überreste zurück in ihr Heimatdorf gebracht, wo ihnen eine Begräbnisfeier mit allen militärischen Ehren zugestanden wurde.

Sources - Quellen:

American Air Museum. « 41-29132B-21 LIBERATOR ». American Air Museum in Britain. [online], http://www.americanairmuseum.com/aircraft/858 (27 septembre 2014, 17 février 2019).

Informations gained from correspondance and research done by Mr Pierry Frère, former member of CEBA.

DERNEDEN (John). Crash : Absturz und Notlandungen von alliierten und deutschen Flugzeugen in Luxemburg 1940-45, Band 1. Luxembourg: GREC, 2004.

KNEPPER (Aimé). Les réfractaires dans les bunkers Luxembourg : Édition Saint-Paul, 2003.

KREINS (Jean-Marie). Histoire du Luxembourg : des origines à nos jours. Paris : Presses Universitaires de France, 2016 (Que sais-je ?).

PAULY (Michel). Histoire du Luxembourg. Bruxelles : Éditions de l'Université de Bruxelles, 2017.

The military fire-toll of the five patriots rendered by the occupiers.
Les funérailles avec honneurs militaires des cinq patriotes rendues par les occupants.
Die Beisetzung mit militärischen Ehren der fünf, von den Besatzern rendierten Patrioten.

For more information
Pour en savoir plus
Für weitere Informationen

Hesperange

In Memory of Corporal Lewis W. Meade, Corporal James G. Russ, and Tec4 Isidore M. Vasko, B Company, 11th Tank Battalion, 10th Armored Division, who died when their tank fell off the bridge into the Alzette River on December 26, 1944.

Route de Thionville 434

5886 Hesperange

Latitude: 49.5728164

Longitude: 6.1568230

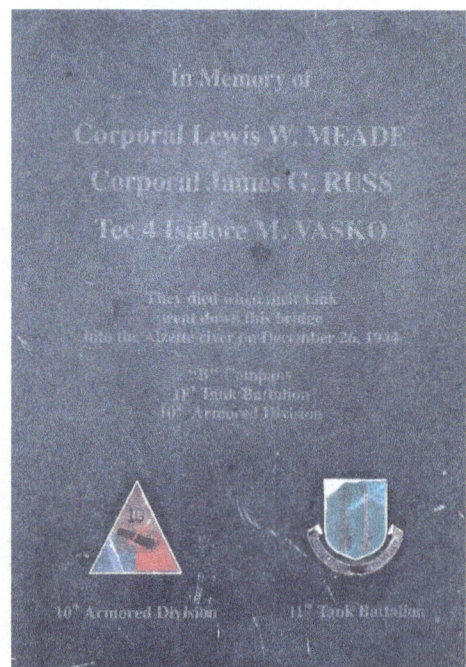

Photo of tank accident in Hesperange courtesy of the U.S. Embassy Luxembourg (retrieved from Wikimedia Commons)

Hesperange

In memory of 8 members of the 84th Infantry Division, 333rd Infantry Regiment, who died on October 22, 1945 in a tragic traffic accident when a military truck loaded with fuel cans hit a small train called 'Jangeli' and exploded.

Route de Thionville 334

5884 Hesperange

Latitude: 49.57828

Longitude: 6.15347

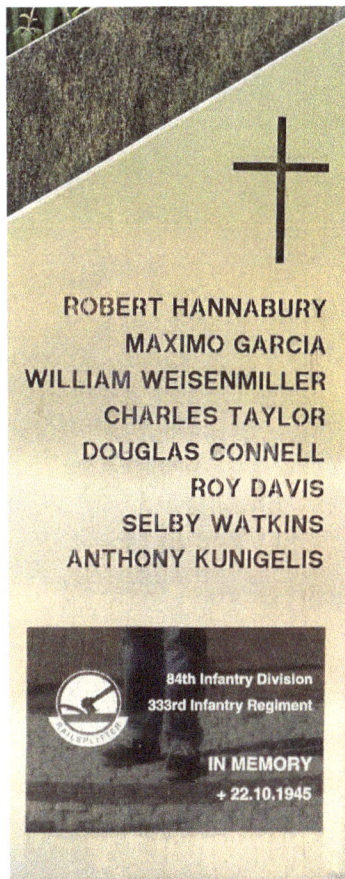

The eight killed in the accident are: Robert Hannabury, Maximo Garcia, William Weisenmiller, Charles Taylor, Douglas Connell, Roy Davis, Selby Watkins, and Anthony Kunigelis

Hoesdorf (Reisdorf)

Dedicated to the 109th Infantry Regiment, 28th Infantry Division. This is also the starting point for the Hoesdorf-Bettendorf-Wallendorf Trail (See trails section).

Route 10

6841 Reisdorf

Latitude: 49.880378

Longitude: 6.256175

While the Monument was presented by the *City of Reisdorf*, the actual location is just at the entrance of Hoesdorf when coming from Reisdorf.

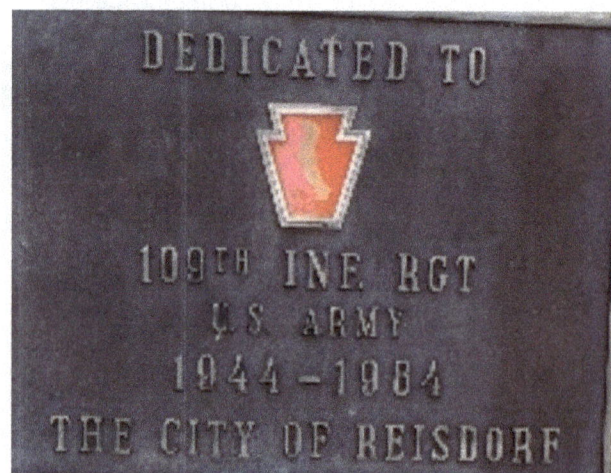

Information Panel Hoesdorf

Promenade du Souvenir Hoesdorf

Circuit autopédestre historique Hoesdorf-Bettendorf

Panneau synoptique (Château d'eau à Hoesdorf)

Welcome to the "Hoesdorf Plateau" historical circuit related to the dramatic events that took place in this area in 1944/45. What is now a peaceful and pleasant landscape, was once the site of fierce fighting and human suffering. The small town of **Wallendorf** (on the right wing), located across the Luxembourg-German border on the confluence point of the Sûre (Sauer) and Our rivers, made the headlines in the wire services and the press, when shortly after the first U.S. troops had reached the borders of the "Third Reich" on **September 11, 1944**, strong elements of **CCR (Combat Command R)** of the **5th U.S. Armored Division**, supported by sub units of the **112th Infantry Regiment (28th U.S. Infantry Division)**, pierced the "Siegfried" line and pushed in direction of Bitburg (Germany), capturing a number of villages, as of **September 14, 1944**. This "high impact" action, initially crowned by success, met considerable enemy resistance around September 20 and had to be abandoned around the end of the month because of lack of flanking protection and sufficient logistical support. In early **October 1944**, units of the **8th U.S. Infantry Division** took over the former positions of the 5th U.S. Armored Divison, and established a number of observation and listening posts on the highground of the "Hoesdorf Plateau" facing the well-camouflaged German "Westwall" (Siegfried line) fortifications across the Our river. Around Thanksgiving Day **(late November 1944)**, elements of the **109th Infantry Regiment (28th U.S. Infantry Division)** after having suffered sizeable losses in the preceding "Huertgen" forest battle near Aachen, Germany, relieved the units of the 8th U.S. in this same sector and unfolded extensive reconnaissance and patrolling activity. From there on, the thinly-spread U.S. defensive line of the entire 28th U.S. Infantry Division stretched from the **Sauer/Our river confluence point at Wallendorf** to the **Ouren/Burg Reuland** sector, some 35 kilometres north, following the flow (in opposite direction) of the Our river. The 109th Infantry Regiment claimed responsibility of the greater **Hoesdorf-Bettendorf-Longsdorf** plateau with **Wallendorf** and **Vianden/Stolzembourg** as its sector limits. When in the early morning hours of **December 16, 1944**, strong German combat elements of the **352nd Volksgrenadier- and 5th Paratroop Divisions** crossed the Our river after a terrific 30-minute artillery barrage to attack the U.S. defense line, the prementioned sector became the site of deadly (often close combat) fighting with heavy losses on both sides. Making good use of the steep incline of the Our river valley, the outnumbered Americans were able to slow down the German advance until December 18-19, 1944, when after days of exhausting fighting, they were ordered to new defensive positions south-west of Diekirch and Ettelbruck. Although the enemy's advance on the Southern shoulder of the "Bulge" was stopped around Christmas, the German units kept control of the Hoesdorf-Bettendorf-Longsdorf plateau until the **end of January 1945**, when units of General Patton's Third Army (primarily the **4th and 5th U.S. Infantry Divisions**) pushed them back to the initial December 16, 1944 jump off line on the "Westwall". In **early February 1945**, units of the **80th U.S. Infantry Division** crossed the Our river from the Hoesdorf plateau angle, piercing the "Siegfried line" marking the beginning of the "Invasion of Germany" campaign.

The map shows the various "tour stops" with more detailed narrative panels of the circuit. While touring this historical ground, please respectfully remember all those unknown who fought, suffered, and died here – the opposed military and the civilians caught in the crossfire.

Attachment : a) croquis livre R. Gaul "Situation unités opposantes du 16 décembre 1944
             a) copie tracte "Security first"

Hoscheid

Monument and information panel dedicated to the 5th Infantry Division and the liberation of Hoscheid, especially honoring the 38 soldiers of the division's 11th Infantry Regiment..

4 Lisseneck

9377 Hoscheid

Latitude: 49.945836

Longitude: 6.079731

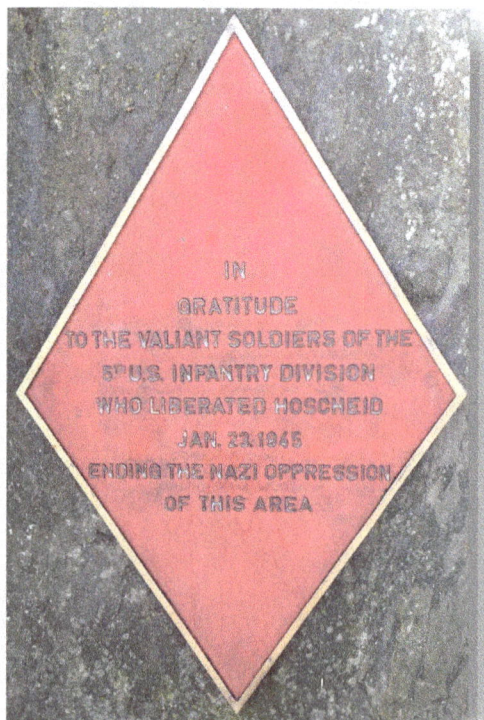

IN
GRATITUDE
TO THE VALIANT SOLDIERS OF THE
5th U.S. INFANTRY DIVISION
WHO LIBERATED HOSCHEID
JAN. 22 1945
ENDING THE NAZI OPPRESSION
OF THIS AREA

★ ★ ★
5 th INFANTRY DIVISION
SQUARE
LIBERATIOUNSPLAAZ
DEDICATED TO THE 38 YOUNG AMERICAN SOLDIERS
WHO GAVE THEIR LIVES FOR THE LIBERATION
OF THE HOSCHEID AREA
WE SHALL ALWAYS REMEMBER
US VETERANS FRIENDS LUXEMBOURG
JAN. 24 th 1999   AND COMMUNE DE HOSCHEID

Information Panel Hoscheid

First Liberation of Hoscheid

While the 109th, 110th and 112th regiments of the **28th US Infantry Division** liberated the north of the Grand Duchy on September 11 - 13, 1944, the **5th US Armored Division (5th AD)** brought freedom back to the south of our country that had greatly suffered under the four-year Nazi-occupation. **Combat Command B** of the **5th US AD** turned north, **liberated Hoscheid on September 11, 1944**, and allied soldiers of the Troop B, 85th Cavalry Reconnaissance Squadron, crossed the German border at Stolzembourg.

Hoscheid in the Battle of the Bulge

On **December 16, 1944**, three guns of 110th Regiment's Anti-tank Company, 28th US Infantry Division, as well as armored support units of the **707th Tank Battalion** defended the village of **Hoscheid**. The German *14th Parachute Infantry Regiment, 5th Parachute Division*, pressed on Hoscheid.

The night of **December 17, 1944**, General Norman D. Cota, commander of the 28th Infantry Division, ordered an armored and infantry platoon in support of the garrison stationed at Hoscheid. However, the *2nd Battalion* of the German *14th Parachute Regiment* stopped the relief force.

On **December 17**, the US garrison repulsed three enemy infantry attacks from the west and assault guns attacking from the north. By defending Hoscheid stubbornly, the US garrison succeeded in delaying the German's advance towards the river Clerf. The unyielding resistance of the **28th US Infantry Division** allowed the Americans to bring reserve units to Bastogne (B) and thus severely disrupt the German advance. As the US tanks ran out of ammunition and fuel, the garrison broke through the enemy ring and retreated towards the 687th Field Artillery command post at Lipperscheid. Together with the artillerymen, they retreated to Wiltz where they helped defend the town.

Second Liberation of Hoscheid

On January 18, 1945, the 4th and 5th US Infantry Divisions (ID) crossed the river Sauer east and west of Diekirch. 5th ID continued its advance north of Diekirch against remnants of 352nd Volks-Grenadier-Division (VGD) and Panzer-Lehr-Division (PLD). The Germans defended the line Hoscheid – Vianden tenaciously to allow their retreating troops to reach their homeland.

The American units needed three days of tough fighting to liberate **Hoscheid**. On January 24, the soldiers of 11th Regiment, part of 2nd Regiment, **5th ID** supported by tanks of Company A, **737th Tank Battalion**, attacked **Hoscheid** under heavy German artillery fire. On the German side, elements of the *212th Grenadier Regiment, 79th VGD*, of the *916th GR, 352nd VGD*, and elements of the *PLD* defended the village. After merciless armor and infantry struggle as well as house-to-house fighting, the G.I.s occupied and liberated Hoscheid at 02:30 p.m.

Sources: Pierre Tatsch, L.I. Melchers, Robert Phillips, United States Army in World War II, Jean Milmeister, The Fifth Infantry Division in the ETO

Première libération de Hoscheid

Alors que les 109e, 110e et 112e régiments de la **28e Division d'Infanterie américaine** libérèrent le nord du Grand-Duché du 11 au 13 septembre 1944, la **5e Division Blindée américaine** libéra le sud de notre pays qui avait très souffert pendant quatre années sous l'occupation nazie. **L'Unité de Combat B** de la **5e Division Blindée** tourna vers le nord, **libéra Hoscheid le 11 septembre 1944** et des soldats alliés de la Troop B, 85th Cavalry Reconnaissance Squadron, traversèrent la frontière allemande à Stolzembourg.

Hoscheid pendant la Bataille des Ardennes

Le 16 décembre 1944, trois pièces de la compagnie anti-chars, 110e Régiment, 28e Division d'Infanterie américaine, ainsi que des unités blindées d'appui du « 707e Tank Bataillon » défendaient le village de **Hoscheid**. Le *14e régiment de la 5e division Parachutiste allemande* fit pression sur Hoscheid.

Durant la nuit du **17 décembre**, le général Norman D. Cota, commandant de la 28e US Division d'Infanterie, ordonna un peloton de chars et d'infanterie en renfort de la garnison de Hoscheid. Cependant, le *2e Bataillon du 14e Fallschirmjägerregiments* arrêta la force de secours.

Le **17 décembre**, la garnison américaine repoussait trois attaques d'infanterie ennemies venant de l'ouest et des chars d'assaut attaquant depuis le nord. En défendant obstinément Hoscheid, la garnison américaine réussit à retarder l'avance allemande vers la rivière Clerf. La résistance acharnée de la **28e Division d'Infanterie américaine** permit aux Américains d'amener des unités de réserve à Bastogne (B) et ainsi de perturber sévèrement l'avance allemande. Quand les chars américains manquaient de munitions et de carburant, les défenseurs percèrent l'anneau ennemi et se retirèrent au poste de commandement du 687e Field Artillery Bataillon à Lipperscheid. Avec les artilleurs, ils se replièrent sur Wiltz où ils participaient à la défense de la ville.

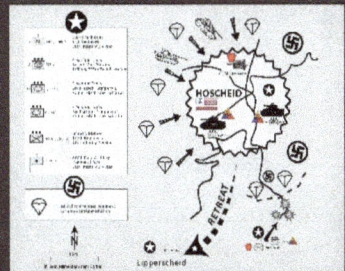

Deuxième libération de Hoscheid

Le 18 janvier 1945, les 4e et 5e divisions d'infanterie américaines (DI) traversèrent la rivière Sûre à l'est et à l'ouest de Diekirch. La 5e DI poursuivit son avance au nord de Diekirch contre les restes de la 352e Volks-Grenadier-Division (VGD) et de la Panzer-Lehr-Division (PLD). Les Allemands défendaient la ligne Hoscheid – Vianden avec ténacité pour permettre à leurs troupes en retraite de rejoindre l'Allemagne.

Les unités américaines eurent besoin de trois jours de durs combats pour libérer **Hoscheid**. Le 24 janvier, les soldats du 11e Régiment, d'une partie du 2e Régiment de la **5e DI**, des chars de la compagnie A du 737e **Bataillon de chars** attaquèrent Hoscheid sous le feu nourri de l'artillerie allemande. Du côté allemand, des éléments du 212e Grenadier Regiment, de la 79e VGD, du 916e GR de la 352e VGD, et des éléments de la PLD défendaient le village. Après d'impitoyables combats de chars, d'infanterie et de maison en maison, les G.I.s occupaient et libéraient Hoscheid à 14h30.

Erste Befreiung von Hoscheid

Während die 109., 110. und 112. Regimenter der **28. US-Infanteriedivision** am 11. - 13. September 1944 den Norden des Großherzogtums befreiten, brachte die **5. US-Panzerdivision** die Freiheit in den Süden des Landes zurück, das vier Jahre schwer unter der Nazi-deutschen Besatzung gelitten hatte. Die **Kampfeinheit B** der **5. US-Panzerdivision** wandte sich nach Norden, befreite **Hoscheid am 11. September 1944** und alliierte Soldaten der Troop B, 85th Cavalry Reconnaissance Squadron, überquerten die deutsche Grenze in Stolzemburg.

Hoscheid in der Ardennenschlacht

Am 16. Dezember 1944 verteidigten drei Geschütze der Anti-Tank-Kompanie des 110. Regiments, 28. US-Infanteriedivision, sowie gepanzerte Unterstützungseinheiten des **707. Panzerbataillons** das Dorf **Hoscheid**. Das deutsche *14. Fallschirmjägerregiment, 5. Fallschirmjägerdivision*, bedrängte Hoscheid.

In der Nacht zum **17. Dezember** befahl General Norman D. Cota, Kommandeur der 28. US-Infanteriedivision, einen Panzer- und Infanteriezug zur Unterstützung der in Hoscheid stationierten Garnison. Das *2. Bataillon des 14. Fallschirmjägerregiments* stoppte jedoch die Hilfstruppe.

Am **17. Dezember** wehrte die US-Garnison drei feindliche Infanterieangriffe aus dem Westen und Sturmgeschützangriffe aus dem Norden ab. Es gelang ihr, den Vormarsch der Deutschen zum Fluss Clerf zu verzögern. Der hartnäckige Widerstand der **28. US-Infanteriedivision** erlaubte es den Amerikanern, Reserveeinheiten nach Bastogne (B) zu bringen und so den Vormarsch der Deutschen empfindlich zu stören. Als den US-Panzern Munition und Treibstoff ausgingen, durchbrachen die Verteidiger den feindlichen Ring und zogen sich zum Gefechtsstand des 687. US-Feldartillerie-Bataillons nach Lipperscheid zurück. Zusammen mit den Artilleristen gelangten sie nach Wiltz, wo sie die Stadt verteidigen halfen.

Zweite Befreiung von Hoscheid

Am 18. Januar 1945 überquerten die 4. und 5. US-Infanteriedivisionen (ID) den Fluss Sauer östlich und westlich von Diekirch. Die 5. ID setzte ihren Vormarsch nördlich von Diekirch gegen Reste der 352. Volks-Grenadier-Division (VGD) und der Panzer-Lehr-Division (PLD) fort. Die Deutschen verteidigten hartnäckig die Linie Hoscheid – Vianden, um ihren Einheiten den Rückzug in die Heimat zu erlauben.

Die amerikanischen Verbände brauchten drei Tage harter Kämpfe, um **Hoscheid** zu befreien. Am 24. Januar griffen die Soldaten des 11. Regiments, Teile des 2. Regiments der 5. ID, sowie Panzer der Kompanie A des 737. **Panzerbataillons**, Hoscheid unter schwerem deutschen Artilleriefeuer an. Auf der deutschen Seite verteidigten Teile des 212. Grenadierregiments, 79. VGD, des 916. GR, 352. VGD, und Teile des Panzer Lehr Division das Dorf. Nach erbarmungslosen Panzer-, Infanterie- und Häuserkämpfen besetzten und befreiten die G.I.s Hoscheid um 14:30 Uhr.

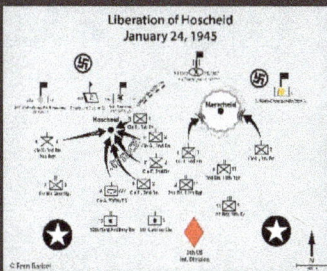

Liberation of Hoscheid
January 24, 1945

© Fern Barbel

For more information
Pour en savoir plus
Für weitere Informationen

Hoscheid (Peace Garden)

Remembering all the young men, resistance fighters, civilians, and American GIs who died during the war from 1940-1945. There is also a panel that lists the names of 38 GIs who died in the area in January 1945, and 38 trees have been planted in their honor.

Geisseck

9378 Parc Hosingen

Latitude: 49.95185

Longitude: 6.07097

Information Panel Peace Garden

Fir dénen 38 G'is hirer ze gedenken, hu mir fir Sie 38 Routeechen geplanzt a wöllen Hir Affer heimat an Erennerung halen.

Ballard	William	23.01.1945		Melillo	Dominik	22.01.1945
Botelho	Antoine	22.01.1945		Menkin	Abraham	24.01.1945
Clancy	James	23.01.1945		Mertz	Albert	24.01.1945
Clarke	Louis	23.01.1945		Mitchell	Clyde	25.01.1945
Corcoran	John	23.01.1945		Morel	Edward	26.01.1945
Curdy	Robert	23.01.1945		Piekielniak	Jos	25.01.1945
Daley	Nickolas	23.01.1945		Postma	John	27.01.1945
Eaton	Woodron	23.01.1945		Potts	Leonhard	23.01.1945
Elliott	Glenn	23.01.1945		Prior	Keith	22.01.1945
Fountain	Walter	23.01.1945		Ring	Roy	23.01.1945
Gaesser	Raymond	22.01.1945		Rusnak	John	24.01.1945
Hall	James	23.01.1945		Rutkowski	Frank	24.01.1945
Heilmann	Daniel	23.01.1945		Salsman	Ben	24.01.1945
Herbert	Arthur	25.01.1945		Schaefer	Elwood	22.01.1945
Hildebrand	Albert	27.01.1945		Scharp	James	23.01.1945
Jeffries	Robert	23.01.1945		Sullivan	Therman	23.01.1945
Jones	James	22.01.1945		Switzer	Estill	25.01.1945
King	Estle	22.01.1945		Tarant	Eschol	22.01.1945
Mc Dermott	Edward	23.01.1945		Wilson	Cecil	25.01.1945

E grousse Merci un d'Donnateuren vun de Beem fir d'Erennnerungsplatz.

Chorale St. Cécile		Hoscheid	Lucas	Judith	Hoscheid	Schmit-Liefgen	Fernand		Hoscheid
Coremans	Antoine	Wahlhausen	Lucas-Wagener	Nico	Hoscheid	Schmitz-Kiesch	Felix		Wolper/Consdorf
Emringer	Guy	Hoscheid-Dickt	Maisch	Barbe	Hosingen	Schroeder-Serres	Helène		Unterschlinder
Faber-Welbes	Lydia	Hoscheid	Mergen	Horst	Hoscheid	Syndicat d'Initiative et de Tourisme			Hoscheid
Friends of Pattons		Merzig	Reuter-Kohl	Joseph	Schlindermanderscheid	Thill	Denise		Bergem
Gales-Wagner	Paul	Hoscheid	Reuter-Rollgen	Joël	Hoscheid-Dickt	Thiry-Schmit	Rudy		Niederkorn
Gaul	Roland	Diekirch	Richarts	Jean-Pierre	Bettange/Mess	Trausch-Turmes	Joseph		Hoscheid
Gleis	Albert	Gralingen	Sapeurs-Pompiers		Hoscheid	U.S. Veterans Friends Luxembourg			Contern
Goedert	Marc	Hoscheid	Scheidweiler	Marcel	Weiler/Pütscheid	Wagener-Schroeder	Carlo		Hoscheid
Gregorius-Leclere	Lucien	Diekirch	Schloesser	Renée	Dudelange	Wagener-Schuller	Valerie		Hoscheid
Heirens	Michel	Luxembourg	Schmit Nico.	Chauffage-Sanitaire	Hoscheid-Dickt	Wagener-Schuller	Philippe		Hoscheid
Heirens-Reimen	Henri	Hoscheid	Schmit-Koenig	Eduard	Bettembourg	Weis-Blau	Edmond		Dondelange
Lakaff	Thilly	Bettembourg							

In order to remember these 38 GIs we have planted 38 trees to memorize their sacrifice.

Ballard, William	Postma, John	Herbert, Arthur
Melillo, Dominik	Eaton, Woodron	Schaefer, Elwood
Botelho, Antoine	Potts, Leonhard	Hildebrand, Albert
Menkin, Abraham	Elliott, Glenn	Scharp, James
Clancy, James	Prior, Keith	Jeffries, Robert
Mertz, Albert	Fountain, Walter	Sullivan, Therman
Clarke, Louis	Ring, Roy	Jones, James
Mitchell, Clyde	Gaesser, Raymond	Switzer, Estill
Corcoran, John	Rusnak, John	King, Estle
Morel, Edward	Hall, James	Tarant, Eschol
Curdy, Robert	Rutkowski, Frank	McDermott, Edward
Piekielniak, Jos	Heilmann, Daniel	Wilson, Cecil
Daley, Nickolas	Salsman, Ben	

Information Panel Peace Garden

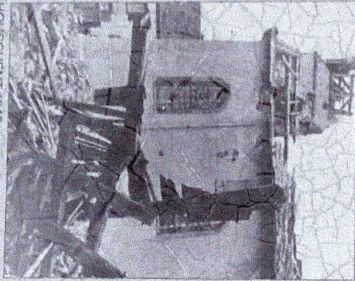

Houschent 1945

Hosingen

Combined memorial honoring Ralph R. Wardle and John W. Kelly of D Company 702nd Tank Battalion. The 17th US Airborne Division, B Company 103rd Engineer Combat Battalion, and K Company, 110th Infantry Regiment, 28th Division.

These monuments are due to be moved to the Hosingen Memorial Park (Water Tower), see next monument for more.

Boukelzerstrooss 1

9807 Parc Hosingen

Latitude: 50.012401

Longitude: 6.089429

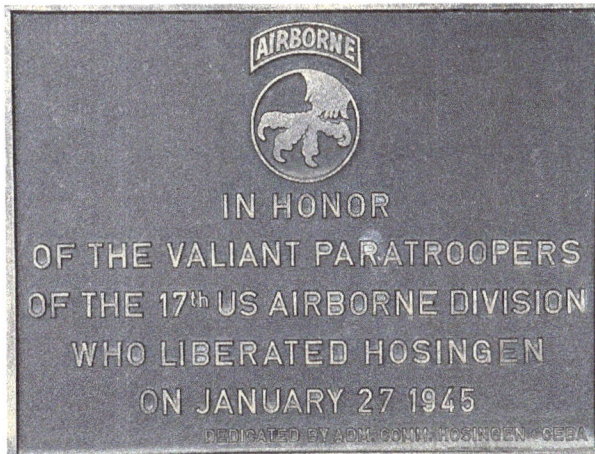

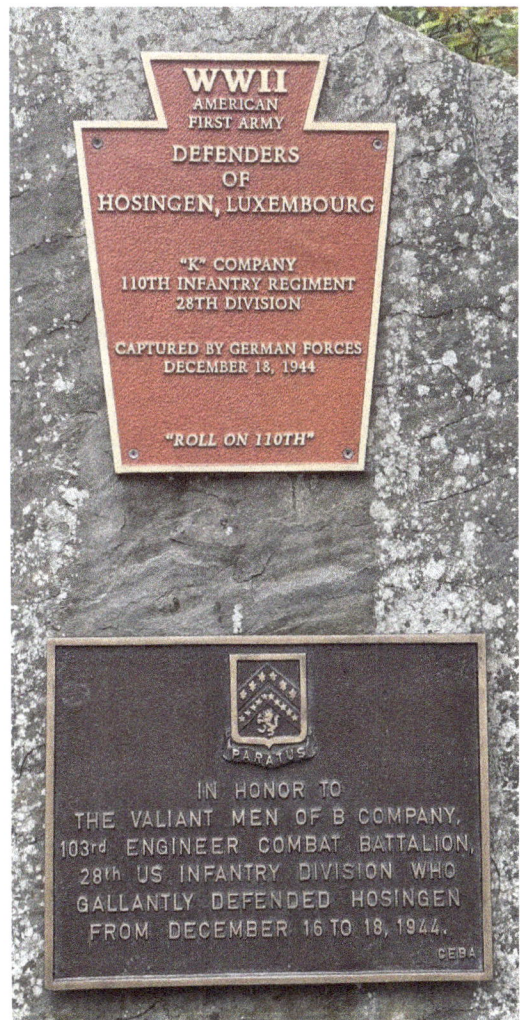

IN HONOR
OF THE VALIANT PARATROOPERS
OF THE 17th US AIRBORNE DIVISION
WHO LIBERATED HOSINGEN
ON JANUARY 27 1945
DEDICATED BY ADM. COMM. HOSINGEN - CEBA

WWII
AMERICAN
FIRST ARMY
DEFENDERS
OF
HOSINGEN, LUXEMBOURG

"K" COMPANY
110TH INFANTRY REGIMENT
28TH DIVISION

CAPTURED BY GERMAN FORCES
DECEMBER 18, 1944

"ROLL ON 110TH"

HONORING
RALPH R. WARDLE
& JOHN W. KELLY
D CO 702nd TANK BN.
WHO GAVE THEIR LIVES
ON 27 JANUARY 1945
TO LIBERATE HOSINGEN
DEDICATED BY THEIR COMRADES
AND TANK COMMANDER LT. MILTON STILL

RED DEVILS

PARATUS

IN HONOR TO
THE VALIANT MEN OF B COMPANY,
103rd ENGINEER COMBAT BATTALION,
28th US INFANTRY DIVISION WHO
GALLANTLY DEFENDED HOSINGEN
FROM DECEMBER 16 TO 18, 1944.
CEBA

Information Panel Hosingen

REMEMBER US HOSINGEN

THE DEFENSE OF HOSINGEN

On December 16, 1944, Hitler launched a surprise attack on the front extending from Monschau (Germany) to Echternach (Luxembourg), with three German armies totalling 240,000 men against 83,000 American soldiers holding the front line. This became known as the "Battle of the Bulge". It was intended to reach Antwerp (Belgium) within less than a week, split the US- and British troops, urge the Allies towards a cease-fire and secure the port of Antwerp, Belgium, to capture its huge dump of supplies.

Battle of the Bulge - Bataille des Ardennes - Ardennenoffensive
16.12.1944 - 28.01.1945

Hitler's plan for the Offensive in the Ardennes "Wacht am Rhein"
Le plan d'Hitler de l'offensive à l'ouest « Wacht am Rhein » (« La Garde au Rhin »)
Hitlers Plan der Offensive im Westen "Wacht am Rhein", und der reelle deutsche Vorstoß

Despite of being outnumbered in the different strong points, the US soldiers held their positions as long as the stocks of their ammunition enabled a defense. Thanks to the stiff resistance of the initial defenders of the 28th US Infantry Division, the 101st US Airborne Division and elements of the 9th and 10th US Armored Divisions reached the town of Bastogne (Belgium) before the Germans. The delaying actions of the defenders contributed to thwart Hitler's plan. General George S. Patton Jr. attacking the enemy with his 5th Army from the south drove the German troops back to their initial "jump off" lines after a six-weeks-battle.

In Hosingen, at the beginning of the German offensive at 5:30 a.m., Company K (160 men), 3rd Battalion, 110th Regiment, 28th US Infantry Division, under the command of Captain Frederic FEIKER, defended the north, south and southeast of the town. Captain William JARRETT's Company B of 103rd Engineer Combat Battalion (125 men) joined in the defense and set up its .50-caliber-machine-guns on the west flank. Three 57-mm antitank guns from the 630th Tank Destroyer Battalion guarded the crossroads in the south of the locality. About 4:00 p.m. on December 16, a platoon of five Sherman tanks of the 707th Tank Battalion came to the rescue of the Hosingen garrison.

The entrenched soldiers in Hosingen impeded the German march westwards by well-directed mortar, machine gun and small arms fire. The observation post on top of the water tower provided an excellent field of observation for "zeroing in" on the attackers.

After two days and nights of heavy fighting to hold the town at all costs, completely surrounded by infantry and tanks, the garrison was running out of ammunition. The G.I.s defended themselves by hand grenades up to the point when the officers decided to surrender.

On December 18, about 300 men and 8 officers became prisoners of war after having inflicted considerable losses to the Germans and after having suffered 10 killed and 12 wounded soldiers.

LA DÉFENSE DE HOSINGEN

Le 16 décembre 1944, Hitler lança une attaque surprise sur le front s'étendant de Montjoie (Monschau D) à Echternach (L) avec trois armées totalisant 240.000 hommes contre 83.000 Américains défendant le front. Cette offensive est connue sous le nom de « Bataille des Ardennes ». Elle avait pour but d'atteindre le port d'Anvers (Belgique) en moins d'une semaine, de diviser les troupes britanniques et américaines, de les forcer à un armistice, de prendre le port et de s'approprier les ravitaillements à Anvers.

En dépit du manque de soldats dans les différents points forts, les soldats américains résistaient aussi longtemps que le stock de leurs munitions le leur permettait. Grâce à la défense acharnée des défenseurs, la 101e US Division Aéroportée et des unités de la 9e et 10e Division Blindée atteignirent la ville de Bastogne (Belgique) avant les Allemands. Les actions retardatrices des défenseurs contribuèrent à avorter le plan d'Hitler. Le Général George S. Patton Jr attaquant les Allemands avec sa 3e Armée par le sud, repoussait les Allemands vers leur point de départ après un combat acharné de six semaines.

À Hosingen, au commencement de l'offensive allemande, la Compagnie K (160 hommes), 3e Bataillon, 110e Régiment, 28e US Division d'Infanterie, sous le commandement du Capitaine Frédéric FEIKER, défendait le nord, le sud et le sud-est de la localité. La Compagnie B du 103e Bataillon de Génie (125 hommes) du Capitaine William JARRETT, se joignit à la défense et installa ses mitrailleuses lourdes, calibre .50, au flanc ouest. Trois canons anti-char 57 mm du 630e Bataillon de Chasseurs de Chars gardaient le croisement de routes au sud de la localité. Le 16 décembre vers 16.00 heures, un peloton de cinq chars Sherman du 707e Bataillon Blindé venait à la rescousse de la garnison de Hosingen.

Defense of Hosingen - 16.- 18.12.1944

HOSINGEN

The defense map of Hosingen - Le plan de défense de Hosingen
Der Verteidigungsplan von Hosingen

Les soldats bien retranchés à Hosingen entravaient fortement l'avance allemande vers l'ouest par des tirs de mortiers, de mitrailleuses et d'armes légères bien ajustés. L'observateur en haut du château d'eau indiquant les objectifs ennemis à bombarder aux équipes de mortiers.

Après deux jours et deux nuits de combats acharnés pour tenir la localité à tout prix, la garnison américaine, complètement encerclée, manquait de munitions. Les soldats se défendaient à l'aide de grenades à main jusqu'au moment où les officiers décidèrent de se rendre.

Le 18 décembre 1944, environ 300 soldats et 8 officiers furent faits prisonniers après avoir infligé de considérables pertes aux Allemands et avoir subi 10 morts et 12 blessés.

DIE VERTEIDIGUNG VON HOSINGEN

Am 16. Dezember 1944 startete Hitler einen Überraschungsangriff mit drei deutschen Armeen, insgesamt 240.000 Soldaten, gegen 83.000 amerikanische Verteidiger, die von Monschau (D) bis Echternach (L) die Front hielten. Diese Offensive ist unter dem Namen „Ardennenoffensive" bekannt. Sie hatte zum Ziel, den Hafen von Antwerpen (B) in weniger als einer Woche zu erreichen, die amerikanischen und britischen Truppen zu spalten, die Alliierten zu einem Waffenstillstand zu zwingen und sich der im Hafen gelagerten Versorgungsgütern zu bemächtigen.

Obschon die befestigten Posten entlang der Frontlinie unterbesetzt waren, hielten die amerikanischen Soldaten ihre Stellungen so lange wie ihre Munitionsreserven einen Widerstand erlaubten. Dank dieser Entschlossenheit erreichten die 101. US Luftlandedivision und Teile der 9. und 10. US-Panzerdivision die Ortschaft Bastogne (B) vor den Deutschen. Die Verzögerungstaktik der Verteidiger trug dazu bei, dass Hitlers Plan zunichte gemacht wurde. General George S. Patton Jr. griff mit seiner 3. Armee von Süden an und trieb die Deutschen nach einem 6-wöchigen Kampf auf ihre Ausgangspositionen zurück.

In Hosingen, am Beginn der deutschen Offensive um 05:30 Uhr, verteidigte die Kompanie K (160 Soldaten), 3. Bataillon, 110. Regiment, 28. US Infanteriedivision, unter dem Kommando von Kapitän Frederic FEIKER den Norden, Süden und Südosten der Ortschaft. Kapitän William JARRETT's Kompanie B des 103. Pionierkampfbataillons (125 Mann) schloss sich der Verteidigung an und ließ seine schweren .50-Kaliber-Maschinengewehre an der Westflanke aufstellen. Drei 57mm-Panzerabwehrgeschütze des 630. Panzerjägerbataillons bewachten die Straßenkreuzung im Süden des Ortes. Gegen 16.00 Uhr kam ein Zug von fünf Sherman Panzern des 707. Panzerbataillons der Hosinger Garnison zu Hilfe.

Die in Hosingen verschanzten Soldaten behinderten den deutschen Vormarsch nach Westen durch gut gezieltes Granatwerfer-, Maschinengewehr- und Gewehrfeuer. Der Beobachtungsposten im Wasserturm gab den Granatwerfer-Mannschaften wertvolle Zielanweisungen auf die Gegner an.

Nach schweren Kämpfen mit dem Ziel, die Ortschaft trotz vollständiger Einkesselung durch Infanterie und Panzer um jeden Preis zu halten und die zwei Tage und zwei Nächte dauerten, ging der amerikanischen Garnison die Munition aus. Die Amerikaner verteidigten sich mit Handgranaten, bis die Offiziere entschieden zu kapitulieren.

Am 18. Dezember wurden etwa 300 Mann und 8 Offiziere zu Kriegsgefangenen, nachdem sie den Deutschen beträchtliche Verluste zugefügt und selbst nur 10 Tote und 12 Verwundete zu beklagen hatten.

Hosingen church, now basic visual, coffee, maison des religieuses, école
Kirche, Schwesternhaus, Schule · Photo: Archives of Hosingen

Sources · Quellen:
To Save Bastogne, Robert F. Phillips, Borodino Books, Burke, USA
Kriegsgeschehen 1939 bis 1945 von Michel Schiltzgebuger in
130 Joër Pompjeën Hosen, Imprimerie Saint-Paul, Lu, Luxembourg, 1987
After Action Report of F. Lt Thomas J. FLYNN, Executive officer, Co K
The Heroes of Hosingen; their untold Story by Alan M. Flynn
Sky Blue Publishing, LLC, Charleston, USA
Schwere Zeiten im Oesrel, Yves Pasqui, ISBN 978-3-00-048836-4,
Verein Historik für Freiburg, C. D-51465 Freiburg
US ARMY HANDBOOK 1939-1945, George Forty, Alan Sutton Publishing, Limited, GB Butler and Tanner, Frome, Somerset

For more information
Pour en savoir plus
Für weitere Informationen

Information Panel Hosingen

THE LIBERATION OF HOSINGEN ON JANUARY 27, 1945

Starting at 01:30 a.m. on January 27, 1945, 1st Battalion, 317th Regiment, 80th US Infantry Division (ID), captured Dorscheid and occupied Neidhausen together with 2nd Platoon of Company A (Co), 702nd Tank Battalion (TB). 3rd Battalion, 317th Regt., with 3rd Platoon, Co A, 702nd TB, liberated Bockholtz. 1st Battalion, 318th Regt., together with 1st Platoon, Co D, 702nd TB and 1st Platoon, Co B, 702nd TB advanced towards the hills west of Hosingen ("Uecht" 503 m).

The first attack on Hosingen started at "Uecht" on January 27 at 07:00 a.m. Five Stuart tanks of 1st Platoon, Co D, 702nd TB, five Sherman tanks of 1st Plat., Co B, 702nd TB, together with 1st Battalion, 318th Regt., attacked the town in the direction of road CR324. The infantrymen advanced with two companies of the battalion, the third company stayed in reserve. However, intense enemy mortar and machine gun fire forced the attackers to retreat to "Schmitzberg".

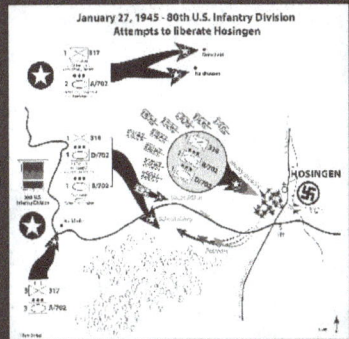

January 27, 1945 - 80th U.S. Infantry Division Attempts to liberate Hosingen

80th US Infantry Division attacked Hosingen three times
La 80e Division d'Infanterie américaine attaqua Hosingen trois fois
Die 80. US Infanteriedivision griff Hosingen dreimal an.

A second attempt to attack Hosingen had the same result. Under the cover of artillery fire, the attack was resumed for a third time. A Sherman and two Stuart tanks were hit, and the infantry had many casualties. Again, the units retreated towards "Schmitzberg".

Ralph R. Wardle and John W. Kelly, Company D, 702nd Tank Battalion, lost their lives when their tank was hit.

The Germans set the ambulance that left Bockholtz to rescue the wounded soldiers, afire. The wounded soldiers were loaded on the remaining three light tanks that brought them to the aid station at Bockholtz.

During this operation, the Americans suffered approximately 200 casualties.

As soon as the last wounded man was back to safety, paratroopers of the 17th US Airborne Division took over the positions of the 80th US Inf. Div. The Germans shelled them with mortar fire. Soon after the shelling, patrols of the 17th Airborne Div. advanced towards Hosingen and noticed that the Germans had left the town. Company A, 139th Airborne Engineer Battalion, removed the numerous booby traps that made the advance most difficult. After the clean-up operations, units of the 17th Airborne Division occupied the battered town. Again and again German artillery fire from the Siegfried Line covered the town with violent shelling.

LA LIBÉRATION DE HOSINGEN LE 27 JANVIER 1945

Le 27 janvier 1945 à partir de 01:30 heures, le 1er Bataillon, 317e Régiment (Régt.), de la 80e US Division d'Infanterie (DI) prit le village de Dorscheid et occupa Neidhausen avec le 2e Peloton de la Compagnie A (Co) du 702e Bataillon Blindé (BB). Le 3e Bataillon, 317e Régt., 80e DI, appuyé par le 3e Peloton, Co A, 702e BB, libérèrent Bockholtz. Entretemps, le 1er Bataillon, 318e Régt., avança avec le 1er Peloton, Co D, 702e BB, et le 1er Peloton, Co B, 702e BB vers les hauteurs à l'ouest de Hosingen (Uecht 503 m).

La première attaque sur Hosingen commença au point « Uecht » le 27 janvier à 07:00 heures. Cinq chars Stuart du 1er Peloton de la Co D du 702e Bataillon Blindé, cinq chars Sherman du 1er Peloton de la Co B du 702e Bataillon Blindé ensemble avec le 1er Bataillon du 318e Régt. de la 80e US DI attaquèrent la localité dans la direction de la route CR324. L'infanterie avança avec deux compagnies du bataillon, la troisième compagnie restant en réserve. Toutefois, un tir intense de mortiers et de mitrailleuses obligea les attaquants à se retirer au « Schmitzberg ».

January 27 - 28, 1945 - 17th U.S. Airborne Division Final Liberation of Hosingen

17th US Airborne Division liberated Hosingen - La 17e Division Aéroportée libéra Hosingen
Die 17. US Luftlandedivision befreit Hosingen

Une deuxième tentative d'attaque connut le même résultat. Avec le support de l'artillerie, l'attaque fut répétée pour la troisième fois. Un char Sherman et deux chars Stuart furent perdus et l'infanterie avait de fortes pertes. De nouveau, les unités se replièrent au « Schmitzberg ».

Ralph R. Wardle et John W. Kelly, Compagnie D, 702e Bataillon Blindé, perdirent leur vie quand leur char fut touché.

Les Allemands tirèrent sur l'ambulance qui partait de Bockholtz pour récupérer les blessés. Ceux-ci furent chargés sur les trois chars légers restants qui les transportèrent au poste de secours établi à Bockholtz.

En tout, quelque 200 pertes américaines étaient à déplorer lors des trois attaques.

Après l'évacuation de tous les blessés, les soldats de la 17e Division Aéroportée reprirent les positions de la 80e DI sous le feu des mortiers allemands. Peu après, des patrouilles de la 17e Division Aéroportée avancèrent vers Hosingen pour constater que les Allemands avaient évacué la localité. La Compagnie A, 139e Bataillon du Génie Aéroporté, désamorça les nombreux objets piégés qui entravèrent l'avance. Après les travaux de déblaiement, la 17e Division Aéroportée occupa la localité en ruines malgré l'artillerie allemande, qui ne cessait de bombarder Hosingen à partir de la Ligne Siegfried.

DIE BEFREIUNG VON HOSINGEN AM 27. JANUAR 1945

Am 27. Januar 1945 begann um 01:30 Uhr der Angriff auf die Dörfer Dorscheid und Neidhausen. Im Laufe der Nacht besetzten das 1. Bataillon, 317. Regiment, 80. US Infanteriedivision (ID), und der 2. Zug der Kompanie A des 702. Panzerbataillons (PB) die beiden Ortschaften. Das 3. Bataillon, 317. Regt., 80. ID und der 3. Zug, Kompanie A, 702. PB befreiten Bockholtz. Das 1. Bataillon, 318. Regt., rückte zusammen mit dem 1. Zug, Kompanie D, 702. PB und dem 1. Zug, Kompanie B, 702. PB, auf die Hügel westlich von Hosingen (Uecht 503 m) vor.

Der erste Angriff auf Hosingen begann am 27. Januar um 07:00 Uhr in „Uecht". Fünf Stuart Panzer des 1. Zuges der Kompanie D, 702. PB, fünf Sherman Panzer des 1. Zuges der Kompanie B, 702. PB und das 1. Bataillon des 318. Regiments, 80. US ID, griffen die Ortschaft entlang der CR324 an. Die Infanterie ging mit zwei Bataillonen vor, das dritte Bataillon blieb in der Reserve. Starkes Granatwerfer- und MG-Feuer zwang die Angreifer, sich zum „Schmitzberg" zurückzuziehen.

Ein zweiter Versuch Hosingen anzugreifen, hatte dasselbe Resultat. Unter Artilleriefeuerschutz wurde der Angriff ein drittes Mal wiederholt. Ein Sherman und zwei Stuart Panzer wurden abgeschossen und die Infanterie erlitt zahlreiche Verluste. Wieder mussten sich die Einheiten auf den „Schmitzberg" zurückziehen. Ralph R. Wardle und John W. Kelly, Kompanie D, 702. Panzerbataillon, verloren ihr Leben, als ihr Panzer einen Volltreffer erhielt. Die Deutschen zerstörten den Rettungswagen, der von Bockholtz aus den Verwundeten zu Hilfe kommen wollte. Die Verletzten wurden auf die drei letzten Stuart Panzer geladen, die sie zum Lazarett nach Bockholtz brachten.

Die Amerikaner hatten bei diesem Einsatz einen Verlust von rund 200 Mann zu beklagen.

Medium tank M4A3E8 "Sherman"
Char moyen M4A3E8 „Sherman"
Mittelschwerer M4A3E8 „Sherman"

Light US tank M5A1 "Stuart"
char léger américain M5A1 „Stuart"
leichter amerikanischer M5A1 „Stuart" Panzer

Sobald die letzten Verwundeten in Sicherheit waren, übernahmen die Soldaten der 17. US Luftlandedivision die Stellungen der 80. US ID. Die Deutschen belegten sie mit Granatwerferfeuer. Wenig später wagten sich Patrouillen der 17. Luftlandedivision nach Hosingen vor. Sie stellten fest, dass die Deutschen den Ort verlassen hatten. Die A-Kompanie des 139. Fallschirmpionierbataillons entfernte die vielen Sprengfallen, die den Vormarsch erschwerten. Nach den Räumungsarbeiten besetzten Einheiten der 17. Luftlandedivision die zerstörte Ortschaft. Immer wieder nahm die deutsche Artillerie aus den Stellungen des Westwalls Hosingen unter heftigen Beschuss.

Sources - Quellen:
- Schwere Zeiten im Ourtal, Yves Raugel, ISBN 978-3-00-033386-1
- Verein Historikum, Dasburg e.V. D-54689 Dasburg
- After Action Report of 702nd Tank Battalion in the Siebert Combined Arms Research Library (CARL) Digital Library
- Kriegsgeschichten 1939 bis 1945 von Michel Schardenberg u. 130 Joer Fräiwëllege Hëseri, Imprimerie Saint-Paul, 20, Luxembourg, 1987 (Seite 292 - 296)
- Die Ardennenschlacht 1944-1945 in Luxemburg von Jean Milmeister, Editions Saint-Paul, 1971 Luxembourg, (Seite 600, 601 - 673, 786 - 788)
- Die 17th US Airborne Division im Raume Hosingen von Jean Milmeister, Luxemburger Wort, 27. Oktober 1993
- Amerikanische Kriegsdenkmäler in Luxemburg, Nelson Hill, Heintz und Milmeister Imprimerie Saint-Paul, Luxembourg, 1995
- Panzerdarstellungen: Achim Beck, Luxus-Sergy flyzfate und -, 1945-TOM -

For more information
Pour en savoir plus
Für weitere Informationen

Information Panel Hosingen

REMEMBER US HOSINGEN

THE WATER TOWER OF HOSINGEN

The water tower of Hosingen was built in 1931.

During the period of September 25 to December 17, 1944, the Americans used the water tower as an observation post, as it provided a wide panoramic 360-degrees view.

The Germans had promised the "Knight's Cross", a six-week-furlough and a reward of 20,000 Reichsmark to any soldier who could infiltrate the town and blow up the tower by explosives. However, the Americans denied any enemy to approach their observation post.

On December 16, 1944, at 5:30 a.m., the observer in the water tower of Hosingen of Company K, 3rd Battalion, 110th Regiment, 28th US Infantry Division, reported: "I see countless pinpoints of light in direction of the German lines!" Seconds later, the first shells exploded in Hosingen setting several houses on fire and lighting up the town. The Battle of the Bulge had begun.

From the top of the water tower, Company K Weapon Platoon's Staff Sergeant gave his mortar crews directions as to where to aim their mortars along the Skyline Drive. That way, the US defenders shattered three German assaults and stopped the attacks.

The water tower of Hosingen in the year 1931 - Le château d'eau de Hosingen en 1931
Der Hosinger Wasserturm im Jahre 1931 - Photo Courtesy of Yves BASTOT

At 01:00 p.m. on December 17, a German tank opened fire on the water tower. The strong construction of the building was able to support several direct hits. The outside walls were of thick masonry and a steel shaft supported the centre. A steel winding-staircase offered the possibility to the observers to leave safely their post on top of the tower. The observers abandoned their post only when six more enemy tanks appeared later in the day and fired on the tower.

The Germans blew up the water tower around the end of January 1945.

LE CHÂTEAU D'EAU DE HOSINGEN

Le château d'eau de Hosingen fut construit en 1931.

Pendant la période du 25 septembre au 17 décembre 1944, les Américains utilisaient le château d'eau en poste d'observation, étant donné qu'il offrait une vue panoramique de 360°.

Les Allemands avaient promis la croix de chevalier, un congé de six semaines et une récompense de 20.000 Reichsmark à quiconque qui se serait infiltré dans la localité et aurait dynamité la tour. Mais les Américains ne permettaient à nul ennemi de s'approcher de leur poste d'observation.

Le 16 décembre 1944, à 05:30 heures, l'observateur en haut du château d'eau de Hosingen, Compagnie K, 3e Bataillon, 110e Régiment, 28e US Division d'Infanterie, rapporta : « Je vois d'innombrables points de lumière dans la direction des lignes allemandes! » Quelques secondes plus tard, les premiers obus explosaient dans Hosingen mettant plusieurs maisons en feu qui illuminaient la localité. La Bataille des Ardennes avait commencé.

Du haut du château d'eau, le sergent du Peloton Armes de la Compagnie K, communiquait les coordonnées des objectifs sur le Skyline Drive à ses soldats qui opéraient les mortiers. De cette façon, les défenseurs brisèrent trois assauts allemands et arrêtèrent les attaques.

Le 17 décembre à 13:00 heures, un char allemand ouvrit le feu sur le château d'eau. La forte construction du mur pouvait supporter quelques coups directs. Les murs extérieurs étaient en béton et une cage en acier supportait l'intérieur de la construction. Un escalier en acier en colimaçon permettait aux observateurs de quitter en sécurité leur poste en haut du château. Les observateurs quittèrent leur poste uniquement quand six chars en plus apparurent plus tard dans la journée et firent feu sur la tour.

Les Allemands démolirent le château d'eau vers la fin du mois de janvier 1945.

DER WASSERTURM VON HOSINGEN

Der Hosinger Wasserturm wurde im Jahr 1931 erbaut. In der Zeit vom 25. September bis zum 17. Dezember 1944 benutzten die Amerikaner den Wasserturm als Beobachtungsposten, da er eine Panoramasicht von 360° bot.

Die Befehlshaber der deutschen Truppen hatten den Soldaten, die sich in Hosingen einschleichen und den Turm sprengen würden, das Ritterkreuz, sechs Wochen Urlaub und eine Belohnung von 20.000 Reichsmark versprochen. Doch die Amerikaner verhinderten jegliche Annäherung des Feindes an ihren Beobachtungsposten.

Am 16. Dezember 1944 um 05:30 Uhr meldete der Beobachter im Hosinger Wasserturm, K-Kompanie, 3. Bataillon, 110. Regiment, 28. US Infanterie Division: "Ich sehe unzählige helle Punkte auf der Seite der deutschen Linien!" Sekunden später schlugen die ersten Granaten in Hosingen ein, entzündeten mehrere Häuser, die dadurch die Ortschaft hell erleuchteten. Die Ardennenoffensive hatte begonnen.

Vom Wasserturm oben gab der Oberfeldwebel des Waffenzuges der K-Kompanie seinen Granatwerfer-Mannschaften Zielangaben entlang des Skyline Drives an. So konnten die Amerikaner drei Anstürme abschmettern und die Angriffe stoppen.

Am 17. Dezember um 13:00 Uhr eröffnete ein deutscher Panzer das Feuer auf den Wasserturm. Die solide Konstruktion des Gebäudes konnte einige direkte Einschüsse verkraften. Die Außenwände waren aus dickem Beton und mit Stahlarmierung. Eine stählerne Wendeltreppe gab den Beobachtern die Möglichkeit, ihren Posten oben im Turm sicher zu verlassen. Die Beobachter verließen ihren Posten erst, als sechs weitere Panzer später am Tag erschienen und auf den Turm feuerten.

Die Deutschen sprengten den Wasserturm Ende Januar 1945.

The water tower of Hosingen after its demolition in 1945 - Le château d'eau après sa démolition en 1945 - Die Wasserturm nach seiner Sprengung in 1945
Photo Courtesy of Archives of the Commune Parc Hosingen

Sources - Quellen:
"To Save Bastogne", Robert F. Phillips, Borodino Books, Barks, USA
"Kriegsgeschichten 1939 bis 1945", von Michel Schoukmelzer, in 130 Iris Pompge-ju-Haven", Imprimerie Saint-Paul s.a., Luxembourg, 1987
"The Heroes of Hosingen Their untold Story" by Alice M. Flynn, Sky Blue Publishing, LLC, Charleston, USA, 2015, ISBN-10: 151-2685313, ISBN-13: 978181/268436
"Schwere Zeiten im Oestal", Yves Racqué, Veräin Historesche Dsiburg eV., Druckerei Anders Prüm, 2015, ISBN 978-3-00-051536-1
"ALAMO in the Ardennes", by John C. McManus, John Wiley & Sons, Inc.

For more information
Pour en savoir plus
Für weitere Informationen

Hosingen Memorial Park (Water Tower)

Planned to open December 2024

Memorial site for the 110th Infantry Regiment of the 28th Infantry Division. An exhibition about the Battle of the Bulge will be located inside the tower and the monuments at the Hosingen crossroads at Boukelzerstrooss 1 (see previous monument) will be relocated next to the water tower.

33 Um Knupp

9808 Hosingen

Latitude: 50.02057

Longitude: 6.09318

Huldange

Monument to the 87th US Infantry Division, 3rd US Army. Units of the 87th liberated the border town of Wasserbillig and subsequently the village of Huldange. In doing so, they liberated the highest and lowest points in Luxembourg.

Op d'Burrigplatz

Huldange

Latitude: 50.16148

Longitude: 6.02398

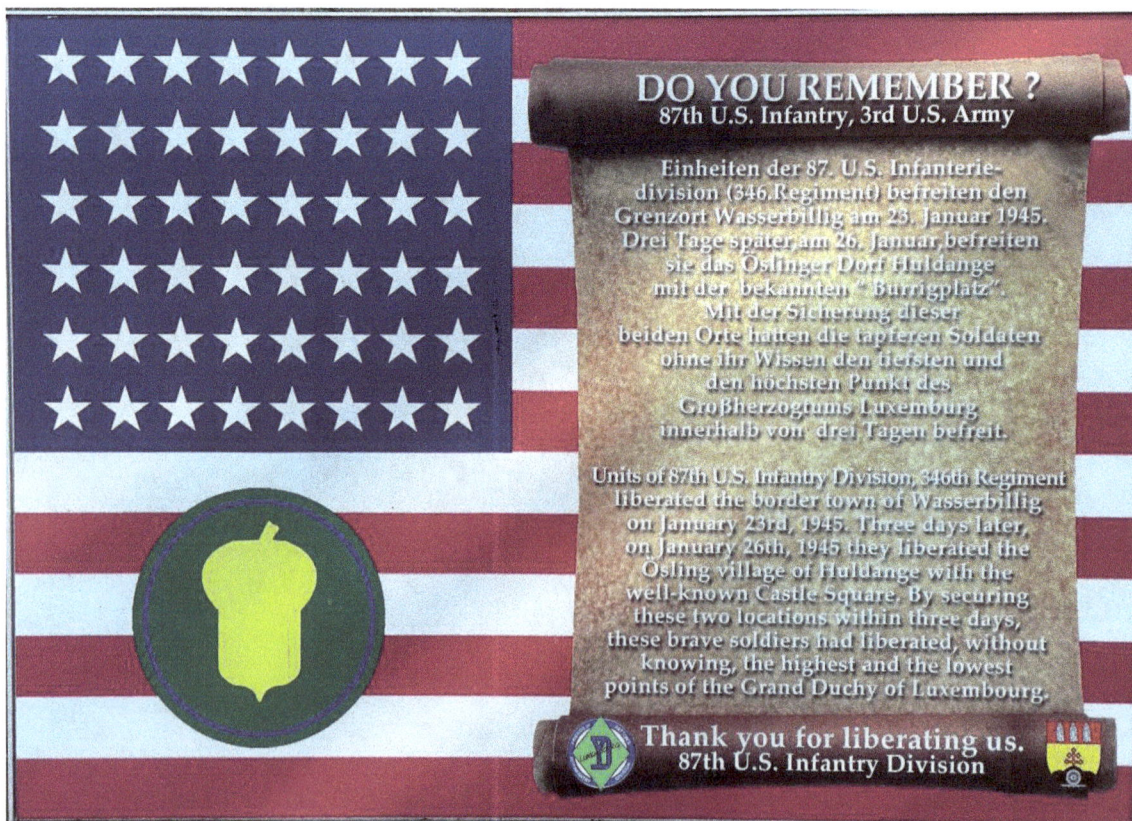

DO YOU REMEMBER ?
87th U.S. Infantry, 3rd U.S. Army

Einheiten der 87. U.S. Infanterie-division (346.Regiment) befreiten den Grenzort Wasserbillig am 23. Januar 1945. Drei Tage später am 26. Januar, befreiten sie das Oslinger Dorf Huldange mit der bekannten "Burrigplatz". Mit der Sicherung dieser beiden Orte hatten die tapferen Soldaten ohne ihr Wissen den tiefsten und den höchsten Punkt des Großherzogtums Luxemburg innerhalb von drei Tagen befreit.

Units of 87th U.S. Infantry Division, 346th Regiment liberated the border town of Wasserbillig on January 23rd, 1945. Three days later, on January 26th, 1945 they liberated the Osling village of Huldange with the well-known Castle Square. By securing these two locations within three days, these brave soldiers had liberated, without knowing, the highest and the lowest points of the Grand Duchy of Luxembourg.

Thank you for liberating us.
87th U.S. Infantry Division

Huldange

Monument to the 26th Infantry Division.

Route 336

Huldange

Latitude: 50.16309

Longitude: 6.01353

Inscription

Do you remember?

Jo, mir denken nach u sie, mir hu sie nёt vergiess!

(Yes, we still remember them, and have not forgotten them)

Friends of Patton's 26th Infantry Division

Hupperdange

Monument to Flt Officer Leo A. Carey of 445th Bomb Group, US Eighth Air Force, killed in crash of B-24 Liberator. Two of the crew of the B-24 that survived the crash hid in the "Bunker" at Heinerscheid (see Heinerscheid bunker page).

CR 339A

Between Hupperdange and Grindhausen

Latitude: 50.089042

Longitude: 6.059836

WE REMEMBER
FLT O LEO A. CAREY
PENNSYLVANIA

KILLED NEAR HUPPERDANGE, APRIL 13TH 1944
IN THE CRASH OF B-24 LIBERATOR BOMBER 41-29132
BURIED AT THE HENRI-CHAPELLE AMERICAN CEMETERY, BELGIUM

WE DEDICATE THIS MONUMENT TO THE HEROIC

445TH BOMB GROUP (H)
703D SQUADRON
U.S. EIGHT AIR FORCE

U.S. VETERANS FRIENDS LUXEMBOURG
COMMUNE DE HEINERSCHEID   FILIERISTES ET PASSEURS L.P.L.

Kalborn

Information panels located on the back wall of the church cover the combat operations in Kalborn, on September 22, 1944, and the murder of seven local men by the German military.

17 Hauptstross

9757 Kalborn

Latitude: 50.10209

Longitude: 6.11182

Kalborn

Kalborn Information Panel

The events of September 22. 1944 in Kalborn

On September 10. 1944, US Army forces liberated Luxembourg City and in the following days, they chased the German occupiers from the rest of the country. The Grand-Duchy, which had been conquered by the Wehrmacht on May 10, 1940 and had been under German occupation since then, was finally free again. The German army forces had left Luxembourg without putting up a fight and had retreated across the Our river into the "Westwall". The "Westwall", called "Siegfriedline" by the Allies, was a dense web of inter-connected pillboxes, tunnels and trenches, built to defend the western border of the "Reich". The US Army was hampered by the lack of supplies, they were not able to bring enough fuel, food, ammunition and equipment to the front lines, to enable them a swift follow-up attack across the Our river into Germany. Thus they dug in along the N7 highway leading from Weiswampach to Diekirch. The inhabitants of the villages in between the N7 and the border with Germany suddenly found themselves living in the "no man's land".

Kalborn was one of these villages. From their defensive positions in Heinerscheid, the GIs were able to overlook the approaches to the village, but during night time, German patrols were able to cross the Our river and sneak into the small hamlet. The situation was tense and potentially dangerous for the villagers, so that, a few days before September 22, the US Army ordered the people of Kalborn to be evacuated to safer villages in the rear. As a consequence only two houses were still occupied by their inhabitants. In the cellars of the Hoelpes and the Freichel homes, the owners and a few neighbors were huddled in fear. In the cellar of the Hoelpes home eleven people and sought refuge. The witnesses can not remember the exact number of villagers who were in the Freichel house, but they are sure that there weren't least six.

On the early morning of September 22, under the cloak of a dense layer of fog, soldiers of the "Kampfgruppe Wegelein" silently entered the village. They possibly wanted to check if there was anybody left in the village and tried to collect some intelligence on the US troop dispositions in the area. The "Landsers" broke into the houses and searched them from top to bottom. Finally they came to the Hoelpes house where they discovered the frightened civilians hiding in the cellar of the family home. They were all rounded up and brought outside where the Germans separated the men form the women and children. Fuming mad the commanding officer of the German patrol asked the civilians if they were hiding any weapons, they denied but this lead to the "Leutnant" getting more and more enraged. He kept on repeating his inquisitional questions, yelling and screaming at the top of his lungs. Behind his back the women and children were making signs to their men that they should keep on denying! Because there was a shotgun in the cellar! But, unbeknownst to the men, the women had been able to hide the hunting gun beneath a pile of straw. Finally one of the men broke under the increasing pressure of the officer and confessed to having a gun. Soon enough the Germans discovered the old and rusty shotgun that hadn't been in use for many years. But this didn't budge "Leutnant" Schneider who had already made up his mind: all the men were to be executed! Even the begging and arguing of his men was to no avail. The six men, who had been hiding in the cellars were marched off to the village pond near the Hoelpes house and, when all the German soldiers refused Schneider's order to open fire, the officer himself butchered the innocent men with his submachine gun. The victims were the four Hoelpes brothers, "Nikela" (Nicolas), a father of four children, Bernard, "Josy" (Joseph) and Emile as well as "Nikela" Peiffer, also father of four kids and "Michi" (Michael) Holper. The six men were aged in-between 27 and 46 years.

Les événements tragiques du 22 septembre 1944 à Kalborn

Le 10 septembre 1944, les troupes américaines libérèrent la ville de Luxembourg. Le Grand-Duché, occupé par la Wehrmacht allemande depuis le 10 mai 1940, fut enfin libéré. Les troupes allemandes quittèrent le pays sans combattre et se replièrent au-delà de l'Our dans les fortifications du Westwall, à savoir un dense réseau de bunkers, de tunnels et de tranchées, qui devaient servir à défendre la frontière occidentale du Reich allemand. Cependant, les forces américaines ne recevaient pas assez de ravitaillements, de munitions, d'essence et de nourriture, leur empêchant ainsi toute avancée directe sur la rive droite de l'Our. Pour cette raison, ils furent contraints de garder leur position le long de la Route Nationale 7, menant de Weiswampach à Diekirch. Les villages situés entre cette ligne américaine et la frontière du Reich allemand se trouvèrent ainsi dans un soi-disant "No Man's Land".

L'un de ces villages fut celui de Kalborn. Même si les troupes américaines surveillèrent la localité depuis Heinerscheid, des soldats allemands continuèrent à traverser la frontière et à pénétrer sur le territoire Luxembourgeois pendant la nuit en franchissant l'Our. La situation était tendue, à tel point que peu avant le 22 septembre 1944, les Américains demandèrent à tous les habitants du village de quitter leurs habitations. Il n'en restèrent que deux maisons habitées : celle de la famille Hoelpes et celle de la famille Freichel. Les derniers habitants restants se réfugièrent dans les caves des deux maisons. Dans la cave de la famille Hoelpes il y en avait onze personnes, on ne sait pas exactement combien de personnes s'abritèrent dans la maison de la famille Freichel, mais on est sure qu'il y'en avait au moins six.

C'est ainsi qu'au petit matin du 22 septembre, des soldats allemands de la "Kampfgruppe Wegelein" profitèrent d'un épais brouillard pour se faufiler à l'intérieur du village sans se faire repérer par les Américains. Il est probable que les Allemands aient voulu découvrir s'il restaient encore des habitants dans le village. Ils pénétrèrent donc dans les caves, fouillèrent les maison et finirent par trouver les occupants de la cave de la maison Hoelpes. Les personnes se trouvant au sous-sol furent rassemblées et emmenées à l'extérieur, où les hommes furent séparés des femmes et des enfants. L'officier en charge de la mission leur demanda aigrement s'il y avait des armes dans la cave, sur ce les hommes répondirent que non. Le jeune Lieutenant répéta sa question encore plus fort qu'auparavant. Derrière lui se trouvèrent les femmes et les enfants qui par des signes de main affirmèrent à nouveau clairement qu'il n'y avait pas d'armes dans la cave. En réalité cependant, il y avait un fusil que les femmes avaient réussi à cacher à temps sous un tas de paille. Néanmoins, l'un des hommes n'ayant pas été au courant de cela et n'ayant aussi probablement pas remarqué les signes de main, s'effondra sous la pression et avoua la présence d'une arme au sous-sol. Il s'agissait d'un fusil rouillé et inutilisable. Mais la sentence avait déjà été prononcée : "Les hommes seront fusillés", ordonna le Lieutenant Schneider. Les six hommes furent amenés à l'étang, où ils furent exécutés. Il s'agissait des quatre frères Hoelpes, à savoir Nikela, père de quatre enfants, Bernard, Josy et Emile, ainsi que Nikela Peiffer, également père de quatre enfants, et Michi Holper. Ils avaient tous entre 27 et 46 ans.

Die tragischen Ereignisse des 22. September 1944 in Kalborn

Am 10. September 1944 befreiten amerikanische Truppen Luxemburg-Stadt, und in den darauffolgenden Tagen wurden die deutschen Besatzer auch aus dem Rest des Landes vertrieben. Das Großherzogtum, welches am 10. Mai 1940, im Zuge des Westfeldzuges, durch die deutsche Wehrmacht besetzt wurde und seitdem unter Deutscher Besatzung gestanden hatte, war endlich wieder frei. Die deutschen Truppen verließen das Land kampflos und zogen sich jenseits der Our in den Westwall, ein dichtes Netz aus Bunkern, Stollen und Gräben, das zur Verteidigung der Westgrenze des Deutschen Reiches dienen sollte, zurück. Die Amerikaner erhielten jedoch nicht genug Nachschub an Nahrung, Munition und Benzin, so dass ein direktes Nachrücken über die Our nicht stattfinden konnte. Stattdessen hielten sie ihre Position entlang der Nationalstraße N7, welche von Weiswampach nach Diekirch führte. Dörfer, die sich zwischen den amerikanischen Linien und der Grenze zum Deutschen Reich befanden, lagen im „No man's land", dem Niemandsland.

Eines dieser Dörfer war Kalborn. Von Heinerscheid aus kontrollierten die amerikanischen Truppen das Dorf, während, von der Our herkommend, nachts immer wieder deutsche Soldaten über die Grenze bis zur Ortschaft vorstießen. Die Situation war angespannt, so dass kurz vor dem 22. September 1944 die Amerikaner alle Einwohner des Dorfes aufforderten, ihre Häuser zu verlassen. Schlussendlich blieben nur noch zwei Häuser bewohnt, das Haus der Familie Hoelpes, sowie das Haus der Familie Freichel. In den Kellern der beiden Häuser harrten die letzten verbliebenen Einwohner aus. Im Keller der Familie Hoelpes waren es elf Personen, wie viele Menschen sich genau im Hause Freichel befanden ist nicht bekannt, es waren jedoch mindestens sechs.

So kam es, dass sich am frühen Morgen des 22. Septembers deutsche Soldaten der "Kampfgruppe Wegelein", vom dichten Nebel vor den Amerikanern geschützt, ins Dorf schlichen. Womöglich ging es den Deutschen darum, herauszufinden, ob sich noch Einwohner in dem Dorf aufhalten würden und an Informationen über die amerikanischen Truppen in der Umgebung zu gelangen. Sie drangen in die Keller ein, durchsuchten die Häuser und stießen schließlich im Keller des Hauses Hoelpes auf noch verbliebene Einwohner. Die Insassen des Kellers wurden zusammengetrieben und nach draußen gebracht. Die Männer wurden von den Frauen und Kindern getrennt. Der oberste Offizier fragte sie wutschnaubend, ob Waffen im Keller versteckt wären. Die Männer verneinten dies. Der Befehlshaber wiederholte seine Frage noch lauter als zuvor. Hinter ihm standen die Frauen und Mädchen, welche mit Handzeichen nochmal versuchten zu verdeutlichen, dass keine Waffen im Keller wären. In Wahrheit jedoch befand sich ein Gewehr im Keller, welches die Frauen noch rechtzeitig unter einem Haufen Stroh hatten verstecken können. Einer der Männer, der dies nicht wusste und die Zeichen womöglich auch nicht sah, brach unter dem Druck zusammen und antwortete schließlich mit einem leisen „Ja". Dabei handelte es sich bei der Waffe um ein altes, verrostetes und untaugliches Gewehr.

Doch das Urteil war bereits gefällt: „Die Männer werden erschossen!", befahl Leutnant Schneider. Die sechs Männer aus dem Keller wurden hinunter zum Weiher gebracht und dort hingerichtet. Es handelte sich um die vier Brüder Hoelpes, Nikela, Vater von vier Kindern, Bernard, Josy und Emile, sowie Nikela Peiffer, der ebenfalls vier Kinder zurückließ, und Michi Holper. Alle sechs waren zwischen 27 und 46 Jahre alt.

For more information
Pour en savoir plus
Für weitere Informationen

Kalborn Information Panel

THE EVENTS OF SEPTEMBER 22, 1944 IN KALBORN

About the same time, two young men were about to enter the hamlet, coming from the neighbouring village of Heinerscheid. They were Benny Hoffmann, 32 years old, and a teenage boy who was a relative of Hoffmann's. As they came into the village they were immediately picked up and searched by the Wehrmacht soldiers. To his downfall, Benny Hoffmann was carrying as well a revolver as a badge of the Belgian "Armée Secrète" resistance movement. The Germans immediately dragged him to the village pond and shot him dead. Due to his young age, the murderers spared the life of the teenage boy who was with Benny Hoffmann. Thus the hamlet of Kalborn lost seven innocent men, who, in a tragic way, fell victims to a savage warcrime.

LES ÉVÉNEMENTS TRAGIQUES DU 22 SEPTEMBRE 1944 À KALBORN

À peu près au même moment, deux jeunes hommes, en venant de Heinerscheid, arrivèrent dans le village de Kalborn. Il s'agissait de Benny Hoffmann, 32 ans, et d'un garçon de sa famille. Ceux-ci furent repérés par des soldats allemands qui les fouillèrent. Lorsqu'ils remarquèrent que Benny Hoffmann portait une arme et l'insigne du Maquis belge, ils l'amenèrent aussitôt à l'étang où il fut abattu. L'autre garçon fut épargné à cause de son jeune âge. Le 22 septembre 1944, le village de Kalborn perdit sept hommes innocents qui furent tragiquement victimes d'un crime de guerre.

Les personnes réfugiées dans le sous-sol de la maison Freichel furent heureusement épargnées. Peu avant que les Allemands ne

DIE TRAGISCHEN EREIGNISSE DES 22. SEPTEMBER 1944 IN KALBORN

Ungefähr zur selben Zeit näherten sich, von Heinerscheid kommend, zwei junge Männer dem Dorf. Es handelte sich um Benny Hoffmann, 32 Jahre alt, und um einen Jungen aus der Familie seines Bruders. Als sie Kalborn betraten, wurden sie von deutschen Soldaten festgehalten und durchsucht. Dabei fanden die Wehrmachtsangehörigen bei Benny Hoffman, der ein Abzeichen der Belgischen "Armée Secrète" trug, eine Waffe. Auch er wurde hinunter zum Weiher geschleppt und dort erschossen. Der andere Junge wurde auf Grund seines noch jungen Alters verschont. Das Dorf Kalborn verlor somit am 22. September 1944 sieben unschuldige Männer, welche auf tragische Art und Weise dem Krieg und den damit verbundenen Verbrechen zum Opfer fielen.

Bernhard Hoelpes
36 years

Emile Hoelpes
32 years

Joey Hoelpes
34 years

Nikola Hoelpes
46 years

Mathi Holper
27 years

Benny Hoffmann
32 years

Nikla Peiffer
39 years

Due to incredible luck, and the sacrifice of US Army GIs, the lives of the men, women and children, having sought shelter in the cellar of the Freichel house, were spared. Just a few moments before the Germans were to storm the Freichel house a US Army patrol, mounted on a Jeep, entered the village. The men from "Kampfgruppe Wegelein" immediately opened fire and killed the driver, taking the other two American soldiers prisoner. Worried by the sudden arrival of the GIs and fearing US reinforcements the German soldiers decided to withdraw from the village. But before fleeing across the border into the "Reich" they abducted the surviving women and children from the Hoelpes house, who until they were able to return to their homes in the spring of 1945, were forced to work as slave labourers on German farms.

In 1976 the Criminal Investigation Branch of Luxembourg Federal Police had started an investigation into this heinous crime. But alas, there were no judicial consequences whatsoever for the accused former Lieutenant Schneider. After 32 years none of the surviving eye witnesses was able to identify him, without any doubt, as the perpetrator of the murders of September 1944. The survivors and the inhabitants of Kalborn were always convinced that the perpetrators of the atrocity were members of the feared "Waffen-SS". But recent researches proved without any possible doubt that the men were members of the "Kampfgruppe Wegelein", this unit had been recently raised with very young soldiers, many of whom had no frontline experience. The best part of these men had come from the "Heeres-Unteroffizier-Schule 2/XII Saarlautern", a German Non-Commissioned-Officers academy.

Today a chapel, built by the surviving members of the Hoelpes family, stands at the place where the killings took place, keeping alive the memory of the seven patriots murdered by the Germans.

Text: Tom Scholtes

parvienrent à prendre d'assaut cette cave, une jeep américaine atteignit le village. Les soldats allemands prirent la fuite tout en faisant feu sur les Américains. Ils touchèrent le conducteur de la jeep à plusieurs reprises. Celui-ci succomba à ses blessures. Après ces événements, les femmes et les enfants furent amenés de l'autre côté de la frontière, où ils ont dû effectuer des travaux forcés pour le compte d'agriculteurs et d'artisans allemands. Au printemps 1945, ils retournèrent finalement dans leur chère patrie.

Une enquête criminelle, diligentée par la sûreté luxembourgeoise en 1976, resta sans suites judiciaires. Ceci comme les témoins survivants, après 32 années, ne se voyaient pas à même d'identifier avec certitude absolue le suspect, le Lieutenant Schneider. Les survivants et les habitants de Kalborn étaient toujours convaincus que les coupables étaient des membres de l'infâme "Waffen-SS". Néanmoins des recherches historiques récentes ont prouvé qu'il s'agissait de membres de la "Kampfgruppe Wegelein". Cette unité, récemment créée, consistait majoritairement de soldats très jeunes, sans aucune expérience du front, un grand nombre des rangs parvenait de la "Heeres-Unteroffizier-Schule 2/XII Saarlautern", une école pour sous-officiers de la Wehrmacht.

Aujourd'hui une chapelle, érigée après la guerre par les soins de la famille Hoelpes, rappelle le sort des sept patriotes luxembourgeois, fusillés le 22 septembre 1944 à Kalborn.

Die Insassen des Kellers im Hause Freichel blieben glücklicherweise verschont. Kurz bevor es den Deutschen gelang, auch diesen Keller zu stürmen, erreichte ein Jeep mit einer amerikanischen Patrouille das Dorf. Die Deutschen eröffneten sofort das Feuer und der Fahrer des Jeeps wurde tödlich getroffen, die zwei weiteren Insassen wurden gefangengenommen. Durch die Ankunft der Patrouille und wohl auch aus Sorge vor weiterer amerikanischer Verstärkung, beschlossen die Mörder sich zurückzuziehen. Jedoch nicht ohne die überlebenden Frauen und Kinder über die Grenze ins Reich zu verschleppen. Diese mussten fortan bei deutschen Bauern und Handwerkern Zwangsarbeit verrichten und konnten erst im Frühjahr 1945 in ihre Heimat zurückkehren.

Ein Ermittlungsverfahren, das im Jahre 1976 von der Luxemburger Kriminalpolizei eröffnet wurde, blieb ohne jurisdische Folgen, da es den überlebenden Zeugen nicht möglich war, den verdächtigten Schneider zweifelsfrei als Täter zu identifizieren. Die Zeugen des schrecklichen Geschehens wie die Einwohner Kalborns sind überzeugt, dass es sich bei dem Mörder um ein Mitglied der Waffen-SS handelte. Rezente Recherchen haben jedoch ergeben, dass es sich bei den deutschen Soldaten um Mitglieder der "Kampfgruppe Wegelein" gehandelt haben muss. Dies war eine Einheit, aufgestellt aus sehr jungen, kampfunerfahreneren Soldaten, von denen die meisten von der Heeres-Unteroffizier-Schule 2/XII Saarlautern stammten.

Heute erinnert eine Gedenkkapelle, nach dem Krieg von der Familie Hoelpes errichtet, an die sieben unschuldigen Opfer von Kalborn.

For more information
Pour en savoir plus
Für weitere Informationen

gemeng CLIARREF

Kaundorf

Um Knupp 10

9662 Bavigne

Latitude: 49.91826

Longitude: 5.90700

Monument with the bomb :

The bomb is in memory of the inhabitants and the allied troops who lived through the horror of the Battle of the Ardennes. It was dropped by a US aircraft during a flight on Christmas Day 1944. It was found in a wood near the monument. The memorial was inaugurated on 28 September 1980.

From this same location you can walk to a bunker system that was used by the resistance during Nazi occupation to hide draft dodgers from the required German forced conscription.

Kehlen

Dedicated to 1st Lt. Harold G. Stalnaker, 358th Fighter Group, 366th Fighter Squadron, 9th US Air Force, killed in action on December 23, 1944 when his aircraft was shot down over Kehlen and crashed.

Rue de Mamer 1

8280 Kehlen

Latitude: 49.66851

Longitude: 6.0345

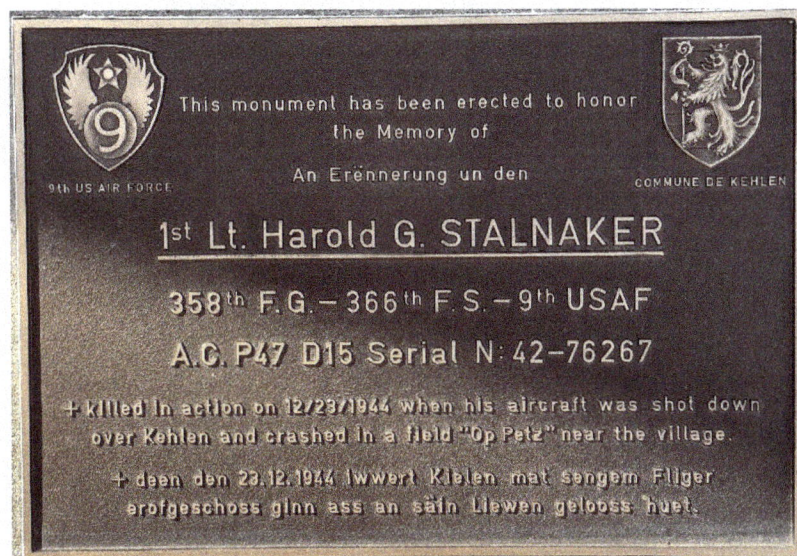

This monument has been erected to honor the Memory of

An Erënnerung un den

9th US AIR FORCE

COMMUNE DE KEHLEN

1st Lt. Harold G. STALNAKER

358th F.G. – 366th F.S. – 9th USAF

A.C. P47 D15 Serial N: 42–76267

+ killed in action on 12/23/1944 when his aircraft was shot down over Kehlen and crashed in a field "Op Petz" near the village.

+ deen den 23.12.1944 Iwwert Kielen mat sengem Flieger erofgeschoss ginn ass an säin Liewen geloss huet.

Larochette

On the 50th anniversary of the liberation 1944-1945. The grateful town of Larochette recognizes the valiant American Army which stopped the Ardennes Offensive in front of their homes.

Rue Osterbour 50

7622 Larochette

Latitude: 49.7896

Longitude: 6.2193

Liberty Road (Voie de la Liberté or Road of Freedom)

Throughout Luxembourg there are *mile marker* monuments on the roadside that mark the route that the 3rd US Army took in its liberation of Europe. The road starts in Normandy at Sainte-Mère-Église (Utah beach) and crosses France, Luxembourg and ends in Bastogne, Belgium. The entire route is marked by a milestone at each kilometer, for a total of 1,147 markers. In Luxembourg, the road generally crosses from Frisange on the southern border and exits near Steinfort on Luxembourg's western border.

(Basemap ©OpenStreetMap.org)

Liefrange

In memory of William D. Rappleye and William H. Goode, 26th Infantry Division, who were killed on December 27, 1944 in Liefrange. The parents of William Rappleye paid most of the cost to install the plaque.

Kirewee 3

9665 Bavigne

Located on side of church in Liefrange

Latitude: 49.90934

Longitude: 5.87620

Lieler

Monument located in 'Place Griffin' named after 1st LT Ray Griffin, commander of 712th Tank Battalion, 90th Infantry Division who used Lieler as his command post in January 1945. Monument dedicated to the 712th Tank Battalion, 8th US Infantry Division, 28th US Infantry Division and 90th US Infantry Division.

CR 338

Lieler

(Beside Church in Lieler)

Latitude: 50.12488

Longitude: 6.1111

Inscription

SIE HANN ÄIS D'FRÄIHEET REMBREIT

(They have restored our freedom)

In honor to the 712th US Tank Battalion and the 90th US Infantry Division liberating units of Lieler on January 26th 1945 and to the 28th and 8th U.S. Infantry Divisions
1944

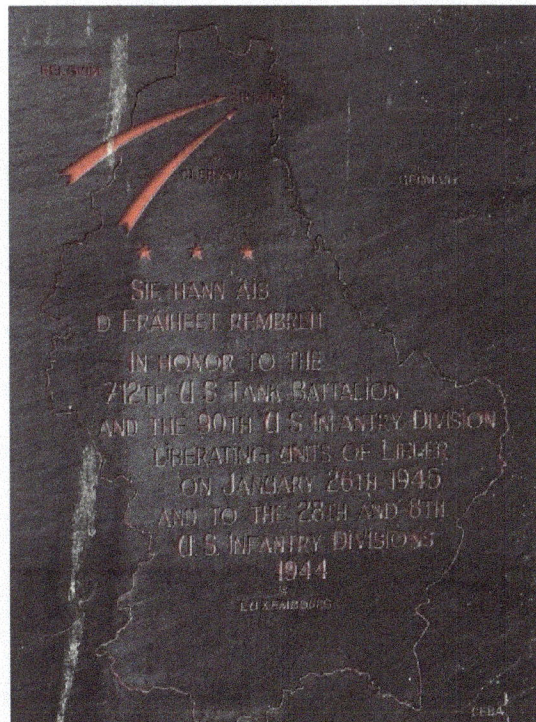

Livange

50th Anniversary of Liberation 1944-1994. There is an information panel located nearby.

Rue Geespelt

3378 Roeser

Latitude: 49.5305

Longitude: 6.1217

Lullange

Dedicated to Saint Barbara, the patron saint of artillerymen, the 6th Armored Division, and the 26th Infantry Division.

Kapellestrooss 36

9762 Wincrange

Latitude: 50.0585

Longitude: 5.9425

Inscription on Lullange Monument

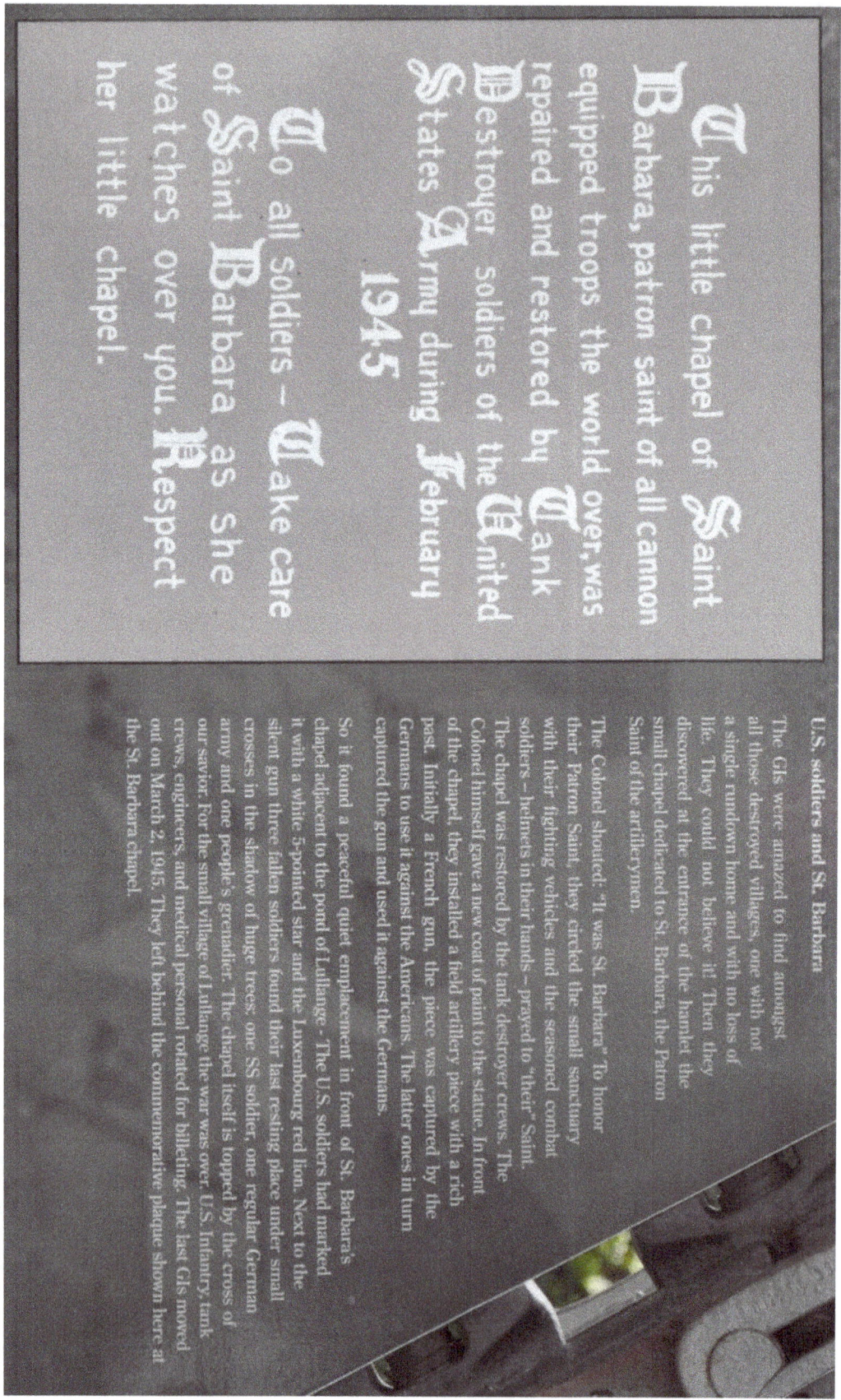

This little chapel of Saint Barbara, patron saint of all cannon equipped troops the world over, was repaired and restored by Tank Destroyer soldiers of the United States Army during February 1945

To all soldiers – Take care of Saint Barbara as She watches over you. Respect her little chapel.

U.S. soldiers and St. Barbara

The GIs were amazed to find amongst all these destroyed villages, one with not a single rundown home and with no loss of life. They could not believe it! Then they discovered at the entrance of the hamlet the small chapel dedicated to St. Barbara, the Patron Saint of the artillerymen.

The Colonel shouted: "It was St. Barbara" To honor their Patron Saint, they circled the small sanctuary with their fighting vehicles and the seasoned combat soldiers – helmets in their hands – prayed to "their" Saint. The chapel was restored by the tank destroyer crews. The Colonel himself gave a new coat of paint to the statue. In front of the chapel, they installed a field artillery piece with a rich past. Initially a French gun, the piece was captured by the Germans to use it against the Americans. The latter ones in turn captured the gun and used it against the Germans.

So it found a peaceful quiet emplacement in front of St. Barbara's chapel adjacent to the pond of Lullange - The U.S. soldiers had marked it with a white 5-pointed star and the Luxembourg red lion. Next to the silent gun three fallen soldiers found their last resting place under small crosses in the shadow of huge trees: one SS soldier, one regular German army and one people's grenadier. The chapel itself is topped by the cross of our savior. For the small village of Lullange the war was over, U.S. Infantry, tank crews, engineers, and medical personal rotated for billeting. The last GIs moved out on March 2, 1945. They left behind the commemorative plaque shown here at the St. Barbara chapel.

Luxembourg – Pescatore Foundation

In this building General George S. Patton, Jr. had his headquarters from December 21, 1944 to March 27, 1945. In the Chapel there is a photo of Patton with a Prayer of General George S. Patton, Jr. *(Please note that the Patton photo and prayer are not in an area that is open to the public)*

Avenue Jean-Pierre Pescatore 13

2324 Luxembourg

Latitude: 49.61603

Longitude: 6.12817

Photo by Tom Scholtes

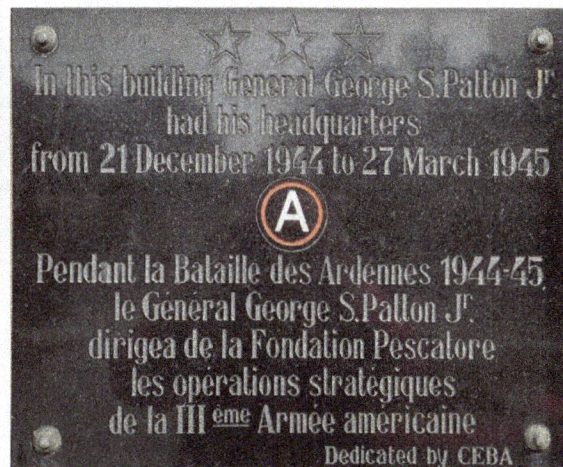

French inscription reads: During the Battle of the Ardennes 1944-45 General George S. Patton Jr. directed operations of the American 3rd Army from the Pescatore Foundation.

Luxembourg

WWII monument for The Battle of the Bulge December 16, 1944 to January 25, 1945.

Boulevard du Prince Henri 1

1724 Luxembourg

Latitude: 49.6153404

Longitude: 6.1279050

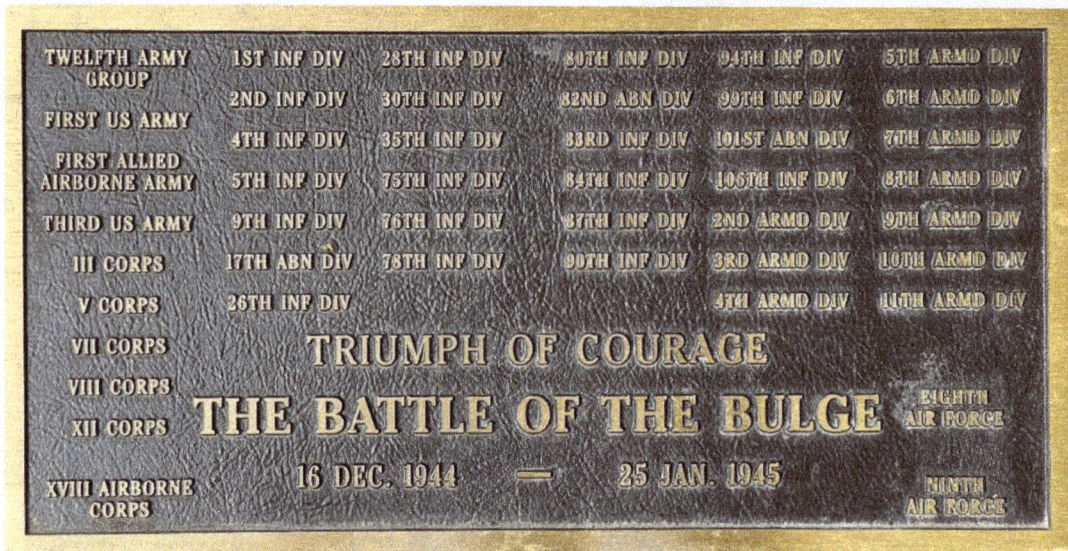

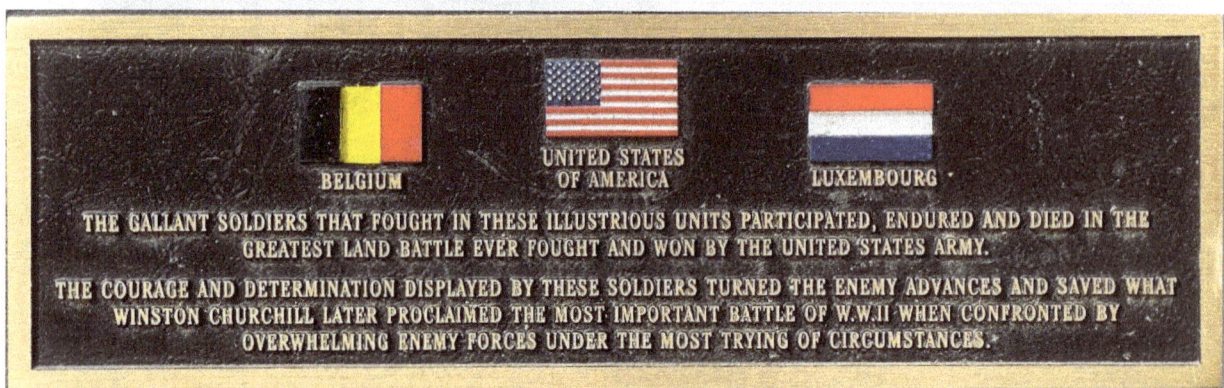

TWELFTH ARMY GROUP	1ST INF DIV	28TH INF DIV	80TH INF DIV	94TH INF DIV	5TH ARMD DIV
	2ND INF DIV	30TH INF DIV	82ND ABN DIV	99TH INF DIV	6TH ARMD DIV
FIRST US ARMY					
	4TH INF DIV	35TH INF DIV	83RD INF DIV	101ST ABN DIV	7TH ARMD DIV
FIRST ALLIED AIRBORNE ARMY	5TH INF DIV	75TH INF DIV	84TH INF DIV	106TH INF DIV	8TH ARMD DIV
THIRD US ARMY	9TH INF DIV	76TH INF DIV	87TH INF DIV	2ND ARMD DIV	9TH ARMD DIV
III CORPS	17TH ABN DIV	78TH INF DIV	90TH INF DIV	3RD ARMD DIV	10TH ARMD DIV
V CORPS	26TH INF DIV			4TH ARMD DIV	11TH ARMD DIV
VII CORPS					
VIII CORPS					EIGHTH AIR FORCE
XII CORPS					

TRIUMPH OF COURAGE
THE BATTLE OF THE BULGE
16 DEC. 1944 — 25 JAN. 1945

XVIII AIRBORNE CORPS

NINTH AIR FORCE

BELGIUM   UNITED STATES OF AMERICA   LUXEMBOURG

THE GALLANT SOLDIERS THAT FOUGHT IN THESE ILLUSTRIOUS UNITS PARTICIPATED, ENDURED AND DIED IN THE GREATEST LAND BATTLE EVER FOUGHT AND WON BY THE UNITED STATES ARMY.

THE COURAGE AND DETERMINATION DISPLAYED BY THESE SOLDIERS TURNED THE ENEMY ADVANCES AND SAVED WHAT WINSTON CHURCHILL LATER PROCLAIMED THE MOST IMPORTANT BATTLE OF W.W.II WHEN CONFRONTED BY OVERWHELMING ENEMY FORCES UNDER THE MOST TRYING OF CIRCUMSTANCES.

Luxembourg

Monument to the 5th US Armored Division, September 10, 1944.

Place d'Armes 2

1136 Luxembourg

Latitude: 49.61145

Longitude: 6.12989

ON THIS SQUARE, ON 10TH SEPTEMBER 1944
THE PEOPLE OF LUXEMBOURG WARMLY WELCOMED ITS LIBERATORS,
THE VALIANT SOLDIERS OF THE US 5TH ARMORED DIVISION
AND THEIR ROYAL HIGHNESSES PRINCE FELIX OF LUXEMBOURG
AND PRINCE JOHN, HEREDITARY GRAND DUKE OF LUXEMBOURG.

Luxembourg

To the memory to General Omar N. Bradley, Commander in Chief of the 12th US Army Group, whose headquarters (Eagle TAC) were located in this building during the liberation of the Luxembourg 1944 - 1945

Avenue de la Liberté 4

1930 Luxembourg

Latitude: 49.607063

Longitude: 6.1278605

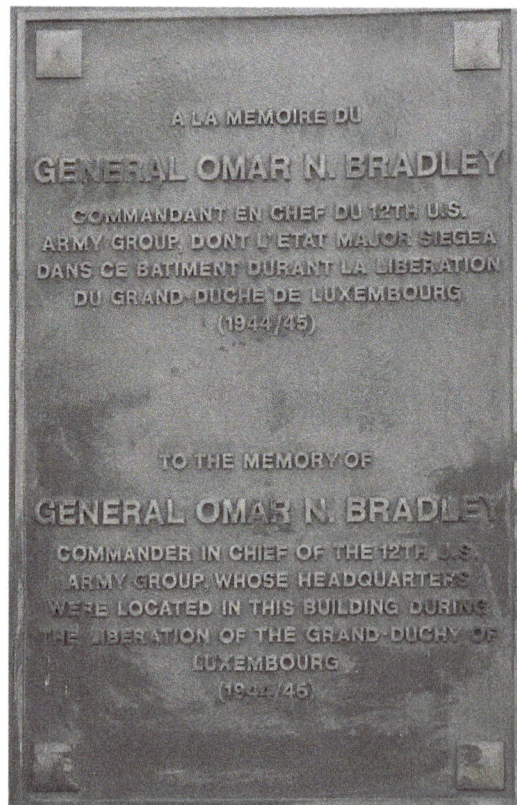

Luxembourg

Information panel to the 23rd Headquarters Special Troops, also known as the *Ghost Army*. The Ghost Army was a deception unit that used inflatable tanks and artillery, sounds effects, impersonation, and radio deception to misled German troops. The panel was inaugurated on 7 September 2023 and is located on the University of Luxembourg, Limpertsberg campus.

162a, avenue de la Faïencerie

1511 Luxembourg

Latitude: 49.623638

Longitude: 6.11284

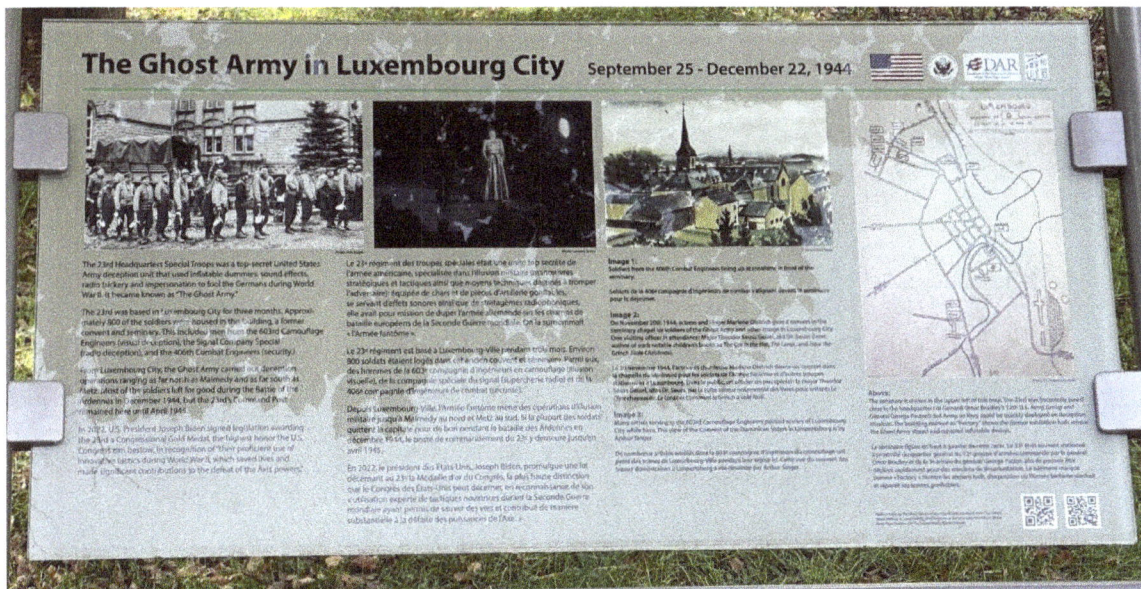

Thanks to ghostarmy.org for providing a high-resolution version of the information board (see next page).

The Ghost Army in Luxembourg City

September 25 - December 22, 1944

The 23rd Headquarters, Special Troops was a top-secret United States Army deception unit that used inflatable dummies, sound effects, radio trickery and impersonation to fool the Germans during World War II. It became known as "The Ghost Army."

The 23rd was based in Luxembourg City for three months. Approximately 800 of the soldiers were housed in this building, a former convent and seminary. This included men from the 603rd Camouflage Engineers (visual deception), the Signal Company Special (radio deception), and the 406th Combat Engineers (security)

From Luxembourg City, the Ghost Army carried out deception operations ranging as far north as Malmedy and as far south as Metz. Most of the soldiers left for good during the Battle of the Ardennes in December 1944, but the 23rd's Command Post remained here until April 1945.

In 2022, U.S. President Joseph Biden signed legislation awarding the 23rd a Congressional Gold Medal, the highest honor the U.S. Congress can bestow, in recognition of "their proficient use of innovative tactics during World War II, which saved lives and made significant contributions to the defeat of the Axis powers."

Le 23e régiment des troupes spéciales était une unité top secrète de l'armée américaine, spécialisée dans l'illusion militaire (manœuvres stratégiques et tactiques ainsi que moyens techniques destinés à tromper l'adversaire), équipée de chars et de pièces d'artillerie gonflables, se servant d'effets sonores ainsi que de stratagèmes radiophoniques, elle avait pour mission de duper l'armée allemande sur les champs de bataille européens de la Seconde Guerre mondiale. On la surnommait « l'Armée fantôme ».

Le 23e régiment est basé à Luxembourg-Ville pendant trois mois. Environ 800 soldats étaient logés dans cet ancien couvent et séminaire. Parmi eux, des hommes de la 603e compagnie d'ingénieurs en camouflage (illusion visuelle), de la compagnie spéciale du signal (supercherie radio) et de la 406e compagnie d'ingénieurs de combat (sécurité).

Depuis Luxembourg-Ville, l'Armée fantôme mène des opérations d'illusion militaire jusqu'à Malmedy au nord et Metz au sud. Si la plupart des soldats quittent la capitale pour de bon pendant la bataille des Ardennes en décembre 1944, le poste de commandement du 23e y demeure jusqu'en avril 1945.

En 2022, le président des États-Unis, Joseph Biden, promulgue une loi décernant au 23e la Médaille d'or du Congrès, la plus haute distinction que le Congrès des États-Unis peut décerner, en reconnaissance de son « utilisation experte de tactiques novatrices durant la Seconde Guerre mondiale ayant permis de sauver des vies et contribué de manière substantielle à la défaite des puissances de l'Axe. »

Image 1:

Soldiers from the 106th Combat Engineers lining up at mealtime in front of the seminary.

Soldats de la 106e compagnie d'ingénieurs de combat s'alignant devant le séminaire pour le déjeuner.

Image 2:

On November 20th 1944, actress and singer Marlene Dietrich gave a concert in the seminary chapel for soldiers of the Ghost Army and other troops in Luxembourg City. One visiting officer in attendance Major Theodor Seuss Geisel, aka Dr. Seuss (later, author of such notable children's books as The Cat in the Hat, The Lorax, and How the Grinch Stole Christmas).

Le 20 novembre 1944, l'actrice et chanteuse Marlene Dietrich donne un concert dans la chapelle du séminaire pour les soldats de l'Armée fantôme et d'autres troupes stationnées à Luxembourg. Dans le public, un officier un peu spécial : le major Theodor Seuss Geisel, alias Dr. Seuss, par la suite auteur notamment des livres pour enfants Le Chat chapeauté, Le Lorax et Comment le Grinch a volé Noël.

Image 3:

Many artists serving in the 603rd Camouflage Engineers painted scenes of Luxembourg City while here. This view of the Convent of the Dominican Sisters in Limpertsberg is by Arthur Singer.

De nombreux artistes enrôlés dans la 503e compagnie d'ingénieurs du camouflage ont peintes scènes de Luxembourg-Ville pendant leur séjour ici. Cette vue du couvent des Sœurs dominicaines à Limpertsberg a été dessinée par Arthur Singer.

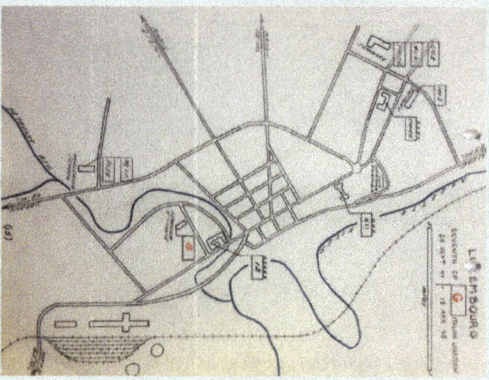

Above:

The seminary is shown in the upper left of this map. The 23rd was frequently based close to the headquarters of General Omar Bradley's 12th US Army Group and General George Patton's 3rd Army, so they could be quickly deployed on deception missions. The building marked as "Factory" shows the former exhibition halls where the Ghost Army stored and repaired inflatable decoys.

Le séminaire figure en haut à gauche de cette carte. Le 23e était souvent stationné à proximité du quartier général du 12e groupe d'armées commandé par le général Omar Bradley et de la 3e armée du général George Patton afin de pouvoir être déployé rapidement pour des missions de décamouflage. Le bâtiment marqué comme « Factory » montre les anciens halls d'exposition où l'Armée fantôme stockait et réparait ses leurres gonflables.

Luxembourg - Hamm

Monument to Dwight D. Eisenhower.

Allée Dwight David Eisenhower

2517 Luxembourg

Latitude: 49.612948

Longitude: 6.187614

DWIGHT D. EISENHOWER

1890 – 1969

CENTENNIAL

1990

Luxembourg - Hamm

Luxembourg - Hamm

To the parish of Hamm and the American Soldiers who gave their lives for the freedom of Luxembourg

Rue des Peupliers 2A

2328 Luxembourg

Latitude: 49.61014

Longitude: 6.16767

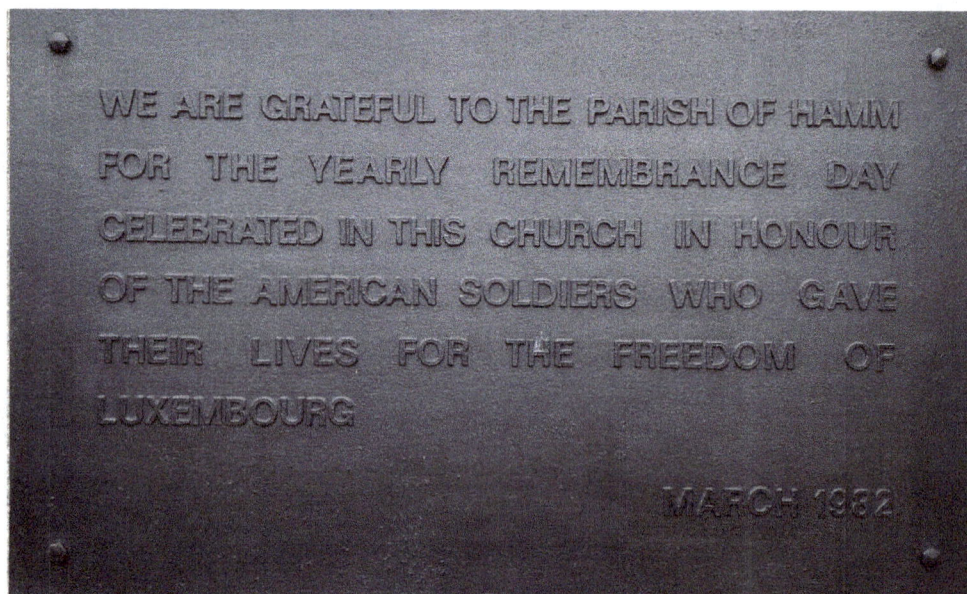

WE ARE GRATEFUL TO THE PARISH OF HAMM FOR THE YEARLY REMEMBRANCE DAY CELEBRATED IN THIS CHURCH IN HONOUR OF THE AMERICAN SOLDIERS WHO GAVE THEIR LIVES FOR THE FREEDOM OF LUXEMBOURG

MARCH 1982

Marnach

Monument to the 707th Tank Battalion and the 28th US Infantry Division with three information panels.

See Information panel section

Schullstrooss 2

Marnach

Latitude: 50.0541

Longitude: 6.0624

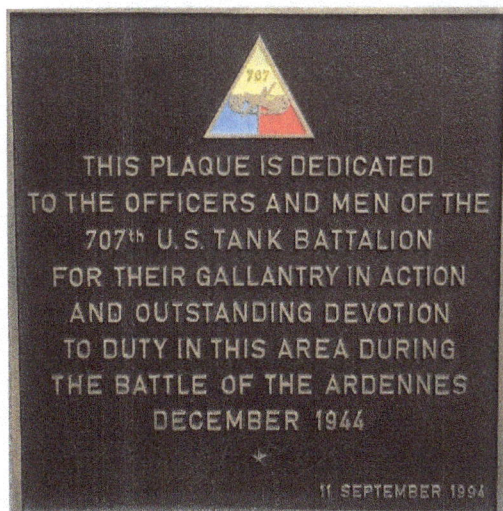

THIS PLAQUE IS DEDICATED
TO THE OFFICERS AND MEN OF THE
707th U.S. TANK BATTALION
FOR THEIR GALLANTRY IN ACTION
AND OUTSTANDING DEVOTION
TO DUTY IN THIS AREA DURING
THE BATTLE OF THE ARDENNES
DECEMBER 1944

11 SEPTEMBER 1994

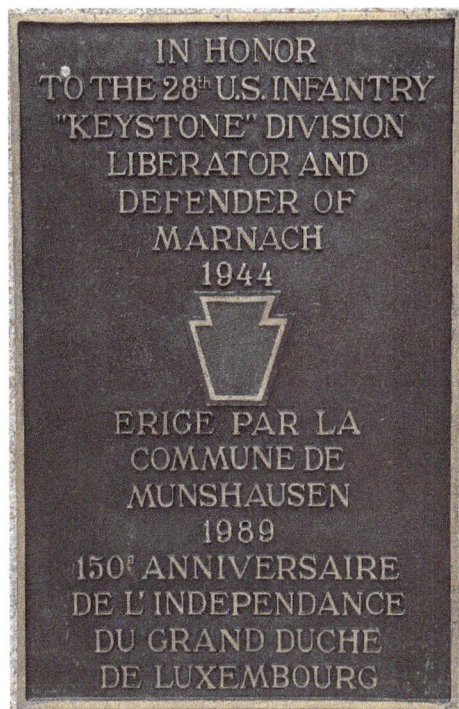

IN HONOR
TO THE 28th U.S. INFANTRY
"KEYSTONE" DIVISION
LIBERATOR AND
DEFENDER OF
MARNACH
1944

ERIGE PAR LA
COMMUNE DE
MUNSHAUSEN
1989
150e ANNIVERSAIRE
DE L'INDEPENDANCE
DU GRAND DUCHE
DE LUXEMBOURG

Information Panel Marnach

THE BATTLE AROUND MARNACH

Early in the morning of Dec 16, 1944, in his headquarters at the Hotel Claravallis, Clervaux, **Colonel Hurley E. FULLER**, commander of the 110th Regiment, 28th US Infantry Division, organized the defense, the support and the counterattacks of his battalions and front-line companies.

Major General Norman D. COTA, Commander of the 28th US Infantry Division, ordered the 110th Regiment to **hold the positions at all costs**. Time had to be bought to move up troops into positions to contain the German thrust.

All day long on December 16, **Company B**, 1st Battalion, 110th Regiment, successfully resisted several German attacks in Marnach. At midnight, tanks and self-propelled guns of the *304th Panzer Regiment* of the *2nd Panzer Division* overran Marnach.

At 02.30 on December 17, a US counterattack from Munshausen towards Marnach with 5 tanks and one infantry platoon was stopped by heavy enemy tank fire. panzerfausts (recoilless one-man antitank hallow-charge bomb discharger) and small arms after a 30-minute fight.

The 2nd Battalion of the 110th Regiment was the divisional reserve. Major General COTA, ordered Company G to Wiltz. Company E & F returned to Col FULLER's command.

At 07.30 on Dec 17, **Company E and F**, attacked east towards **Marnach** supported by the light tanks of **Company D, 707th Tank Battalion**, proceeding from the north. However, German tanks stopped the attack and the US soldiers retreated north-west of Reuler. Spearheads of the *116th Panzerdivision* knocked out 11 from the 17 light tanks around **Fischbach**. Marnach was definitely lost.

At **Hupperdange**, 6 guns (of Battery A, 109th Field Artillery), two 40mm Bofors guns and a quad 50 were knocked out between 01.30 and 07.30 of December 17 by elements of the *560th Volksgrenadierdivision*.

The spearheads of the *116th Panzerdivision* and a regiment of the *560th Volksgrenadierdivision* overran **Heinerscheid** in the afternoon of December 17.

All day long on Dec 17, in **Munshausen**, **Company C**, 1st Battalion, 110th Regiment, and **Regimental Cannon Company** resisted the attacks of the reconnaissance troop of the *2nd Panzerdivision*. At nightfall, the Germans concluded their attacks. At the same time, the exhausted defenders left their position in Munshausen and retreated to the west. All their guns were lost.

On the morning of Dec 17, Major General Norman Daniel "Dutch" COTA sent **Company B** of the 2nd **Tank Battalion** of Combat Command R, 9th **Armored Division**, to support the 110th Infantry Regiment. Colonel Hurley E. FULLER ordered the 1st Platoon to **Heinerscheid**, the 2nd Platoon to **Reuler** and the 3rd Platoon to throw out the Germans from the southern part of **Clervaux**.

After a 24 hours' fight with the German tanks, not a single of the 18 medium Sherman tanks survived.

LA BATAILLE AUTOUR DE MARNACH

Tôt le matin du 16 décembre 1944, à son poste de commandement à l'Hôtel Claravallis, Clervaux, le **Colonel Hurley E. FULLER**, commandant du 110e Régiment de la 28e Division d'Infanterie, organisa la défense, le soutien et les contre-attaques de ses bataillons et compagnies.

Major Général Norman D. COTA, commandant de la 28e Division d'Infanterie américaine, ordonna au 110e Régiment de **tenir toutes les positions à tout prix**. Il fallait gagner du temps pour mettre en place des troupes pour contenir l'avance des Allemands.

Pendant toute la journée du 16 décembre, la **Compagnie B**, 1er Bataillon, 110e Régiment, résista à plusieurs attaques allemandes à **Marnach**. À minuit, des chars et des canons automoteurs du *304e Régiment de Chars de la 2e Panzerdivision* envahirent le village de Marnach.

Le 17 décembre, à 02.30 heures, une contre-attaque américaine de **Munshausen vers Marnach** avec 5 chars et un peloton d'infanterie fut arrêtée après une demi-heure par le feu nourri de chars, de bazookas (lance bombe antichar sans recul de bombe antichar individuel) et d'armes légères ennemies.

Le 2e Bataillon du 110e Régiment était la réserve de la division. Major Général Norman D. COTA ordonna la Compagnie G à Wiltz. Les Compagnies E et F retournèrent sous les ordres du Colonel Hurley E. FULLER.

Le 17 décembre, à 07.30, les **Compagnies E et F** attaquèrent **Marnach** à l'est. Elles furent soutenues par la **Compagnie D, 707e Tank Battalion** venant du nord. Toutefois, des chars allemands arrêtèrent l'attaque et les soldats américains se retirèrent au nord-ouest de Reuler. Des unités de choc de la *116e Panzerdivision* abattirent 11 des 17 chars légers autour de **Fischbach**. Marnach était définitivement perdu.

À **Hupperdange**, le 17.12.44, entre 01.30 et 07.30, des éléments de la *560e Volksgrenadierdivision* détruisirent six obusiers (de la batterie A, 109e Artillerie de Campagne), deux canons anti-aériens Bofors et une pièce à quatre mitrailleuses 50.

Les avant-gardes de la *116e Panzerdivision* et un régiment de la *560e Volksgrenadierdivision* envahirent **Heinerscheid** pendant l'après-midi du 17 décembre.

Tout au long de la journée du 17 décembre, à **Munshausen**, la **Compagnie C**, 1er Bataillon, 110e Régiment, et la **Compagnie d'Artillerie du régiment** résistaient aux attaques des troupes de reconnaissance de la *2e Panzerdivision*. À la tombée de la nuit, les Allemands cessèrent leurs attaques. Au même instant, les défenseurs exténués quittèrent leurs positions à Munshausen et se retirèrent vers l'ouest. Toute leur artillerie était perdue.

Le matin du 17 décembre, Major Général Norman Daniel « Dutch » COTA envoya la **Compagnie B** du 2e **Bataillon Blindé** du Groupement de Combat R de la 9e **Division Blindée** à la rescousse du 110e Régiment. Le Colonel Hurley E. FULLER dirigea le premier peloton à **Heinerscheid**, le deuxième à **Reuler** et le 3e à **Clervaux** pour y chasser les attaquants du sud de la localité.

Après un combat de 24 heures avec les chars allemands, tous les 18 blindés américains étaient hors de combat.

DER KAMPF UM MARNACH

Früh am Morgen des 16. Dezember 1944 organisierte **Oberst Hurley E. FULLER**, Kommandeur des 110. Regiments der 28. US Infanteriedivision, die Verteidigung, die Unterstützung und die Gegenangriffe seiner Einheiten in seinem Hauptquartier im Hotel Claravallis in Clerf.

Generalmajor Norman D. COTA, Kommandeur der 28. US Infanteriedivision, befahl dem 110. Regiment: „Halten um jeden Preis". Es musste Zeit gewonnen werden, um Truppen in Stellung zu bringen, um den deutschen Vormarsch zu stoppen.

Am 16. Dezember hielt **Kompanie B**, 1. Bataillon, 110. Regiment, während des ganzen Tages den deutschen Angriffen in Marnach stand. Um Mitternacht überrannten Panzer und Sturmgeschütze des *304. Panzerregiments* der *2. Panzerdivision* die Ortschaft Marnach.

Am 17. Dezember um 02.30 Uhr, wurde ein amerikanischer **Gegenangriff von Munshausen nach Marnach** mit 5 Panzern und einem Zug Infanterie durch heftiges deutsches Panzer-, Panzerfaust- und Gewehrfeuer nach einem 30-minütigen Kampf zerschlagen.

Das 2. Bataillon des 110. Regimentes war die Divisionsreserve. Generalmajor COTA verlegte die Kompanie G nach Wiltz. Kompanie E und F kamen wieder unter den Befehl von Oberst FULLER.

Am 17. Dezember um 07.30, griffen die **Kompanien E und F. Marnach** im Osten an. Die leichten Panzer der **Kompanie D, 707. Panzerbataillon**, unterstützen die Infanterie indem sie von Norden angriffen. Deutsche Panzer jedoch stoppten den Angriff und die amerikanischen Soldaten zogen sich nordwestlich von Reuler zurück. Die Vorhut der *116. Panzerdivision* schoss 11 von den 17 leichten Panzern in **Fischbach** ab. Marnach war endgültig verloren.

Am 17.12.44 zwischen 01.30 und 07.30 zerstörten Einheiten der *560. Volksgrenadierdivision* sechs Kanonen (der Batterie A, 109. Feldartillerie), zwei Bofors-Kanonen und ein Vierlingsgeschütz in **Hüppderdingen**.

Die Vorhut der *116. Panzerdivision* und ein Regiment der *560. Volksgrenadierdivision* überrannten **Heinerscheid** am Nachmittag des 17. Dezember.

Am 17. Dezember, schlugen **Kompanie C**, 1. Bataillon, 110. Regiment, und **Artilleriekompanie des Regimentes in Munshausen** alle Angriffe der Aufklärungstruppen der *2. Panzerdivision* zurück. Bei Anbruch der Nacht stoppten die Deutschen die Angriffe. Zur gleichen Zeit zogen sich die erschöpften Verteidiger aus Munshausen nach Westen zurück. Ihre Geschütze waren alle verloren.

Am Morgen des 17.12.44 befahl General Norman Daniel „Dutch" COTA der **Kompanie B, 2. Panzerbataillon** Kampfgruppe R, 9. **Panzerdivision**, dem 110. Regiment zu Hilfe zu kommen. Oberst Hurley E. FULLER leitete den ersten Zug nach Heinerscheid, den zweiten nach Reuler und den dritten nach **Clerf**, um dort die Angreifer aus dem südlichen Teil der Ortschaft hinauszuwerfen.

Nach einem 24-stündigen Kampf mit den deutschen Panzern, waren alle 18 amerikanischen Panzer außer Gefecht.

Sources - Quellen:
- To Save Bastogne, Robert F. Phillips, Borodino Books, USA
  US ARMY HANDBOOK 1939-1945, George Forty,
  Alan Sutton Publishing Limited, ISBN 0-7509-1078-X

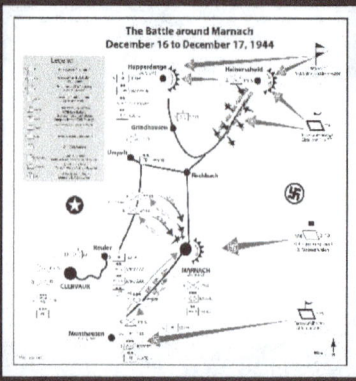

For more information
Pour en savoir plus
Für weitere Informationen

Information Panel Marnach

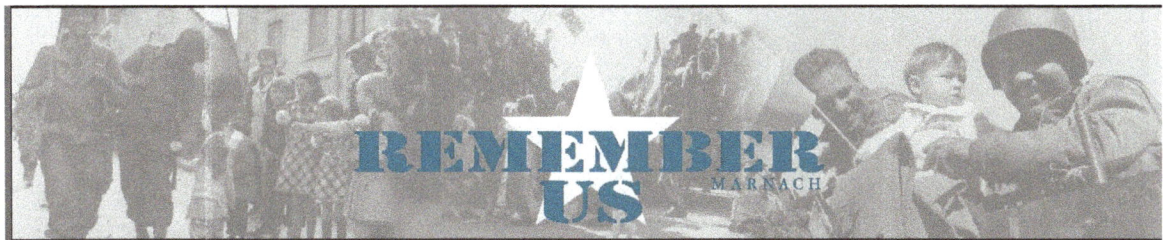

REMEMBER US
MARNACH

THE 707TH TANK BATTALION IN SUPPORT OF THE 110TH REGIMENT, 28TH US INFANTRY DIVISION

In December 1944, the 707th Tank Battalion was attached to the 28th US Infantry Division and was in the Division's reserve. On December 16, 1944, at the beginning of the Battle of the Bulge, the 707th Tank Battalion was alerted at 07.00 a.m. and partitioned as follows:

- Companies A and B were assigned to the 110th Regiment (Clervaux)
- Company C came under the orders of the 109th Regiment (Diekirch)
- Company D (light tanks) was attached to the 112th Regiment (Weiswampach).

For four days, the tanks of Companies A and B, together with the defenders of the 110th Regiment and its attached units, offered stiff resistance to the attacking German units in Marnach, Munshausen, Clervaux, Hosingen, Bockholtz, Consthum, Hoscheid and Wiltz. Gradually, however, the stronger German "Panther" tanks eliminated many of the American tanks.

The remaining tanks concentrated in Wiltz and heavily defended the town. When the Germans encircled the town, and the Americans had run out of ammunition and fuel, the crews destroyed their remaining tanks and fled to Belgium.

Total losses: 36 tanks from Company A and B, 17 light tanks of Company D and 18 tanks of Company B, Combat Command R, 9th US Armored Division that had fought in Clervaux, Reuler and Heinerscheid.

LE 707E BATAILLON DE CHARS EN SUPPORT DU 110E RÉGIMENT, 28E DIVISION D'INFANTERIE AMÉRICAINE

Au mois de décembre 1944, le 707e Bataillon de chars (707th Tank Battalion) était rattaché à la 28e Division d'Infanterie américaine et faisait partie de la réserve de la division. Au matin du 16 décembre 1944, au commencement de la Bataille des Ardennes, le 707e Bataillon de chars fut alerté à 07.00 heures et réparti de la façon suivante :

- Les Compagnies A et B furent attribuées au 110e Régiment (Clervaux),
- la Compagnie C passa sous les ordres du 109e Régiment (Diekirch).
- la Compagnie D (chars légers) rejoignit le 112e Régiment (Weiswampach)

Pendant quatre jours, les chars des compagnies A et B ainsi que les défenseurs du 110e Régiment et les unités qui y étaient rattachées opposèrent une résistance farouche aux attaquants allemands dans les localités de Marnach, Munshausen, Clervaux, Hosingen, Bockholtz, Hoscheid et Wiltz. Toutefois, peu à peu, les chars allemands « Panther » plus forts éliminèrent de nombreux chars américains. Les chars restants se concentrèrent à Wiltz et défendirent la ville avec acharnement. La ville étant encerclée, les Américains manquant de munitions et de carburant, les équipages détruisirent leurs chars et se replièrent sur la Belgique.

Total des pertes : 36 chars des Compagnies A et B, 17 chars légers de la Compagnie D ainsi que 18 chars de la Compagnie B, Combat Command R, de la 9e Division Blindée Américaine qui avaient combattu à Clervaux, Reuler et Heinerscheid.

DAS 707. PANZERBATAILLON IN UNTERSTÜTZUNG DES 110. REGIMENTES DER 28. US INFANTERIEDIVISION

Im Dezember 1944 war das 707. Tank Bataillon der 28. US Infanteriedivision angegliedert und befand sich in der Divisionsreserve. Am 16. Dezember 1944, am Beginn der Ardennenoffensive, wurde das 707. Panzerbataillon um 07.00 Uhr alarmiert und folgendermaßen aufgeteilt:

- Kompanie A und B wurden dem 110. Regiment (Clerf) zugeordnet,
- Kompanie C wurde dem 109. Regiment (Diekirch) zugeteilt,
- Kompanie D (leichte Panzer) wurde dem 112. Regiment (Weiswampach) unterstellt.

Die Panzer der Kompanien A und B leisteten in Marnach, Munshausen, Clerf, Hosingen, Bockholtz, Consthum, Hoscheid und Wiltz mit den Verteidigern des 110. Regimentes und den angegliederten Einheiten den deutschen Angreifern während vier Tagen heftigen Widerstand. Viele Panzer wurden jedoch nach und nach den stärkeren deutschen Panzer Panzern ausgeschaltet.

Die übriggebliebenen Tanks konzentrierten sich in Wiltz und verteidigten die Stadt hartnäckig. Als die Stadt eingekesselt war, den Amerikanern die Munition und den Treibstoff fehlten, zerstörten die Besatzungen ihre Panzer und flüchteten nach Belgien.

Alle 36 Panzer der Kompanien A und B, die 17 leichten Panzer der Kompanie D sowie die 18 Panzer der Kompanie B, Kampfgruppe R (CCR), der 9. US Panzerdivision, die in Clerf, Reuler und Heinerscheid eingesetzt gewesen waren, waren zerstört.

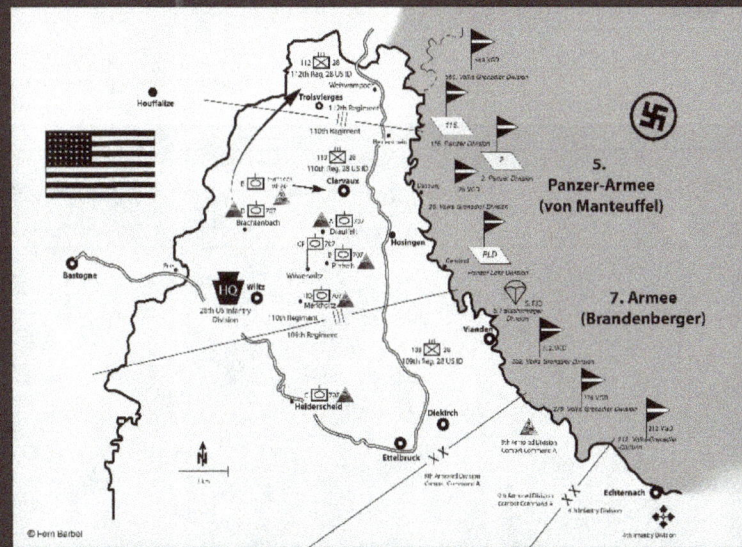

5. Panzer-Armee (von Manteuffel)

7. Armee (Brandenberger)

Medium tank M4A3E8 "Sherman"
Char moyen M4A3E8 "Sherman"
Mittelschwerer M4A3E8 "Sherman"

Light US tank M5A1 "Stuart"
char léger américain M5A1 "Stuart"
leichter amerikanischer M5A1 "Stuart" Panzer

Sources - Quellen:
- The Heroes of Hosingen by Alice M. Flynn Sky Blue Printing, LLC, USA
- Die Ardennenschlacht 1944-1945 in Luxemburg von Jean Milmeister Éditions Saint-Paul Luxembourg, 1994
- U.S.Army (ETO 1944-45) Marquages et organisation von Emile Becker und Jean Milmeister, Imprimerie G. Willems, Dudelange, 1988
- US ARMY HANDBOOK 1939-1945, George Forty, Alan Sutton Publishing Limited, ISBN 0-7509-1078-X
- Panzerdarstellungen: Adobe Stock

For further details about 707th Tank Battalion's operational areas, please consult this QR-CODE.

Pour plus de détails concernant les aires de combat du 707e Bataillon Blindé, veuillez consulter ce QR-CODE.

Für weitere Informationen über die Einsatzgebiete des 707. Panzerbataillons bitte QR-CODE scannen.

Information Panel Marnach

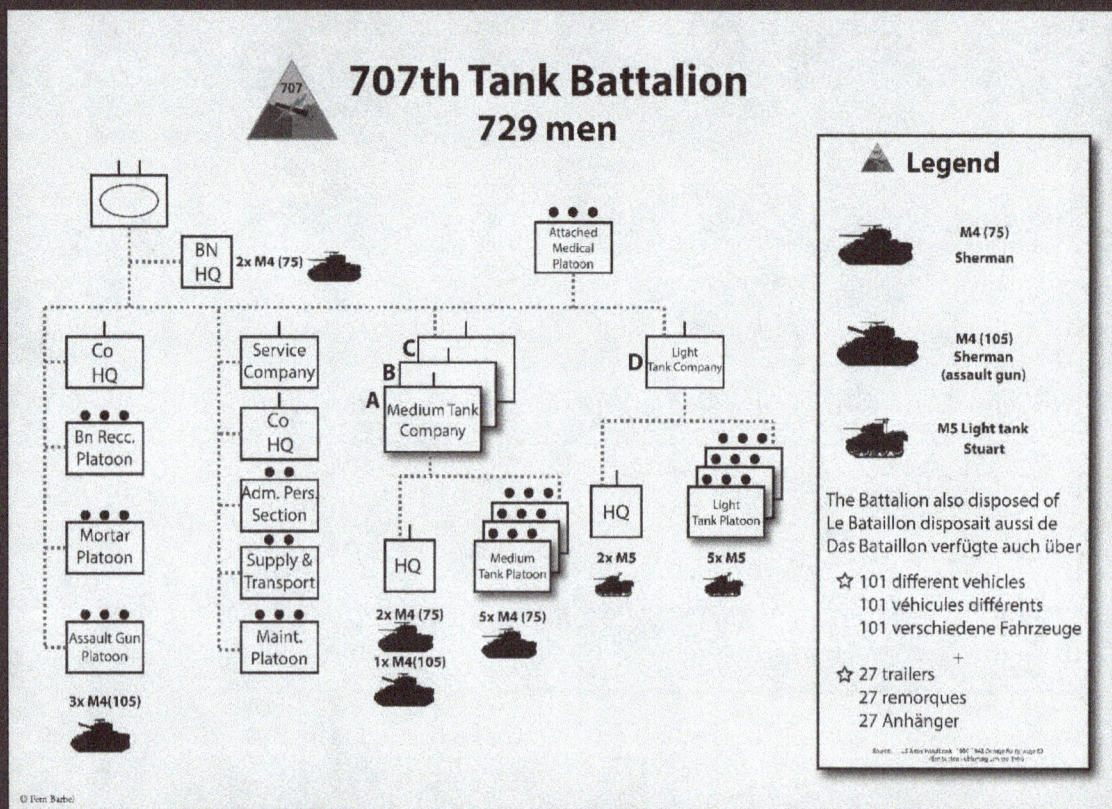

Martelange

In honor of the 4th US Armored Division which liberated Martelange-Rombach on December 23, 1944.

Route d'Arlon

Martelange

Latitude: 49.833253

Longitude: 5.740306

Medernach

Dedicated to the 9th US Armored Division, Commanded by Major General John W. Leonard and its three Combat Commands A, B, and R.

Rue de Larochette

7661 Vallée de l'Ernz

Latitude: 49.8064

Longitude: 6.2136

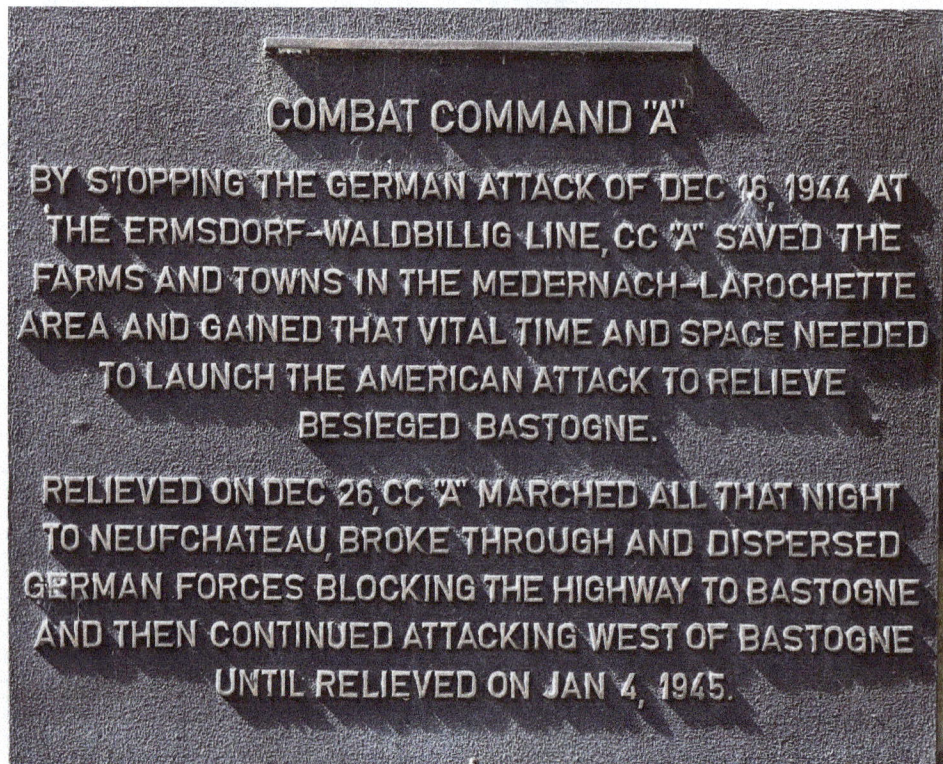

COMBAT COMMAND "A"

BY STOPPING THE GERMAN ATTACK OF DEC 16, 1944 AT THE ERMSDORF-WALDBILLIG LINE, CC "A" SAVED THE FARMS AND TOWNS IN THE MEDERNACH-LAROCHETTE AREA AND GAINED THAT VITAL TIME AND SPACE NEEDED TO LAUNCH THE AMERICAN ATTACK TO RELIEVE BESIEGED BASTOGNE.

RELIEVED ON DEC 26, CC "A" MARCHED ALL THAT NIGHT TO NEUFCHATEAU, BROKE THROUGH AND DISPERSED GERMAN FORCES BLOCKING THE HIGHWAY TO BASTOGNE AND THEN CONTINUED ATTACKING WEST OF BASTOGNE UNTIL RELIEVED ON JAN 4, 1945.

Medernach (Continued)

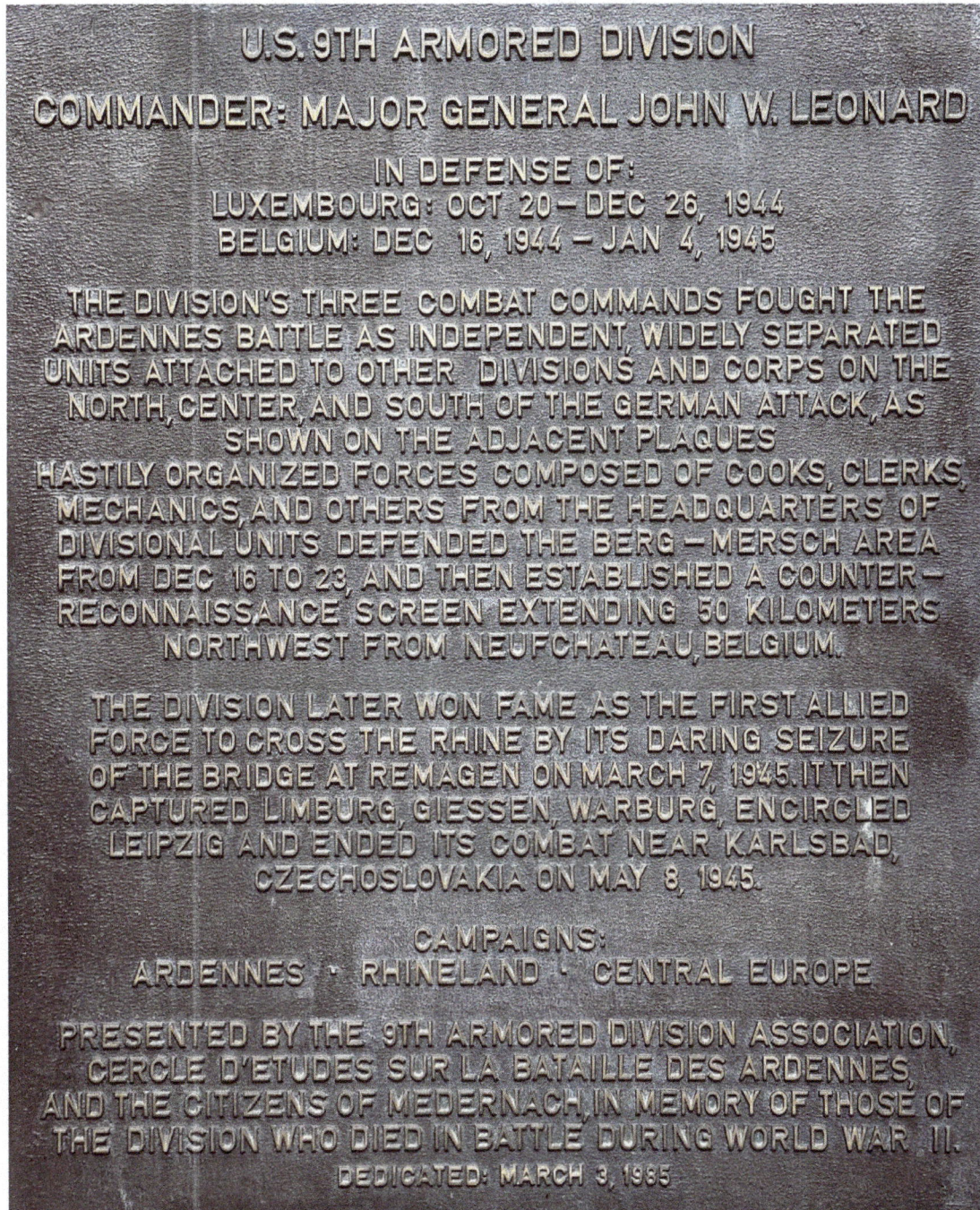

U.S. 9TH ARMORED DIVISION

COMMANDER: MAJOR GENERAL JOHN W. LEONARD

IN DEFENSE OF:
LUXEMBOURG: OCT 20 – DEC 26, 1944
BELGIUM: DEC 16, 1944 – JAN 4, 1945

THE DIVISION'S THREE COMBAT COMMANDS FOUGHT THE ARDENNES BATTLE AS INDEPENDENT, WIDELY SEPARATED UNITS ATTACHED TO OTHER DIVISIONS AND CORPS ON THE NORTH, CENTER, AND SOUTH OF THE GERMAN ATTACK, AS SHOWN ON THE ADJACENT PLAQUES
HASTILY ORGANIZED FORCES COMPOSED OF COOKS, CLERKS, MECHANICS, AND OTHERS FROM THE HEADQUARTERS OF DIVISIONAL UNITS DEFENDED THE BERG – MERSCH AREA FROM DEC 16 TO 23, AND THEN ESTABLISHED A COUNTER – RECONNAISSANCE SCREEN EXTENDING 50 KILOMETERS NORTHWEST FROM NEUFCHATEAU, BELGIUM.

THE DIVISION LATER WON FAME AS THE FIRST ALLIED FORCE TO CROSS THE RHINE BY ITS DARING SEIZURE OF THE BRIDGE AT REMAGEN ON MARCH 7, 1945. IT THEN CAPTURED LIMBURG, GIESSEN, WARBURG, ENCIRCLED LEIPZIG AND ENDED ITS COMBAT NEAR KARLSBAD, CZECHOSLOVAKIA ON MAY 8, 1945.

CAMPAIGNS:
ARDENNES · RHINELAND · CENTRAL EUROPE

PRESENTED BY THE 9TH ARMORED DIVISION ASSOCIATION, CERCLE D'ETUDES SUR LA BATAILLE DES ARDENNES, AND THE CITIZENS OF MEDERNACH, IN MEMORY OF THOSE OF THE DIVISION WHO DIED IN BATTLE DURING WORLD WAR II.
DEDICATED: MARCH 3, 1985

Medernach (Continued)

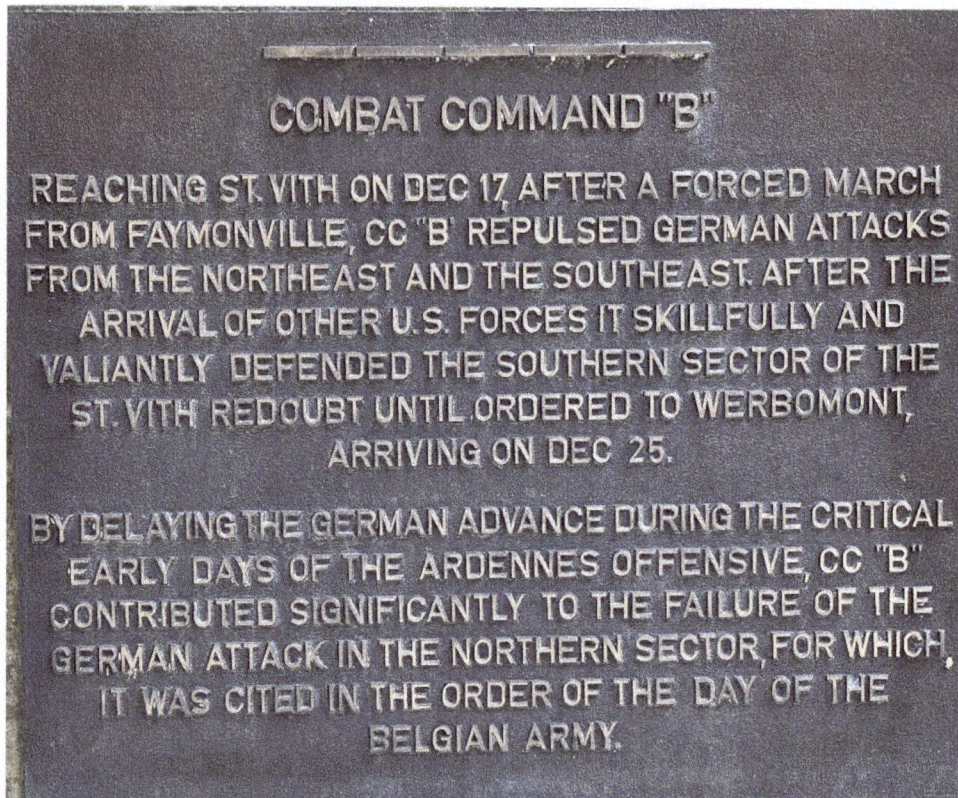

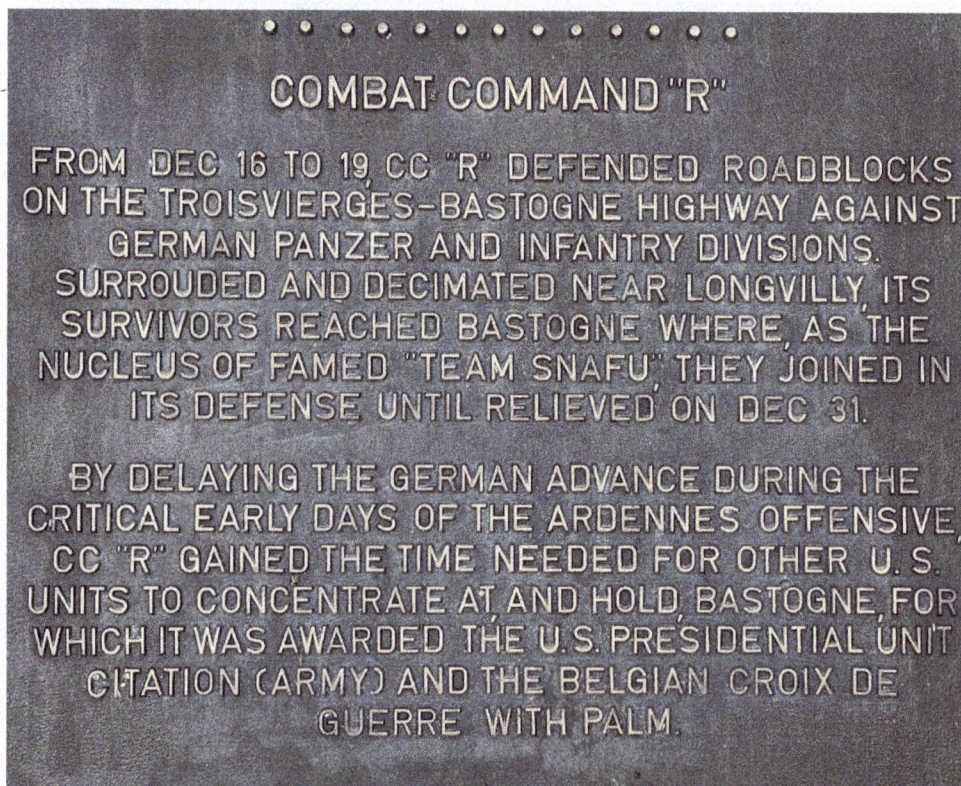

COMBAT COMMAND "B"

REACHING ST. VITH ON DEC 17, AFTER A FORCED MARCH FROM FAYMONVILLE, CC "B" REPULSED GERMAN ATTACKS FROM THE NORTHEAST AND THE SOUTHEAST. AFTER THE ARRIVAL OF OTHER U.S. FORCES IT SKILLFULLY AND VALIANTLY DEFENDED THE SOUTHERN SECTOR OF THE ST. VITH REDOUBT UNTIL ORDERED TO WERBOMONT, ARRIVING ON DEC 25.

BY DELAYING THE GERMAN ADVANCE DURING THE CRITICAL EARLY DAYS OF THE ARDENNES OFFENSIVE, CC "B" CONTRIBUTED SIGNIFICANTLY TO THE FAILURE OF THE GERMAN ATTACK IN THE NORTHERN SECTOR, FOR WHICH, IT WAS CITED IN THE ORDER OF THE DAY OF THE BELGIAN ARMY.

COMBAT COMMAND "R"

FROM DEC 16 TO 19, CC "R" DEFENDED ROADBLOCKS ON THE TROISVIERGES–BASTOGNE HIGHWAY AGAINST GERMAN PANZER AND INFANTRY DIVISIONS. SURROUDED AND DECIMATED NEAR LONGVILLY, ITS SURVIVORS REACHED BASTOGNE WHERE, AS THE NUCLEUS OF FAMED "TEAM SNAFU", THEY JOINED IN ITS DEFENSE UNTIL RELIEVED ON DEC 31.

BY DELAYING THE GERMAN ADVANCE DURING THE CRITICAL EARLY DAYS OF THE ARDENNES OFFENSIVE, CC "R" GAINED THE TIME NEEDED FOR OTHER U.S. UNITS TO CONCENTRATE AT, AND HOLD, BASTOGNE, FOR WHICH IT WAS AWARDED THE U.S. PRESIDENTIAL UNIT CITATION (ARMY) AND THE BELGIAN CROIX DE GUERRE WITH PALM.

Merscheid-lès-Putscheid

Information panel dedicated to first and second liberation of Merscheid, and the war victims of Merscheid. Merscheid was originally liberated on September 11, 1944 by the 5th US Armored Division.

See Panels Section for details.

2 rue du Puits

Merscheid (Putscheid)

Intersection of Rue de Hoscheid and rue du Puits

Latitude: 49.954147

Longitude: 6.104430

Information Panel 1 at Merscheid-lès-Putscheid

REMEMBER US
MERSCHEID

FIRST LIBERATION AND BEGINNING OF THE "BATTLE OF THE BULGE"

Merscheid was liberated on September 11, 1944, by units of the 5th US Armored Division after a more-than-four-year occupation by the Nazi regime. As the occupants were chased to the other side of the river Our (NB: the "Our" marks the border between Luxembourg and Germany), everything seemed to develop back to normality.

Three months later, however, 250,000 German soldiers started the Battle of the Ardennes, known as the "Battle of the Bulge". Hitler intended to reconquer the port of Antwerp in Belgium, divide the American and British forces, and sue the Allies to the negotiation table towards an armistice. Units of the 5th Paratroop Division of the 7th German Army attacked our region.

The small liaison unit of the 109th Regiment of the 28th US Infantry Division and members of the Luxembourg armed resistance (militiamen) defended Merscheid ferociously with all their means. However, the large number of attackers forced the Americans to surrender. The Germans entered the village, took the villagers hostage, and threatened to shoot them. The militiamen not belonging to a regular army, were considered by the Germans as terrorists. Owing to the courageous engagement of Reverend Antoine RUPPERT, priest of Merscheid, the German officer did not execute his threat.

During the occupation of the village, the Germans installed a field hospital with a pharmacy and surgical station in the houses of the central village. Hundreds of medical interventions were executed here. A provisional cemetery was established in the "Wahlhausen" street where twenty-seven German soldiers were temporarily buried. After the war, their remains were returned to their families in Germany.

PREMIÈRE LIBÉRATION ET COMMENCEMENT DE LA BATAILLE DES ARDENNES

Le village de Merscheid fut libéré le 11 septembre 1944 par des unités de la 5e Division Blindée américaine, ceci après 5 années d'occupation par le régime nazi. Les occupants ayant été chassés de l'autre côté de l'Our (NB: rivière frontalière entre l'Allemagne et le Luxembourg), tout semblait retourner dans la normalité.

Cependant, trois mois plus tard, le 16 décembre 1944, l'Offensive des Ardennes commença avec 250.000 soldats allemands. Hitler voulait reconquérir le port d'Anvers (B), diviser les troupes américaines et britanniques et forcer les Alliés à un armistice. Des unités de la 5. Fallschirmjäger Division (5e Division Parachutiste) de la VIIe Armée attaquèrent notre région.

La petite unité de liaison du 109e Régiment de la 28e Division d'Infanterie américaine et quelques membres de la résistance armée luxembourgeoise (miliciens) défendaient Merscheid de toutes leurs forces. Toutefois, le nombre élevé des attaquants forçait les Américains à se rendre. Les Allemands entrèrent dans le village, prirent les habitants comme otages et menaçaient de les exécuter. En effet, les miliciens n'appartenant pas à une armée régulière, les Allemands les considéraient comme des terroristes. Grâce à l'intervention courageuse du curé de Merscheid, Révérend Antoine RUPPERT, l'officier allemand ne mit pas en exécution sa menace.

Durant l'occupation, les Allemands installèrent un hôpital de campagne avec une pharmacie et une salle d'opération dans les maisons au centre de la localité. Des centaines d'interventions médicales y furent effectuées. Un cimetière improvisé fut aménagé dans la rue de Wahlhausen où vingt-sept soldats allemands furent provisoirement inhumés. Après la guerre, leurs corps furent rapatriés vers leurs proches en Allemagne.

ERSTE BEFREIUNG UND BEGINN DER ARDENNENOFFENSIVE

Merscheid wurde am 11. September 1944 von Einheiten der 5. US-Panzerdivision befreit, nachdem das Land fünf Jahre lang vom Nazi-Regime besetzt war. Da die Besatzer auf die andere Seite der Our (NB: Grenzfluss zwischen Deutschland und Luxemburg) verjagt worden waren, schien alles zur Normalität zu entwickeln.

Drei Monate später startete jedoch die Ardennenoffensive mit 250.000 deutschen Soldaten. Hitler wollte den belgischen Hafen von Antwerpen zurückerobern, die amerikanischen und britischen Truppen teilen und die Alliierten zu einem Waffenstillstand zwingen. Einheiten der 5. Fallschirmjägerdivision der VII. Armee griffen unsere Region an.

Die kleine Verbindungseinheit des 109. Regimentes der 28. US-Infanteriedivision und einige luxemburgische Widerstandskämpfer (Miliz) verteidigten Merscheid mit all ihren Kräften. Die große Anzahl der Angreifer zwang die Amerikaner, sich zu ergeben. Die Deutschen kamen ins Dorf, nahmen die Einwohner als Geiseln und drohten, sie zu erschießen. Da die Widerstandskämpfer keiner regulären Armee angehörten, betrachteten die Deutschen sie als Terroristen. Dank des mutigen Einsatzes des Pfarrers von Merscheid, Hochwürden Antoine RUPPERT, setzte der deutsche Offizier seine Drohung nicht in die Tat um.

Während der Besatzung installierten die Deutschen in den Häusern im Zentrum des Dorfes ein Feldlazarett mit einer Apotheke und einem Operationssaal. Hunderte medizinische Eingriffe wurden vorgenommen. Ein improvisierter Friedhof wurde in der Wahlhausener Straße angelegt, auf dem 27 deutsche Soldaten provisorisch bestattet wurden. Nach dem Krieg wurden ihre Leichname zu ihren Angehörigen nach Deutschland zurückgeführt.

WAR VICTIMS IN MERSCHEID

Meanwhile, the villagers had to suffer a lot during the second occupation. Nine "Merschter" citizens did not survive this decisive battle.

Pierre SCHINNERT, hit by a German bullet on December 16, 1944, while working on his farm. He died two days later, on the way to the hospital.

Félix MARTZEN and **François WEILER**, both residents of Merscheid, as well as **Michel Menster** of Bettendorf, all three men members of the clandestine Luxembourg underground organization LPL, "Lëtzebuerger Patriote Liga", were arrested by the Gestapo and executed in Bauler, near Vianden on January 9, 1945.

Nicolas THEISEN, conscripted by force to serve in the German Wehrmacht, lost his life due to an accident in Gorodok, Russia, on June 12, 1943.

Marcel ROBERT, conscripted by force to serve in the German Wehrmacht, was killed in Nowograd-Wolynsk, Russia, on January 2, 1944.

The two sisters **Marguerite** and **Bernadette TURPEL** were victims of diphtheria on January 26 and February 4, 1945.

Pierre MOLITOR, hit by a piece of shrapnel, was found dead in his bed on January 23, 1945.

Jeanne THELEN, hit by a piece of shrapnel, passed away on January 22, 1945.

Félix MARTZEN

François WEILER

VICTIMES DE GUERRE DE MERSCHEID

Les villageois ont dû souffrir beaucoup sous cette deuxième occupation. Neuf « Merschter » habitants n'ont pas survécu à cette guerre cruciale.

Pierre SCHINNERT, touché par une balle allemande le 16 décembre 1944 en voulant effectuer des travaux dans sa ferme, est décédé deux jours après en route vers l'hôpital.

Félix MARTZEN et **François WEILER**, résidents de Merscheid, ainsi que **Michel Menster** de Bettendorf, tous les trois membres de la LPL (Lëtzebuerger Patriote Liga), une organisation clandestine de résistance luxembourgeoise) furent déportés par la Gestapo et exécutés à Bauler près de Vianden le 9 janvier 1945.

Nicolas THEISEN, enrôlé de force dans l'armée allemande, est décédé à la suite d'un accident à Gorodok, en Russie, le 12 juin 1943.

Marcel ROBERT, enrôlé de force dans l'armée allemande, est décédé à Nowograd-Wolynsk, en Russie, le 2 janvier 1944.

Les deux jeunes sœurs **Marguerite** et **Bernadette TURPEL** furent victimes de la diphtérie le 26 janvier et le 4 février 1945.

Pierre MOLITOR, touché par un éclat d'obus, a été trouvé mort dans son lit le 23 janvier 1945.

Jeanne THELEN, touchée par un éclat d'obus, est décédée le 22 janvier 1945.

Pierre SCHINNERT

Nicolas THEISEN

KRIEGSOPFER IN MERSCHEID

Unterdessen mussten die Dorfbewohner viel unter dieser zweiten Besatzung leiden. Neun „Merschter" überlebten diesen entscheidenden Krieg nicht.

Pierre SCHINNERT wurde am 16. Dezember 1944 von einer deutschen Kugel getroffen, als er Arbeiten auf seinem Bauernhof verrichten wollte. Er erlag seiner Verletzung zwei Tage später auf dem Weg ins Krankenhaus.

Félix MARTZEN, **François WEILER**, beide in Merscheid wohnhaft, sowie **Michel Menster** aus Bettendorf waren Mitglieder der geheimen luxemburgischen Untergrundbewegung LPL (Lëtzebuerger Patriote Liga). Sie wurden von der Gestapo verschleppt und in Bauler nahe Vianden am 9. Januar 1945 erschossen.

Nicolas THEISEN, Zwangsrekrutierter der Wehrmacht, kam infolge eines Unfalls am 12. Juni 1943 in Gorodok in Russland ums Leben.

Marcel ROBERT, Zwangsrekrutierter der Wehrmacht, fiel in Nowograd-Wolynsk, Russland, am 2. Januar 1944.

Die beiden Schwestern **Marguerite** und **Bernadette TURPEL** fielen der Diphtherie am 26. Januar und 4. Februar 1945 zum Opfer.

Pierre MOLITOR, von einem Granatsplitter getroffen, wurde am 23. Januar 1945 tot in seinem Bett aufgefunden.

Jeanne THELEN, von einem Granatsplitter getroffen, starb am 22. Januar 1945.

Photos of the victims of Merscheid: Veiner Geschichtsfrënn, Ons der Veiner Geschicht, Nr. 17, 2019, Gestorwen fir d'Heemecht, fédération des enrôlés de force LIGUE „ONS JONGEN VEIANEN"

For more information
Pour en savoir plus
Für weitere Informationen

Information Panel 2 at Merscheid-lès-Putscheid

REMEMBER US — MERSCHEID

SECOND LIBERATION OF MERSCHEID

Already on December 22, 1944, General George S. PATTON jr. attacked the south flank of the German advance with three divisions. On January 18, 1945, the 4th and 5th US infantry divisions crossed the river Sauer (Sûre) and advanced towards the north. As the Germans intended to keep under their control the road Putscheid-Vianden at all costs to enable their troops to retreat, they defended themselves tenaciously. The snow and the cold as well as the mine fields hidden under a thick snow layer made the American advance exceedingly difficult. The 3rd Battalion, 11th Regiment, 5th US Infantry Division, under the command of Lieutenant-Colonel William H. BIRDSONG jr. encountered heavy resistance against some two hundred entrenched soldiers of the 9th "Volksgrenadierdivision" (9th People's Grenadier Division). Three American tanks coming to support of the American infantrymen, turned turtle on the icy roads. After the engineers attached to the 3rd Battalion had removed the mines, I, K and L Companies succeeded in surrounding the village and liberating it after violent combat.

Merscheid was liberated on January 25, 1945, at 11.40 a.m. Both the Americans as well as the Germans had to deplore a heavy toll on human lives in the Battle of the Bulge.

SECONDE LIBÉRATION DE MERSCHEID

Dès le 22 décembre 1944, le Général George S. PATTON Jr. attaqua le flanc sud de l'avance allemande avec trois divisions. Le 18 janvier 1945, la 5e et 4e Divisions d'Infanterie américaines traversèrent la rivière Sûre et avancèrent vers le nord. Les Allemands voulant maintenir la route Putscheid-Vianden ouverte pour permettre à leurs troupes de se retirer, la résistance allemande fut farouche. La neige et le froid ainsi que les champs de mines dissimulés dans la neige rendaient l'avance des Américains très difficile. Le 3e Bataillon du 11e Régiment de la 5e Division d'Infanterie américaine sous les ordres de Lieutenant-Colonel William H. BIRDSONG, Jr., se heurta à la forte résistance des 200 soldats de la 9e Volksgrenadierdivision (9e Division des grenadiers du peuple) retranchés autour de Merscheid. Trois chars américains venant en aide aux fantassins américains, se renversèrent sur les routes glacées. Après que le peloton de pionniers attaché au 3e Bataillon avait enlevé les mines, les compagnies I, L et K réussirent à contourner le village et à le libérer après de lourds combats.

Merscheid fut libéré le 25 janvier 1945 à 11.40 heures. Les Allemands aussi bien que les Américains avaient à déplorer d'énormes pertes en vies humaines pendant la Bataille des Ardennes.

ZWEITE BEFREIUNG VON MERSCHEID

Schon am 22. Dezember 1944 griff General George S. PATTON Jr. die südliche Flanke des deutschen Vorstoßes mit drei Divisionen an. Am 18. Januar 1945, überquerten die 4. und 5. US-Infanteriedivisionen den Fluss Sauer und rückten gegen Norden vor. Da die Deutschen unbedingt die Straße Pütscheid - Vianden unter ihrer Kontrolle behalten wollten, um ihren Truppen einen Rückzugsweg freizuhalten, verteidigten sie sich erbittert. Der Schnee und die Kälte sowie die im Schnee versteckten Minenfelder machten den Amerikanern das Vorrücken sehr schwer. Das 3. Bataillon des 11. Regimentes der 5. Infanteriedivision unter dem Befehl von Oberstleutnant William H. BIRDSONG Jr. stieß auf den zähen Widerstand der um Merscheid verschanzten zweihundert Soldaten der 9. Volksgrenadierdivision. Drei amerikanische Panzer, die den Infanteristen zu Hilfe kommen wollten, kippten auf den vereisten Straßen um. Nachdem der an das 3. Bataillon angegliederte Zug der Pioniere die Minen entfernt hatte, gelang es den Kompanien I, K und L das Dorf zu umringen und es nach heftigen Kämpfen zu befreien.

Merscheid wurde am 25. Jan. 1945 um 11.40 Uhr befreit. Sowohl die Amerikaner als auch die Deutschen hatten sehr hohe Verluste an Menschenleben während der Ardennenoffensive zu beklagen.

September 11, 1944
US M4 Sherman tanks at the main crossroads of Merscheid · Des chars américains « M4 Sherman » au carrefour principal du village · Amerikanische M4 Sherman Panzer an der Hauptkreuzung des Dorfes
Photo : Archives Jérôme Sinnes

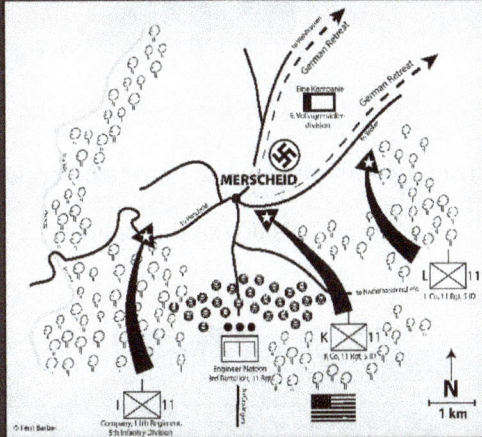

Second liberation of Merscheid on January 25, 1945 · Deuxième libération de Merscheid, le 25 janvier 1945 · Zweite Befreiung von Merscheid am 25. Januar 1945

The church of Merscheid received more than 20 shell impacts · L'église de Merscheid a souffert plus de 20 impacts d'obus · Die Kirche von Merscheid erhielt mehr als 20 Einschläge
Photo : Archives Jérôme Sinnes

03.01.1946: The "Junker-Grill" house, panel "AIR RAID SHELTER" - "ABRI ANTI-AÉRIEN" - "LUFTSCHUTZ-UNTERSTAND"
Photo : Robert LEER

April 1951: The old school building · La vieille école du village · Die alte Schule
Photo : Robert LEER

1946: The "Schmartz-Diener" house · Maison Schmartz-Diener · Haus Schmartz-Diener
Photo : Robert LEER

03.01.1946: Emergency hut on the road to Weiler · Baraque de fortune à la rue de Weiler · Notunterkunft an der Weiler Straße
Photo : Robert LEER

For more information
Pour en savoir plus
Für weitere Informationen

Merscheid-lès-Putscheid

Rue de Wahlhausen 2

9380 Putscheid

Latitude: 49.95415

Longitude: 6.1055

Inscription reads: Zu Ehren vun e'se Patrioten vergiesst sie nett

To our patriots do not forget them

Mertzig

Dedicated to the liberators of the town of Mertzig, the 80th Infantry "Blue Ridge" Division.

Rue de Michelbouch

9170 Mertzig

Latitude: 49.82771

Longitude: 6.01064

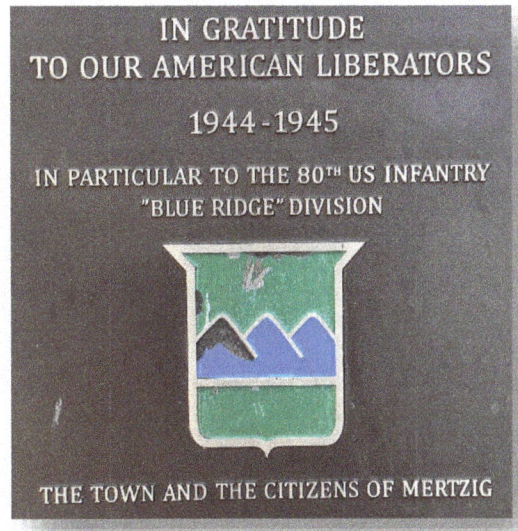

Moestroff

Sauer River Crossing Operations January 18-22, 1945 Bettendorf - Moestroff. Also dedicated to Private first class Vincent J. Festa, of G Company, 8th Infantry Regiment, 4th US Infantry Division, KIA January 20, 1945 in house to house fighting in Moestroff. There is also an information panel nearby.

Rue de la Gare

9382 Bettendorf

Latitude: 49.8688

Longitude: 6.2385

The "Moestroff" bridge is named in honor and remembrance of

Vincent J. FESTA

Private First Class,
"G" company,
8th US Infantry Regiment,
4th US Infantry Division,
KIA(†) on January 20, 1945 in Moestroff

"Lest we forget"

PFC Vincent J. Festa, KIA January 20, 1945

Inaugurated on January 18, 2020

Information Panel Moestroff

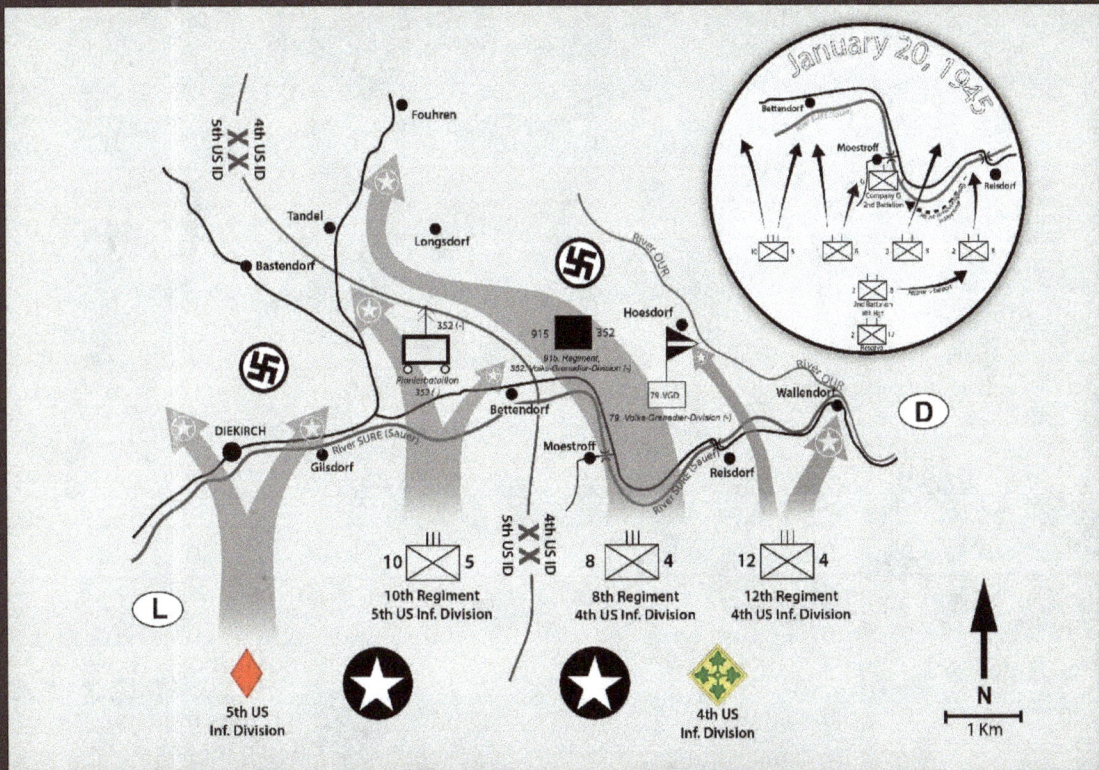

Information Panel Moestroff

REMEMBER US
MOESTROFF

SAUER RIVER CROSSING OPERATIONS
JANUARY 18 - 22, 1945
BETTENDORF – MOESTROFF

After fiercely fighting in late December 1944 in the Echternach sector and „little Switzerland", the two divisions remained in defensive position until mid-January 1945 on the hills of the southern banks of the „Sauer", from where extensive reconnaissance patrolling operations were conducted. One of those patrols, which became later known as the „Bettendorf raid" by a small unit of the 10th US Infantry Regiment, resulted in a German NCO being captured with maps and overlays of their defensive positions in the Bettendorf-Moestroff sector.

The order for the two divisions to carry out an aggressive „river crossing" operation by Third Army's XIIth Corps came on January 17, 1945. Accordingly, both divisions (5th Division on the left and the 4th on the right wing) were ordered to jump off at 3:00 a.m. on January 18, 1945 without initial artillery preparation, supported by their divisional engineer units. The entrance of the village of Moestroff had been fixed as the boundary between the two divisions. The 5th Infantry Division in direction of Diekirch was to attack in the Bettendorf-Diekirch-Ingeldorf sector, the 4th Infantry Division from Moestroff further east in direction of Dillingen.

Moestroff itself lay in the operational sector of the division's own 8th Infantry Regiment, followed in line on the right wing by the 12th and 22nd regiments. Jumping off from the Medernach-Eppeldorf area, where its frontline companies had been billed before the attack, the 8th Infantry Regiment forced the icy Sauer river at Moestroff and got across towards the village and high ground on the north banks. Heavy fighting and house-to-house combat took place until January 21, 1945 against well-dug in German forces. The 8th Infantry Regiment continued its progression towards the Our river and succeeded in capturing Tandel, Fouhren and the outskirts of Vianden by the end of January 1945. In early February 1945, the regiment crossed the Our river into Germany in direction of Prüm.

Portrait of Vincent J. Festa, Private first class and member of "G" company/8th Regt., 4th US Infantry Division, KIA Jan 20, 1945.

Portrait de Vincent J. Festa, soldat 1re classe et membre de la compagnie "G" du 8e Régiment de la 4e Division d'Infanterie US, tombé le 20 janvier 1945.

Porträt des Vincent J. Festa, Soldat 1. Klasse und Angehöriger der "G"-Kompanie des 8. Regiments der 4. US-Infanteriedivision, gefallen am 20. Januar 1945.

(photo coll. R. Gaul).

As one amongst those numerous American casualties of the Sauer river crossing operations, Pfc Vincent FESTA of "G" company, 8th Inf. Rgr. was killed on January 20, 1945 during the house-to-house fighting to recapture Moestroff. Pfc Festa originated from Pennsylvania. He is buried at the Luxembourg-American cemetery in Hamm/Luxbg: Plot H, Row 12, Grave No. 49. The Moestroff bridge was renamed in honor and memory of him in "Vincent Festa" bridge.

Copyright: RG/TPV/SG

LE FRANCHISSEMENT DE LA SÛRE
18 AU 22 JANVIER 1945
BETTENDORF – MOESTROFF

Après les rudes combats de fin décembre 1944 dans le secteur d'Echternach et de la « Petite Suisse », les deux divisions restaient cantonnées jusqu'en janvier 1945 à des positions défensives sur les hauteurs septentrionales de la Sûre, d'où elles déployaient de vives activités de reconnaissance. Une telle patrouille de reconnaissance, composée d'une petite unité du 10e régiment d'infanterie américain, parvint à capturer un sous-officier allemand porteur de cartes et de croquis révélant les positions allemandes dans le secteur Bettendorf-Moestroff. Cette action entra dans l'histoire sous le nom de « coup de main de Bettendorf ».

Soldiers of "G" company, 8th US Infantry Regiment, en route to Reilly leading to take up positions, in the upper village of Moestroff on Jan 21, 1945.

Des soldats américains de la compagnie "G" du 8e Régiment d'infanterie passent en pont de génie pour occuper de nouvelles positions dans le village supérieur de Moestroff, le 21 Jan. 1945.

US Soldaten der "G" Kompanie der 8. Infanterieregiments ziehen über eine Pionierbrücke über die Sauer der 21. Jan. 1945 und beziehen neue Stellungen im oberen Dorf von Moestroff.

(Photo credit: NARA/GNM/M. Aelbrecht, R. Gaul).

L'ordre émanant du XIIe Corps de la 3e Armée d'attaquer au-delà de la Sûre parvint aux deux divisions le 17 janvier 1945. Les deux divisions (la 5e sur l'aile gauche, la 4e sur l'aile droite) étaient supposées lancer l'assaut le 18 janvier 1945 à 3 heures du matin, sans préparation d'artillerie préalable et avec le seul appui de leurs pionniers. Ce fut l'entrée du village de Moestroff qui fut défini comme point de démarcation sectoriel entre les deux divisions. La 5e division devait attaquer vers le nord dans le secteur de Bettendorf-Diekirch-Ingeldorf, tandis que la 4e division avait comme objectif de pénétrer vers l'est à partir de Moestroff, vers Dillingen.

Moestroff se trouvait dans le secteur du 8e régiment d'infanterie, couvert sur son aile droite par les 12e et 22e régiments. L'offensive fut engagée à partir d'Eppeldorf-Medernach, point de ralliement des compagnies du 8e régiment. Ce dernier franchit la Sûre gelée près de Moestroff, se retrancha en bordure de la localité pour enfin s'emparer des hauteurs au nord de la rivière. S'en suivirent d'âpres combats de rue contre des unités allemandes, bien retranchées dans le village, jusqu'au 21 janvier 1945. Le 8e régiment d'infanterie reprit alors l'attaque vers l'Our et il réussit à capturer Tandel, Fouhren et l'entrée de la ville de Vianden jusqu'à la fin du mois. Début février 1945, il franchit l'Our et pénétra en Allemagne en direction de Prüm.

Parmi les nombreuses victimes américaines du franchissement de la Sûre, figure aussi le soldat de première classe Vincent J. Festa, du 8e régiment, tué le 20 janvier 1945 dans des combats de rue à Moestroff. Vincent J. Festa fut originaire de l'État de Pennsylvanie. Il repose sur le cimetière américain de Hamm – Champ H, Rangée 12, Tombe 49. En son honneur et en sa mémoire, le point de Moestroff fut baptisé « Pont Vincent Festa ».

DIE SAUER-ÜBERQUERUNG
18. - 22. JANUAR 1945
BETTENDORF – MOESTROFF

Nach schweren Kämpfen Ende Dezember 1944 im Sektor Echternach und der "kleinen Luxemburger Schweiz" verblieben die zwei Divisionen bis Mitte Januar 1945 in Verteidigungsposition auf dem Höhengelände am südlichen Sauerufer, von wo aus umfangreiche Spähtrupp-Aktionen gesteuert wurden. Einem dieser Spähtrupps, aus einer kleinen Einheit des 10. US-Infanterieregiments bestehend, gelang es, einen deutschen Unteroffizier, der einige Karten und Skizzen der deutschen Verteidigungslinie im Abschnitt Bettendorf - Moestroff bei sich hatte, gefangen zu nehmen. Diese Aktion ging als "Bettendorfer Husarenstück" in die Geschichte ein.

Der Befehl des XII. Korps der 3. US-Armee an die beiden Divisionen, einen Angriff über die Sauer durchzuführen, erging am 17. Januar 1945. Demnach sollten die beiden Divisionen (die 5. US Infanteriedivision am linken und die 4. US-Infanteriedivision am rechten Flügel) am 18. Januar 1945 frühmorgens ab 03:00 Uhr ohne vorherige Artillerievorbereitung, nur durch ihre divisionseigenen Pioniereinheiten unterstützt, den Angriff über die Sauer einleiten. Dabei war der Ortsrand von Moestroff als sektorielle Trennlinie zwischen den beiden Divisionen festgelegt worden. Die 5. US Infanteriedivision sollte dabei in Richtung Diekirch im Abschnitt Bettendorf - Diekirch - Ingeldorf angreifen, die 4. US-Infanteriedivision hauptsächlich von Moestroff aus weiter östlich in Richtung Dillingen.

Moestroff selbst lag im Bereich des divisionseigenen 8. Infanterieregiments, in weiterer Linie am rechten Flügel befanden sich das 12. und 22. Regiment. Ausgangspunkt des Angriffs war der Umkreis Eppeldorf-Medernach, wo vorher die einzelnen Kompanien des 8. Regiments in Bereitschaft lagen. Das 8. US Infanterieregiment setzte über die gefrorene Sauer bei Moestroff und drang an den Ortsrand sowie bis zur Anhöhe am nördlichen Sauerufer vor. Es folgten heftige Straßen- und Häuserkämpfe gegen gut verschanzte deutsche Einheiten; die bis zum 21. Januar 1945 andauerten. Das 8. Infanterieregiment setzte seinen Angriff in Richtung der "Our" fort und es gelang ihm, bis Ende Januar 1945 Tandel, Fouhren sowie den Ortseingang von Vianden einzunehmen. Anfang Februar 1945 überquerte das Regiment die Our und betrat deutsches Gebiet, es bewegte sich daraufhin in Richtung Prüm.

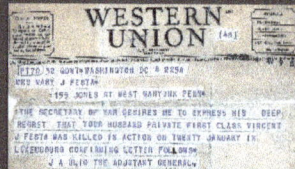

WESTERN UNION

WESTERN UNION Telegram informing the family that Vincent J. Festa was killed on January 20, 1945 in Luxembourg.

Télégramme-type de la Western Union – informant la famille que Vincent J. Festa a été tué en Luxembourg, le 20 janvier 1945.

WESTERN UNION Telegramm an die Familie mit Information, dass Vincent J. Festa am 20. Januar 1945 in Luxemburg gefallen ist.

(document: Festa family).

Unter den zahlreichen Verlusten der "Sauer"-Überquerung auf amerikanischer Seite befand sich auch der Soldat 1. Klasse, Vincent J. FESTA, der "G"-Kompanie/ 8. Infanterieregiment, der am 20. Januar 1945 beim Häuserkampf in Moestroff getötet wurde. Vincent J. Festa stammte aus dem Bundesstaat Pennsylvania. Er fand seine letzte Ruhestätte auf dem US-Friedhof in Hamm, Plot H, Reihe 12, Grab 49. Die Brücke von Moestroff wurde ihm zu Ehren und Gedenken in "Vincent Festa" Brücke umbenannt.

For more information
Pour en savoir plus
Für weitere Informationen

Syndicat d'initiative Bettendorf

Commune du Bettendorf

NORTIC

Moutfort

5th US Armored Division and American Veterans Square.

Rue d'Oetrange 8

5333 Contern

Latitude: 49.58827

Longitude: 6.25521

PLAATZ ONS JONGEN
AMERICAN VETERANS SQUARE
EN SOUVENIR DE LA
LIBERATION
10. 9. 1944

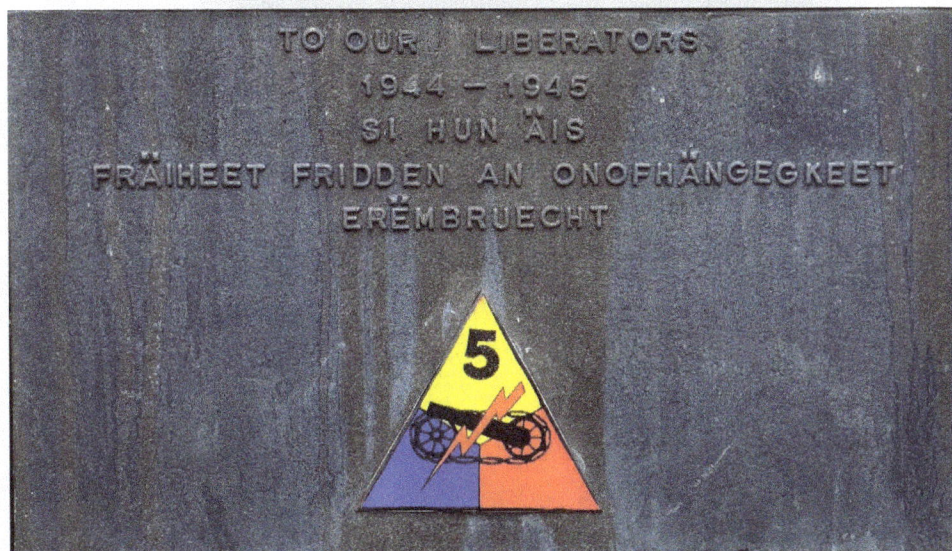

TO OUR LIBERATORS
1944 – 1945
SI HUN AIS
FRÄIHEET FRIDDEN AN ONOFHÄNGEGKEET
ERËMBRUECHT

Inscription reads: To our liberators 1944-1945. They have restored peace and independence to us

Niederfeulen

Dedicated to the 1st Battalion (Reinforced), 319th Infantry Regiment, 80th Infantry Division, Third United States Army who liberated the community of Feulen.

Route de Bastogne 25

9176 Feulen

Latitude: 49.8538

Longitude: 6.0457

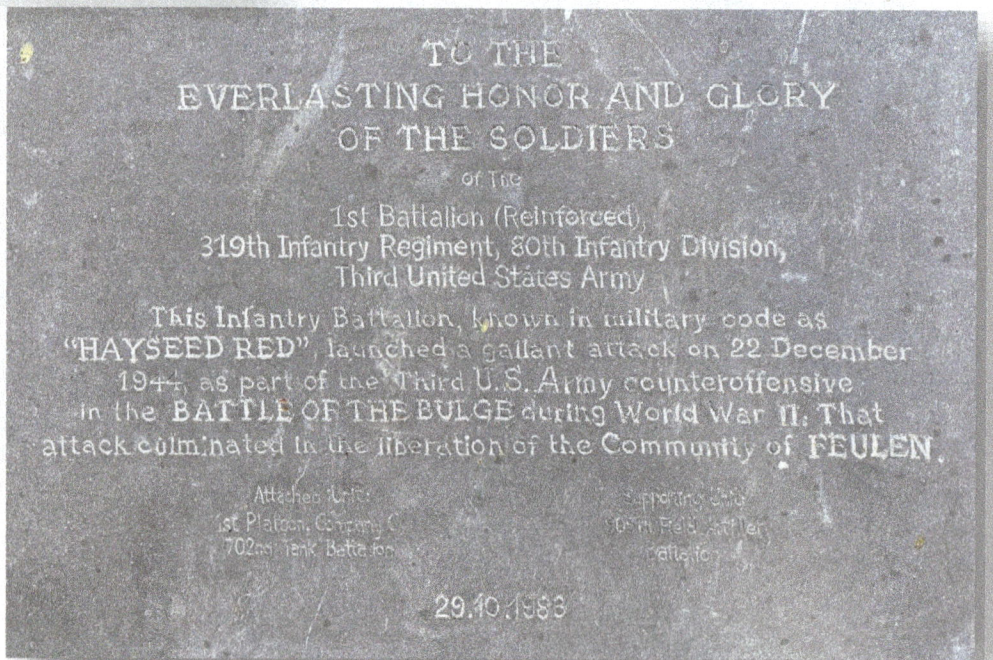

TO THE
EVERLASTING HONOR AND GLORY
OF THE SOLDIERS
of the
1st Battalion (Reinforced),
319th Infantry Regiment, 80th Infantry Division,
Third United States Army

This Infantry Battalion, known in military code as
"HAYSEED RED", launched a gallant attack on 22 December
1944, as part of the Third U.S. Army counteroffensive
in the BATTLE OF THE BULGE during World War II. That
attack culminated in the liberation of the Community of FEULEN.

Attached Units:
1st Platoon, Company C
702nd Tank Battalion

Supporting Units:
10th Field Artillery
Battalion

29.10.1988

Niederfeulen

Dedicated to the children of Feulen by soldiers of the 1st Battalion (Reinforced), 319th Infantry Regiment, 80th Infantry Division, Third United States Army. In Niederfeulen there is also a street named after Major General Dudley Ives (LTC at the time of the Battle of the Bulge), 319th Infantry Regiment whose troops liberated Feulen on December 22, 1944. The correct spelling of his last name is Ives, although the street name and plaque list him as Yves.

Bongerterwee 1

9175 Feulen

Latitude: 49.8550

Longitude: 6.0448

Niederwampach

Inscription: This is an American 105 mm, M2A1 Howitzer, built in 1939.

Om Knupp 3

9672 Wincrange

Latitude: 50.0108

Longitude: 5.8469

Nocher

Dedicated to the 80th "Blue Ridge" Infantry Division, who fought for 16 days to free this area in January 1945.

Um Knupp 12

9674 Goesdorf

Latitude: 49.94753

Longitude: 5.97543

Information Panel Nocher

REMEMBER US
NOCHER

THE LONG BATTLE FOR NOCHER

It didn't take long for the Germans to conquer the high plateau, upon which lie the villages of Goesdorf, Dahl and Nocher, in the opening days of the Battle of the Bulge. The plateau was of a great strategic importance, dominating the transport node of Wiltz and the main road which connected Bastogne to the "Reich".

After the US Army had retaken the village of Heiderscheid, on December 23 1944, they knew as well of the importance of the high plateau as of the difficulties they were facing in trying to retake the three villages. The high ground was separated from their positions by the deep valley of the "Sûre" river. As the roads were covered by ice and a thick layer of snow, the steep slopes of the valley would be difficult to scale for the tanks supporting the attack. On January 6 the GIs of the 3rd battalion, 319th Infantry, 80th Division, stormed up the steep slopes to the west of Goesdorf and, in a "coup-de-main", pushed the German defenders out of the village. Once they had the "Landsers" on the run the GIs didn't stop and drove the enemy out of the neighbouring village of Dahl on the same day. This in spite of being under extreme heavy German artillery fire. The following days the men of the 319th Infantry had to repulse extremely heavy enemy counterattacks. But regardless of the heavy artillery fire and the support by Panther and Jagdpanther tanks the GIs held the line. One of the most famous "Blue Ridgers", Sergeant Day G. Turner was posthumously awarded the Medal of Honor for his actions at the "Aaschterthaff" farm near Dahl, on January 8 1945.

The reconquest of the village of Nocher would be a much tougher nut to crack for the men of the 80th Division. The village was surrounded by extensive minefields and the wooded slopes on the western and eastern edges of the plateau were littered by "Kraut" foxholes and machine-gun nests. After the war the inhabitants of Nocher told about the rare sight the German defenders were: "When they returned to the village, coming from their fighting positions, they reminded one of a carnival parade. The "Krauts" were wearing capes made out of bedlinen and crocheted tablecloths, several of them were wearing the white robes of a priest and one of them had even donned a cut up girls communion dress. Their artillery pieces and vehicles had also been camouflaged by white cloth, all of which had been stolen from the wardrobes of the houses, right under our noses." The Americans were aware that they would have to get into bitter hand-to-hand fighting to retake the village and that they would suffer heavy losses. To soften the German defences and to batter their morale Nocher was subjugated to heavy artillery shelling and air attacks.

LA LONGUE BATAILLE POUR NOCHER

Dès les premiers jours de la bataille des Ardennes, la prise des villages de Goesdorf, Dahl et Nocher, ainsi que du haut plateau sur lequel les villages se trouvent, n'eut pris, ni beaucoup de temps ni de grands efforts aux attaquants allemands. Ce plateau domine le nœud routier de Wiltz et l'axe routier principal, reliant Bastogne à l'Allemagne. Ainsi le contrôle des alentours des trois villages ardennais fut d'une grande importance tactique.

Les forces américaines, qui eurent repris Heiderscheid en date du 23 décembre 1944, furent bien conscientes de cette grande importance du plateau qui se trouvait devant leurs yeux, de l'autre côté de la vallée de la Sûre. Les commandants de l'US Army avaient également réalisé que la reprise du plateau de Goesdorf ne serait pas une opération facile. Déjà, sous conditions normales, la vallée de la Sûre, profonde et à pentes raides, forma un obstacle formidable! Mais avec les routes à fort dénivellement, gelées et couvertes d'une couche de neige épaisse, l'avance des chars serait presque impossible. Au matin du 6 janvier, les soldats américains montèrent à l'assaut. Ils escaladèrent la pente, presque verticale, qui donna accès au côté ouest du village de Goesdorf, et malgré le fait de se trouver sous un bombardement intense de l'artillerie allemande, ils parvinrent à déloger l'ennemi de ses positions. Comme les fantassins de la Wehrmacht se trouvèrent en déroute totale, les GIs du 3ème bataillon, 319ème régiment d'infanterie du 80ème division US, ne traînèrent pas le pas. Ils libérèrent le village avoisinant de Dahl le même jour. Malgré les contre-attaques farouches de l'ennemi au cours des jours suivants, les soldats américains ne lâchèrent pas. Un rôle décisif dans la défense des lignes américaines joua le sergent Day G. Turner. Sa défense obstinée de la ferme "Aaschterthaff" lui a valu d'avoir été décerné, à titre posthume, la "Medal of Honor", la plus haute médaille militaire de l'US Army.

La reprise du village de Nocher alla poser de plus grands problèmes aux hommes du 319ème régiment. Les alentours du village avaient été minés par les allemands. Les forêts, qui se trouvèrent sur les pentes bordant le plateau, furent parsemées de trous de tirailleurs et de nids de mitrailleuses. A la fin de la guerre, les habitants du village relatèrent l'apparence parfois bizarre des troupes allemandes: "Quand ils retournaient de leurs positions on croyait voir une cavalcade de carnaval. Des draps de lits, des nappes crochetées, des robes de prêtre et même des robes de communion de petite fille, découpés en loques, servirent de camouflage aux "Landser". Il fut de même de leurs pièces d'artillerie et de leurs véhicules, qui furent couverts de tissus blancs. Bien sûr, tous ces textiles furent dérobés sous nos nez, des armoires des habitations du village." Les américains eux furent bien conscients du fait que l'attaque sur le petit village ne pourrait se faire qu'au combat rapproché et entraînerait des pertes importantes en vies humaines. Ainsi ils avaient commencé à bombarder Nocher si bien avec l'artillerie qu'avec l'aviation, qui profita de l'amélioration des conditions météorologiques.

DER LANGE KAMPF UM NOCHER

Das Plateau um die Dörfer Goesdorf, Dahl und Nocher war in den ersten Tagen der Ardennenoffensive sehr schnell von den deutschen Streitkräften besetzt worden. Es beherrschte den strategisch wichtigen Knotenpunkt Wiltz mit der Hauptverbindungsstraße zwischen der belgischen Stadt Bastogne und dem Reichsgebiet.

Als die Amerikaner am 23. Dezember 1944 Heiderscheid zurückeroberten, war ihnen die militärische Bedeutsamkeit dieses Plateaus bewusst, welches zudem als Schutzwall für das von den Deutschen besetzte Wiltz fungierte. Die Amerikaner waren sich aber auch den Schwierigkeiten bewusst, die sich ihnen bei der Eroberung der drei Dörfer in den Weg stellen würden. Ein tiefer Einbruchsgraben des Sauer-Flusses trennte sie vom Gebiet und mit ihren Panzern konnten sie die Erhöhung nur sehr schwer erklimmen, denn der Boden war mit einer glatten Eisschicht bedeckt. Am 6. Januar 1945 erstürmte die amerikanische Infanterie die Höhe vom westlichen Steilhang aus. Trotz schwerem Artilleriebeschuss des Gegners konnten die Soldaten des 3. Bataillons, des 319. Infanterieregiments, der 80. US-Division die Dörfer Goesdorf und Dahl befreien. Obwohl die Wehrmacht schwere Gegenangriffe startete um die Ortschaften zurückzugewinnen, gelang es den Amerikanern ihre Position zu halten. Dies ist unter anderem dem Einsatz von Sgt. Day G. Turner zu verdanken, welcher mit seiner kleinen Abteilung, das nahe gelegene "Aastert" Gehöft, allen Widrigkeiten zum Trotz, erfolgreich verteidigen konnte. Für seinen unerschrockenen und heldenhaften Einsatz wurde dieser mit der höchsten Auszeichnung der US-Armee, der "Medal of Honor", ausgezeichnet.

Die Rückeroberung des Nachbardorfes Nocher würde den US-Truppen allerdings noch einiges abverlangen, denn ein Teil des Umlandes war in ein gewaltiges Minenfeld verwandelt worden. Zudem waren die Waldstücke bei Nocher mit zahllosen Schützenlöchern, Gräben und Bunker übersät, in denen sich die Deutschen festgesetzt hatten. Nach dem Krieg berichteten die Einwohner des kleinen Dorfes, wie sich die deutschen Soldaten in langen Reihen im Schnee versteckt hatten. "Kamen sie dann zurück" erzählte ein Einwohner, "so glaubte man, einen Fastnachtszug zu sehen. Leintücher, Kirchenkleider, ja sogar die weißen Kommunionskleider Mädchen dienten zur Tarnung. Auch die Geschütze waren mit Leinen behangen, das man uns vor der Nase wegstahl." Die Amerikaner hingegen waren sich bewusst, dass die Rückeroberung Nochers nur im Nahkampf erfolgen konnten und mit einem hohen Verlust an Menschenleben verbunden war. Zuerst wurde aber noch versucht, die Moral der Deutschen durch tagelanges Bombardement, unter anderem aus der Luft, zu brechen.

For more information
Pour en savoir plus
Für weitere Informationen

Information Panel Nocher

The long battle for Nocher

For the civilian inhabitants of Nocher these were trying times. About 70 women, men and children had not been able to flee from the advancing Germans and had sought refuge in the cellars of the farms. During the bombardment they could not leave their shelters, neither for scavenging for food nor to provide for their cattle. Even while mourning the destruction of their village they found solace in the fact that their rescue, by the hand of the US soldiers, was near. The "Blue Ridgers" had prepared the upcoming attack by very intensive patrol and reconnaissance activity, so the US forces knew every detail of the enemy's defences and positions. Finally the 4 companies of the 3rd Battalion, 319th Infantry, were to attack on the early morning of January 18. For this difficult mission the Battalion had been reinforced with 2nd Battalion's G company, a platoon of Sherman tanks from C company 702nd Tank Battalion, a platoon of A company, 610th Tank Destroyer Battalion and a platoon of C company, 305th Engineer Combat Battalion. The attack jumped off at 07:00 in the morning. In order to surprise the "Krauts" there had been no artillery preparation. The attackers advanced in two columns, along the covered western and eastern edges of the plateau, an advance along the Dahl-Nocher road and over the open ground of the plateau, without any cover, would have been suicidal. K company advanced from the southwest, circumvented the village to the north, and put up a blocking position in order to fend off any German counterattack. I and L companies, with G company in reserve, were to attack from the southeast and gain control of the village. According to the account of the inhabitants of Nocher and the US Army Combat Interviews, there was a brutal house to house fighting, the men of the 319th had to fight for every cellar and barn, using hand grenades and bazookas to drive out the dug-in German defenders. The Wehrmacht answered with extremely heavy artillery fire and the advance of the tanks and tank destroyers was slowed down by heavy direct fire, nonetheless, at 08:20 the GIs had managed to take control of the village.

But this was not the end of the harsh fighting. The Germans had retreated to the north, on the heights around the village of Merkholtz, from where they continued pouring down intensive artillery fire onto the GIs in Nocher. But three days later, on January 21, the Germans, under heavy pressure of the 319th Infantry, had to abandon their positions at Merkholtz. Nocher was finally safe. With the fall of Merkholtz the strategically important high plateau was firmly in US hands and Patton's 3rd Army could venture to retake Wiltz and thus cut off the most important road connecting Bastogne to the German rear bases. In a hurry the US Army assembled heavy artillery batteries on the plateau, in order to shell the road network around the villages of Hoscheid and Hosingen, and thus further hamper the flow of German supplies to their forces around Bastogne.

The combats for the villages of Goesdorf, Dahl and Nocher, and the high plateau they lie on, raged for 16 days. The casualties the 80th Division suffered were heavy, 161 GIs were killed or wounded in action. The village of Nocher had been completely destroyed, most of the cattle had been lost and several civilians lost their lives. After the war many US veterans thought of these 16 days in January 1945 as the worst fighting they had ever seen. Captain Robert J. Bee, 3rd Battalion's S-3 said that Nocher "…was about the most beat up town he had ever seen. There were no houses with roofs remaining…"

La longue bataille pour Nocher

Pour les habitants de Nocher, les bombardements présentèrent un calvaire effroyable. Environ 70 femmes, hommes et enfants , qui n'avaient pas su prendre la fuite lors de l'attaque allemande, se furent refugiés dans les caves du village. Lors des attaques de l'artillerie et de l'aviation américaine, il leur fut impossible de quitter leur refuge. Ainsi ils ne purent ni se ravitailler en nourriture, ni s'occuper de leur bétail. Mais, même s'ils eurent leurs coeurs déchirés par la destruction de leur village, ils surent que leur libération, aux mains des soldats américains, fut imminente, et ils y prirent courage. En préparation de l'attaque sur Nocher les troupes américaines avaient effectués un grand nombre de patrouilles de reconnaissance. Ainsi, ils avaient su se familiariser avec le terrain et l'emplacement des défenses de l'ennemi, ils connurent tous les points faibles de leur adversaire. La mission de reprendre le petit village fut confiée aux 4 compagnies du 3me bataillon du 319me régiment, qui pour l'occasion, avait reçu des renforts non négligeables: la compagnie G du 2me bataillon, ainsi qu'un peloton de chars Sherman, un peloton de canons anti-chars autopropulsés et un peloton des troupes du génie. A 07:00 heures, le matin du 18 janvier, les GIs lancèrent leur attaque, ceci, afin de ne pas alerter la garnison allemande, sans aucune préparation d'artillerie. Les troupes américaines avancèrent en deux colonnes, chacune camouflée par les forêts, se trouvant en haut des pentes abordantes du haut plateau. Une avance par la route qui reliant Dahl à Nocher ainsi que par le plateau dégagé, sans aucun couvert, aurait mené infailliblement à un massacre! La première colonne, qui progressa du sud-ouest, eut comme mission de couper la route qui entra dans Nocher par le nord-est. Leur but fut de repousser une éventuelle contre-attaque des troupes de la Wehrmacht. L'attaque principale fut menée par la deuxième colonne, qui avança par un petit vallon boisé au sud-est du village. Cette route d'approche les mena en plein coeur du village, sans être vu des défenseurs allemands. Les combats s'y jouèrent au corps à corps. A l'aide de grenades à mains et de lance-roquettes, les GIs ont dû déloger les allemands de leurs positions. Le village a été pris maison par maison, cave après cave, sous un tribut en vies humaines effroyable. L'avance des chars fut retardée, à maintes occasions, par des tirs directs des allemands. Mais finalement les américains prirent le dessus et les survivants allemands furent forcés de se rendre.
Mais ceci ne présenta pas la fin des combats acharnés. Les allemands s'étaient repliés sur les hauteurs du village de Merkholtz, au nord de Nocher. De ces positions, les hommes du Führer prirent les GIs du 319me régiment sous un feu nourri de l'artillerie. Ce ne fut que trois jours plus tard, en date du 21 janvier 1945, que les hommes de la 80me division purent déloger les "Boches" de leurs positions autour de Merkholtz. Finalement, le village éprouvé de Nocher eut retrouvé son calme. La redoute sud-est de Wiltz fut tombé et le Général Patton put envisager la prise de cette ville si importante. Tout de suite, la 3me armée US commença à concentrer des batteries d'artillerie lourde sur le plateau reconquis. Ceux-ci prirent sous leur feu le réseau routier dans les parages de Hoscheid et Hosingen, afin d'entraver le flux du ravitaillement vers les troupes allemandes autour de Bastogne.

Les combats pour le haut plateau et les villages de Goesdorf, Dahl et Nocher avait ragé pendant 16 jours. Les pertes des deux côtés furent très lourdes. La US Amy avait perdu 161 hommes, tués ou blessés. Le village de Nocher fut complètement dévasté, les fermes détruites, le bétail tué et plusieurs civils avaient perdu leur vie. Après la guerre, beaucoup de vétérans américains clamèrent que les combats pour Nocher furent les plus âpres auxquels ils avaient dû participer pendant toute la guerre. Le capitaine Bee, officier au 3me bataillon, dit que "… Nocher fut le village le plus ramolli que je n'ai jamais vu. Il n'y avait plus aucune maison qui gardait son toit …"

Der Lange Kampf um Nocher

Für die Bewohner Nochers war dies eine schwere Zeit. Etwa 70 Männer, Frauen und Kinder hatten sich zusammen in einem Keller verschanzt, welchen sie wegen der Bombardements nicht verlassen konnten, nicht einmal, um das Vieh zu füttern. Sie trauerten um die Zerstörung ihres Dorfes, wussten aber auch, dass die Befreiung durch die Amerikaner mit jedem Einschlag näher rücken würde. Um den Angriff auf Nocher vorzubereiten, hatten etliche Amerikanische Späbtrupps und Patrouillen die Gegend um Nocher ausgekundschaftet, um sich mit dem Terrain und den feindlichen Positionen vertraut zu machen. Der Auftrag das Höhendorf einzunehmen, ging schließlich an die vier Kompanien des 3. Bataillons des 319. US Infanterieregiments. An diese Einheit als Verstärkung angegliedert waren zudem die G Kompanie des 2. Bataillons sowie ein Zug Panzer und ein Zug Panzerjäger. Am 18. Januar stürmten die Amerikaner das Dorf dann ohne Artillerievorbereitung, um die Deutschen zu überraschen. Dies geschah in zwei Infanteriekolonnen: Einmal vom Südwesten aus, um den West- und Nordzugang des Dorfes abzuschneiden, und einmal vom Südosten aus, durch das bewaldete Tal östlich der Straße Dahl-Nocher, um das Dorfzentrum einzunehmen. Hätten sie die Hauptstraße von Dahl aus genommen, hätten sie sich dem Feind auf offener Fläche präsentiert und der Angriff hätte in einem Massaker geendet. So konnten sie aber, durch die Wälder geschützt, in das Dorf eindringen und stellten sich den Deutschen in einem brutalen Häuserkampf entgegen: Mit Panzerfaust und Handgranaten musste Keller um Keller, Scheune um Scheune, Schutthaufen um Schutthaufen von deutschen Schützen gesäubert werden, die sich nicht kampflos geschlagen gaben. Amerikanische Panzer versuchten nachzurücken, konnten auf ihrem Weg von den deutschen Kämpfern aber mehrmals gestoppt werden. Schließlich kam das Dorf dann doch unter amerikanische Kontrolle und nur wenigen Deutschen gelang die Flucht.

Aber damit war der Kampf noch nicht vorbei. Deutsche Truppen hatten sich nördlich auf die Höhe von Merkholtz zurückgezogen und begannen nun auf das eh schon zertrümmerte Dorf zu feuern. Die amerikanischen Truppen planten daraufhin einen Fliegerangriff auf Merkholtz, doch bevor es dazu kommen sollte, hatte sich die deutsche Artillerie bereits wieder zurückgezogen und die beiden Dörfer Nocher und Merkholtz wurden am 21. Januar endgültig von den Deutschen aufgegeben. Mit dem Fall von Merkholtz war endlich das ganze Hochplateau von Dahl fest in amerikanischer Hand und somit das süd-östliche Bollwerk des von den Deutschen besetzten Wiltz erobert. Schnell wurde schwere Artillerie herangezogen, um vor allem den wichtigen Höhenzug Hoscheid-Hosingen zu beschießen und den Verkehr auf der Nord-Süd-Straße Hosingen-Diekirch zu behindern.

16 Tage hat es somit gedauert das Dorf Nocher wieder von den Deutschen zu befreien, verbunden mit vielen Verlusten. 161 Amerikaner waren getötet oder verwundet worden und Nocher lag komplett in Trümmern. Die Bauernhöfe waren zerstört, das Vieh war umgekommen und auch etliche Dorfbewohner hatten ihr Leben verloren. Jahre später berichteten viele amerikanische Veteranen, dass diese Schlacht einer der härtesten und blutigsten Kämpfe gewesen war, die sie je erlebt hatten. Und laut Capt. Bee war Nocher "… das verwüstetste Dorf, das er je gesehen hatte. Es gab kein einziges Haus mit Dach mehr …"

For more information
Pour en savoir plus
Für weitere Informationen

Noerdange

Dedicated to the people of Noerdange and to the resistance who held the ground from 19 to 21 December 1944, until the relief by the 328th Regiment, 26th "Yankee" Infantry Division.

Kierchwee 1

8551 Beckerich

Latitude: 49.74075

Longitude: 5.92405

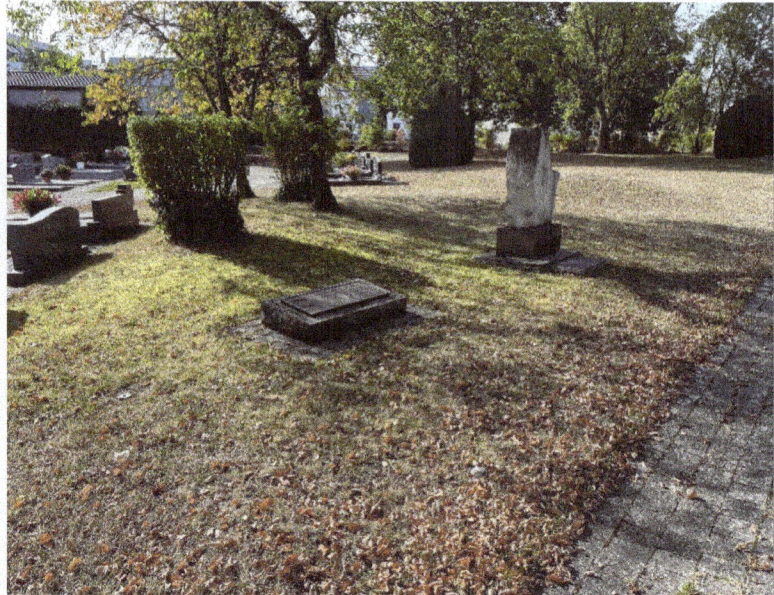

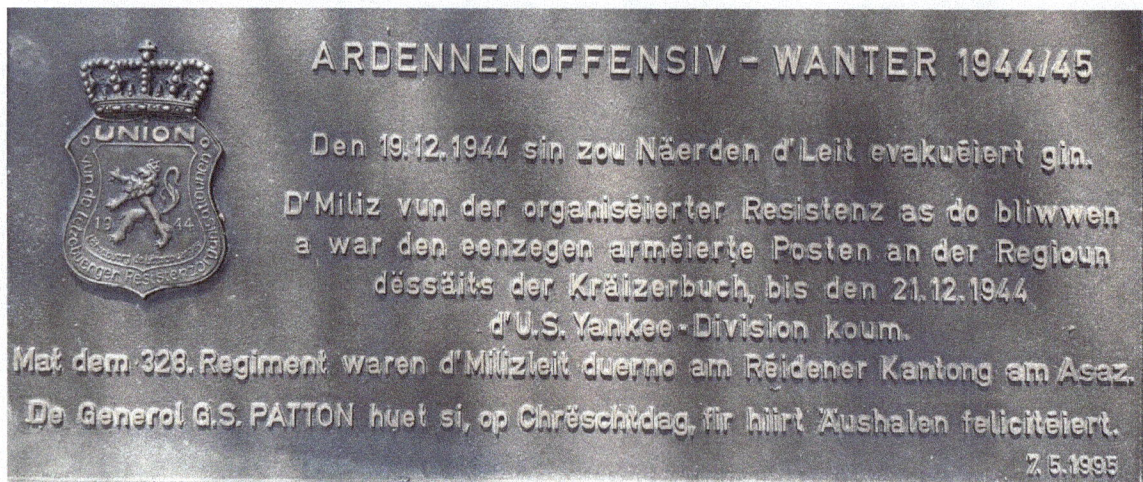

Inscription reads: Ardennens Offensive – Winter 1944/45 On December 19, 1944 the people of Noerdange were evacuated. The Resistance militia stayed and were the only armed forces on this side of the 'Kräizerbuch' until December 21, 1944 when the US Yankee Division arrived. When the 328th Regiment arrived, the militia was deployed in the Rédange district. General G. S. Patton congratulated them for their endurance at Christmas.

Nothum

Located on the side of a private house in Nothum, plaque is in memory of Capt. James D. Richter 915th Field Artillery Battalion (105 mm Howitzer), 90th Infantry Division, who was killed in action on January 9, 1945.

Um Knupp 2

9678 Bavigne

Latitude: 49.93997

Longitude: 5.88286

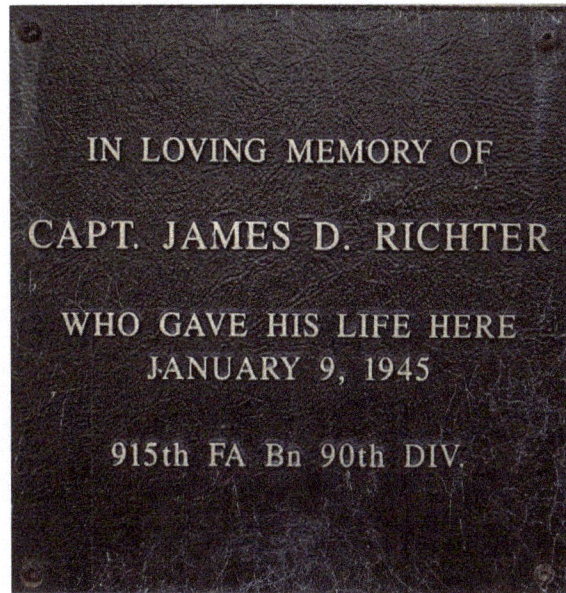

Oberwampach

Dedicated to Sgt. Hassel C. Whitefield, 90th Infantry Division.

Am Wolerech 21

9673 Wincrange

Latitude: 50.0175

Longitude: 5.8579

Inscription reads: In memory of Sgt. Hassel C. Whitefield of the 90th Inf Div who sacrificed his life on 17 January 1945 while trying to save the child Marcel Schilling.

Oetrange

Dedicated to our American friends who liberated us on September 12, 1944.

Rue du Pont

5355 Contern

Latitude: 49.5998

Longitude: 6.2583

PONT DE LA LIBERATION

AN DANKBARER ERËNNERUNG UN
D'LIBERATIOUN VUN EITER
DEN 12. SEPTEMBER 1944
DUERCH EIS AMERIKANESCH FRËNN
WOUBÄI EIS AWUNNER AN EIST DUERF
DUERCH GLËCKLECH ÖMSTÄNN
KEE SCHUED GELIDDEN HUN

D'LEIT VUN EITER
SEPTEMBER 2004

Inscription reads: In grateful remembrance of our liberation on 12 September 1944 by our American friends, whereas our citizens and village did not suffer any damage due to lucky circumstances.

Osweiler

To the soldiers that defended this area from September 1944 to March 1945. Units included 5th Armored Division, 28th Infantry Division, 83rd Infantry Division, 4th Infantry Division, 9th Armored Division, 5th Infantry Division, 87th Infantry Division, Tank Destroyer – Field Artillery, 76th Infantry Division.

Rue Principale 40

6570 Rosport

Latitude: 49.7851

Longitude: 6.4385

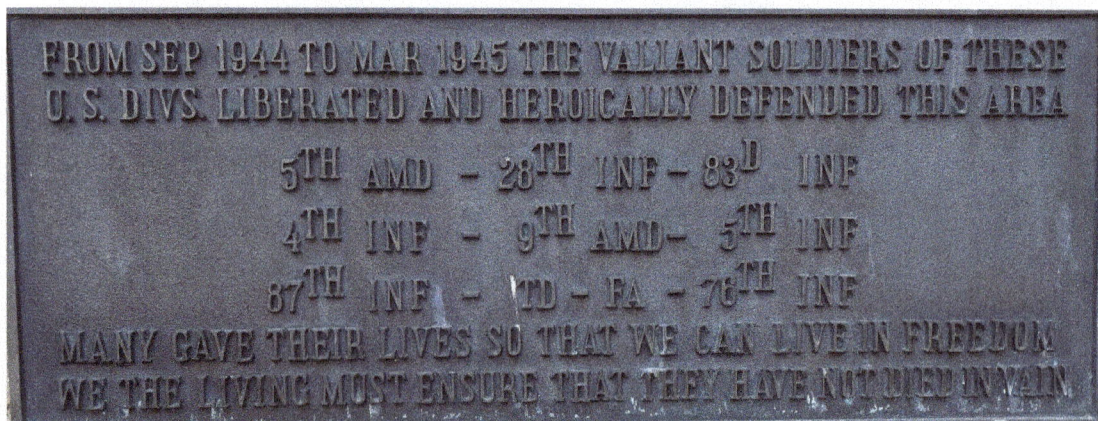

Osweiler

In memoriam to 2nd Battalion, 22nd Infantry Regiment, 4th Infantry Division, commanded by Lieutenant Colonel Thomas A. Kenan.

Rue de l'École 5

6571 Rosport

Latitude: 49.7861

Longitude: 6.4421

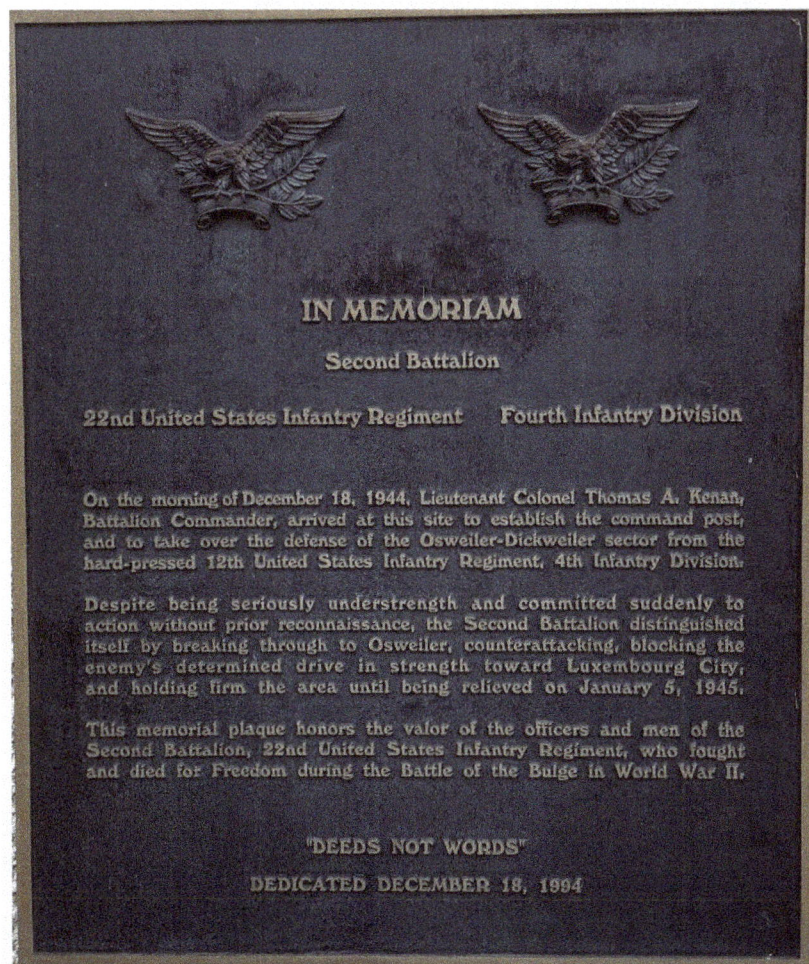

IN MEMORIAM

Second Battalion

22nd United States Infantry Regiment   Fourth Infantry Division

On the morning of December 18, 1944, Lieutenant Colonel Thomas A. Kenan, Battalion Commander, arrived at this site to establish the command post, and to take over the defense of the Osweiler-Dickweiler sector from the hard-pressed 12th United States Infantry Regiment, 4th Infantry Division.

Despite being seriously understrength and committed suddenly to action without prior reconnaissance, the Second Battalion distinguished itself by breaking through to Osweiler, counterattacking, blocking the enemy's determined drive in strength toward Luxembourg City, and holding firm the area until being relieved on January 5, 1945.

This memorial plaque honors the valor of the officers and men of the Second Battalion, 22nd United States Infantry Regiment, who fought and died for Freedom during the Battle of the Bulge in World War II.

"DEEDS NOT WORDS"

DEDICATED DECEMBER 18, 1994

Perlé

Dedicated to the memory of two B-17G bomber crews from the 385th Bomb Group, 550th Squadron, and 551st Squadron who died when their aircrafts collided in mid-air over Perle and Wolwelange on July 12, 1944. Also dedicated to 1st Lt Lewis C. Williams of the 100th Bomb Group, 349th Squadron who died on November 9, 1944.

Rue de l'Eglise

8826 Rambrouch

Latitude: 49.809300

Longitude: 5.765043

Names listed on the monument are: McDonald, Robert L. 1LT; Ryan, Stephen F. 2LT; Henry, William T. 2LT; Chrisman, Francis M. F/O; Hale, Russel S/Sgt; Hefferman, Peter J. S/Sgt; Berosh, Walter R. S/Sgt; Linton, Peter S/Sgt; Brown, George E. T/Sgt; White, Richard B. Capt; Flanagan, Patrick J. 2LT; Gittins, Clarence E. 2LT; Johnston, James W. F/O; Nieman, Marvin, W. S/Sgt; Fitzwater, Harry E. S/Sgt; Lord, William Jr. S/Sgt; Comegys, Homer S/Sgt; Canter, Samuel L. S/Sgt; Williams, Lewis C. 1LT.

Perlé (Field Hospital)

Plaque located on side of house beside church. Inscription reads "This House was used as a Field Hospital under General Patton's command during the 1944 Battle of the Bulge"

13 rue de l'Église

8826 Perlé

Latitude: 49.809778

Longitude: 5.765109

This House was used as a Field Hospital under General Patton's command during the 1944 Battle of the Bulge

Lt Col. A. Burkel

Pétange, Hyman Josefson Square

In Memory of 2LT Hyman Josefson, 1st US Army, 5th Armored Division, the first American soldier killed in action on Luxembourg territory during the liberations of Pétange and the Grand-Duchy of Luxembourg, on September 9, 1944. Also at the site is an M8 "Greyhound" Armored Car.

35 rue de l'Eglise

4732 Pétange

Latitude: 49.560695

Longitude: 5.872327

SQUARE
Hyman JOSEFSON
2Lt, Platoon leader
1st US Army - 5th Arm. Div. - CCA
Troop A 1st Section
85th Cavalery Reconnaissance Squadron (Mecz)

FIRST AMERICAN SOLDIER KILLED IN ACTION FOR
THE LIBERATION OF PETANGE AND THE
GRAND-DUCHY OF LUXEMBOURG
SEPTEMBER 9th 1944

Pintsch

In honor of the soldiers of the 317th Regiment, 80th Infantry Division who liberated the area of "Kiischpelt" in January 1945.

Op der Spier

9767 Kiischpelt

Latitude: 49.992289

Longitude: 6.006006

Putscheid (155-mm Battery)

Dedicated to the 155-mm Howitzer Battery, 21st Field Artillery Battalion, the 3rd Battalion, 11th Regiment, and the 737th Tank Battalion, of the 5th US Infantry Division.

Veinerstrooss (rue de Vianden)

9380 Putscheid

Latitude: 49.96382

Longitude: 6.13117

Information Panel Putscheid (155-mm Battery)

A 155-mm Howitzer Battery in Weiler-Putscheid

When two companies of the 10th Regiment of the 5th US Infantry Division (ID) attacked Putscheid on January 24, 1945, they encountered heavy resistance by units of the "Panzer Lehr Division" entrenched in the village.

After the 3rd Battalion of the 11th Regiment, 5th US ID, had liberated Weiler after a ten-hour-fight, a battery (4 pieces) of the 21st Field Artillery Battalion took its position in "Braetfeld".

On January 28, 1945, four companies of the 10th Regiment occupied Putscheid with the support of the 21st Field Artillery Battalion and of the 737th US Tank Battalion.

Combat for Putscheid, January 24 - January 28, 1945

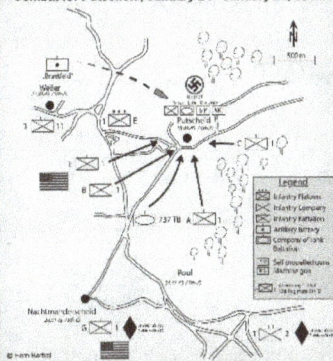

28.01.1945, attack by the 10th Regiment, 5th US ID, on Putscheid with the support of one battery of the 21st Field Artillery Battalion and one company of the 737th Tank Battalion.

28.01.1945, attaque sur Putscheid du 10e Régiment, 5e Division d'Infanterie US, avec le support d'une batterie du 21e Bataillon d'Artillerie de Campagne et d'une compagnie du 737e Bataillon de Chars.

28.01.1945, Angriff des 10. Regimentes, 5. US Infanteriedivision, mit der Unterstützung einer Batterie des 21. Feldartilleriebataillons und einer Kompanie des 737. Panzerbataillons, auf Putscheid.

In 2013, the exact position of the 155-mm Howitzer could be determined by means of discarded percussion primers and lifting plugs that were found on "Braetfeld", north of Weiler.

The first 155-mm Howitzers operated by a 6-man-crew of the U.S. Army were the French models M1917 and M1917A1. The Americans improved these models and constructed the M1918 in the U.S., succeeded and superseded by the M1918M1. This model became the standard of the 155-mm howitzer material. It was a highly successful, accurate weapon, 4 000 of which were produced.

Une batterie de 155-mm obusier à Weiler-Putscheid

Quand deux compagnies du 10e Régiment de la 5e Division d'Infanterie US attaquèrent Putscheid le 24.01.1945, elles furent confrontées à une résistance farouche de la part d'unités de la Division Panzer Lehr qui s'étaient retranchées dans le village.

Après un combat de dix heures, le 3e Bataillon du 11e Régiment, 5e Division d'Infanterie américaine, eut libéré le village de Weiler. Une batterie (4 canons) du 21e Bataillon d'Artillerie de Campagne prit position dans le lieu-dit « Braetfeld ».

Le 28 janvier 1945, quatre compagnies du 10e Régiment avec le support du 21e Bataillon d'Artillerie et du 737e Bataillon de Chars occupèrent Putscheid.

En 2013, la position exacte de la batterie du 155-mm Howitzer au lieu-dit « Braetfeld » au nord de Weiler, a pu être localisée grâce à la découverte d'étoupilles utilisées et d'anneaux de transport trouvés sur le site.

Les premiers 155-mm obusiers desservis par une équipe américaine de six artilleurs étaient les M1917 et les M1917A1 de construction française. Les Américains amélioraient ces modèles et construisaient le M1918 aux États-Unis qui fut succédé et remplacé par le M1918M1. Ce modèle devint le standard des 155-mm obusiers. C'était une arme très efficace et précise qui fut construite en 4.000 exemplaires.

155-mm Howitzer in the yard of the Musée National d'Histoire Militaire in Diekirch (L)

Un 155-mm obusier dans la cour du Musée National d'Histoire Militaire à Diekirch (L)

Eine 155-mm Haubitze im Hof des Musée National d'Histoire Militaire in Diekirch (L)

Photos: Archives MNHM

Sources, Quellen:
- Service de dépannage du l'Armée luxembourgeoise (SEDAL)
- U.S. Army Handbook 1939-1945 by George Forty, Alan Sutton Publishing Limited, Great Britain
- Roland Gaul, 1944-1945 Schicksale zwischen Sauer und Our. Soldaten und Zivilpersonen erzählen, Band II, Sankt-Paulus-Druckerei, 1987.
- Émile Becker, Jean Milmeister, U.S. Army (BTO 1944-45), Marquage et Organisation, Imprimerie G. Willems, Luxelmbourg, 1988.
- The American Arsenal by Ian V. HOGG, The World War II Standard Ordnance Catalog of Small Arms, Tanks, Armored Cars, Artillery, Antiaircraft Guns, Ammunition, Grenades, Mines, etc. etc.

Eine 155-mm Haubitze Batterie in Weiler-Putscheid

Als zwei Kompanien des 10. Regimentes der 5. US Infanteriedivision Pürscheid am 24.01.1945 angriffen, stießen sie auf heftigen Widerstand von Einheiten der Panzer Lehr Division, die sich im Dorf verschanzt hatten.

Nachdem das 3. Bataillon des 11. Regimentes, 5. US Infanteriedivision, nach zehnstündigem Kampf das Dorf Weiler befreit hatte, nahm eine Batterie (4 Geschütze) des 21. Feldartilleriebataillons Stellung in „Braetfeld".

HE Projectile: High Explosive projectile / projectile explosif / Sprenggeschoss

Photo: The American Arsenal by Ian V. HOGG

Am 28. Januar 1945, nahmen vier Kompanien des 10. Regimentes mit der Unterstützung des 21. Feldartilleriebataillons und des 737. US Panzerbataillons Pürscheid ein.

Im Jahre 2013 konnte anhand von vor Ort gefundenen abgeschossenen Zündpatronen und Transportringen die genaue Stellung der 155-mm Haubitze Batterie auf „Braetfeld" nördlich von Weiler festgestellt werden.

Die ersten von sechs amerikanischen Artilleristen bedienten 155-mm Haubitzen waren die französischen M1917 und M1917A1. Die Amerikaner verbesserten diese Modelle und konstruierten in den Vereinigten Staaten das M1918 Modell, gefolgt und ersetzt vom M1918M1. Dieses Modell wurde zum Standard der 155-mm Haubitzen. Es war eine erfolgreiche, präzise Waffe, wovon 4.800 Exemplare hergestellt wurden.

Information Panel Putscheid (155-mm Battery)

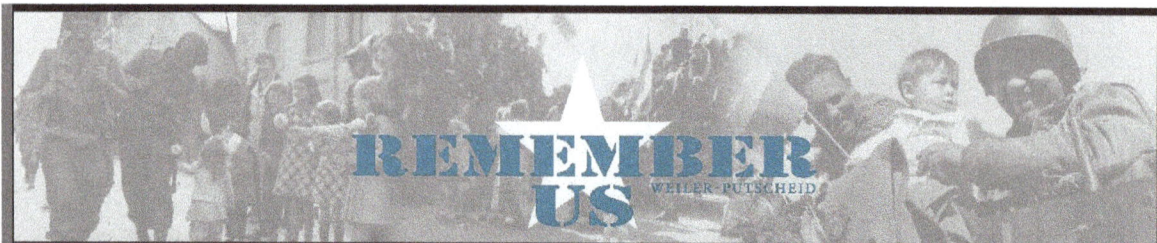

REMEMBER US WEILER-PUTSCHEID

A 155-MM HOWITZER BATTERY IN WEILER-PUTSCHEID

2 squares	weight class of the projectile to calculate the trajectory
2 carrés	classe de poids du projectile pour calculer la trajectoire
2 Quadrate	Gewichtsklasse des Geschosses, um die Flugbahn zu berechnen

155H	155-mm Howitzer

TNT	explosive
	explosif
	Sprengstoff (Trinitrotoluene)

LOT 4T-2-P-A	production number
	numéro de production
	Produktionsnummer

SHELL M107	Projectile model 107
	modèle du projectile 107
	Geschossmodell 107

R2BLA	Ammo Identification Code used in logistics for ordering the ammunition (not always mentioned)
	Code d'identification de la munition utilisé en logistique pour la commande de la munition (le code n'est pas toujours indiqué)
	In der Logistik gebrauchter Identifizierungscode der Munition zur Bestellung der Munition (wird nicht immer vermerkt).

* The Speed and the maximum range vary according to the propellent charge.
La vitesse et la portée varient selon la charge propulsive.
Die Geschwindigkeit und die Reichweite ändern gemäß der Treibladung, gerechnet an der Horizontalen.

Shell, HE, M107 with supplementary charge for 155-mm howitzer

Projectile explosif, M107 avec charge supplémentaire pour 155-mm obusier

Sprenggeschoss M107 mit Zusatzladung für 155-mm Haubitze

Photo: The American Arsenal by Ian V. HOGG

Length / Longueur / Länge:	69 cm

Speed at the end of the barrel	1220 - 1850 fps*
Vitesse à la sortie du canon	372 - 564 m/sec
Geschwindigkeit an der Mündung:	372 - 564 m/sec

W/SUPPL.CHG

With supplementary charge (a small cardboard cylinder in the cavity where the fuze is tightened. This cylinder is taken out, when a VT FUZE M96 (= Variable Time Fuze, a fuze generated by radar) is needed. It is not taken out, when another fuze is needed – usually a PDF M48 = Point Detonationg Fuze model 48).

Avec charge supplémentaire (un petit cylindre en carton dans la cavité où l'on visse la fusée. Ce cylindre est enlevé quand une fusée VT FUZE M96 est utilisée (= fusée variable générée par radar). Ce cylindre n'est pas enlevé quand une autre fusée est utilisée – ordinairement une fusée PDF M48 (une fusée qui explose au contact de l'objectif).

Mit einer Zusatzladung (ein kleiner Pappzylinder im Hohlraum, wo der Zünder aufgeschraubt wird. Dieser Zylinder wird herausgenommen, wenn ein VT FUZE M96 gebraucht wird (ein mit Radar gesteuerter Annäherungszünder). Dieser Zylinder wird nicht herausgenommen, wenn ein Aufschlagzünder verwendet wird. Am häufigsten wird der Zünder PDF M48 aufgesetzt. (Zünder explodiert beim Aufschlag).

Weight	94,52 lb (varies slightly; therefore the indication of the squares)
Poids	43 kg (varie légèrement, pour cette raison l'indication des carrés)
Gewicht	43 kg (ändert leicht, deshalb die Angabe der Quadrate)

Max Range	10780 - 16.355 yd	*
Portée	9.857 m – 14.955 m	
Reichweite	9.857 m – 14.955 m	

Further projectiles / différents projectiles / Weitere Geschosse

AP	Armor Piercing / anti-char / panzerbrechend
M118	Illuminating / projectile éclairant / Leuchtgeschoss
M104	Gas or Smoke HC (Hexachloretane) or WP (White Phosphorus) / Gaz ou brouillard / Gas- oder Nebelgranate
M7	Dummy / projectile d'exercice / Trainingsgeschoss
M107	HE High Explosive projectile; / Projectile explosif / Sprenggeschoss
M102	same as M107 with a smaller copper band / le même que le M107 avec une plus petite bande de forcement en cuivre / wie M107 mit einem kleineren, kupfernen Führungsband

155 mm Howitzer M1918M1

Photo: Artillery in Color 1920 - 1963 by Ian Hogg

Sources, Quellen:
The American Arsenal by Ian V. HOGG, The World War II Standard Ordnance Catalog of Small Arms, Tanks, Armored Cars, Artillery Antiaircraft Guns, Ammunition, Grenades, Mines, etc.

Putscheid (Church)

Information board located at the church on the combat operations for the liberation of Putscheid over the last few days of January 1945. The German *Panzer Lehr Division* held Putscheid until it was defeated by the 10th Regiment, 5th US Infantry Division, supported by the 737th Tank Battalion.

Haaptstrooss (CR 320)

Putscheid

Latitude: 49.9594632

Longitude: 6.1416312

Information Panel at Putscheid Church

REMEMBER US
PÜTSCHEID

THE COMBAT FOR PUTSCHEID

On December 16, 1944, three German armies with 250,000 men attacked the 83,000 US soldiers holding the front line from Echternach (L) to Monschau (Germany) in the west. The stiff resistance of the US defenders thwarted the Germans' plan to reach Antwerp (B).

The Third US Army under the command of General George S. Patton Jr. attacked the Germans at the southern flank on December 22, 1944. The American and the British armies succeeded in throwing back the aggressors to their initial positions in a hard, six-week lasting battle.

Liberation of the northern area of Diekirch, January 18 - February 22, 1945

The map bases on the information of Jean Milmeister's book
„Die Ardennenschlacht 1944-1945 in Luxembourg",
page 447 and 449. Editions Saint-Paul, Luxembourg, 1994.
* Moestroff, the part on the right side of the Sauer River was liberated on Dec 24, 44 by "Task Force Rudder" of 10th US AD. The left side was liberated on Jan 18, 45 by 4th US Inf. Div.

The 5th and 4th US Infantry Divisions (ID) crossed the River Sauer on January 18. They began liberating the villages north of Diekirch on January 19, 1945.

Units of the German "Panzer Lehr Division" held and defended with all their means the village of Putscheid. On January 24, A and B Company, 1st Battalion, 10th Regiment, 5th US ID, attacked Putscheid. In spite of strong artillery, mortar and tank fire, the US troops didn't succeed in liberating the village. On January 27, 1945, the 3rd Battalion, 11th Regiment, 5th US ID, occupied the village of Weiler after a ten-hour fight. An artillery battery took up position in the fields of the so-called "Braetfeld" area. On January 28, the US troops attacked again. However the Germans started a counter-attack with infantry, tanks and assault guns (self-propelled guns) that pinned down the US units. A new attack of A, B, C & E Companies of 1st and 2nd Battalion, 10th Regiment, 5th US ID, with the support of the 737th US Tank Battalion, finally succeeded in driving away the defenders. Putscheid was liberated on January 28, 1945 at 13.40. The village was in ruins.

LA LUTTE POUR PUTSCHEID

Le 16 décembre 1944, trois armées allemandes avec 250.000 soldats attaquèrent à l'ouest les 83.000 troupes américaines qui défendaient le front sur la ligne d'Echternach (L) jusqu'à Monschau (Allemagne). Grâce à la résistance tenace des défenseurs américains, les Allemands ne pouvaient atteindre leur but, Anvers (B).

La 3e Armée Américaine sous le commandement du Général George S. Patton Jr. attaqua les troupes allemandes par le flanc sud. Dans un combat qui durait six semaines, les Américains et les Britanniques réussirent à refouler les agresseurs sur leur ligne de départ.

Les 4e et 5e US Divisions d'Infanterie (ID) traversèrent la rivière Sûre le 18 janvier 1945. Elles commencèrent à libérer les villages au nord de Diekirch à partir du 19 janvier 1945.

Des unités de la « Panzer Lehr Division » occupaient et défendaient farouchement le village de Putscheid. Le 24.01.1945, les compagnies A et B, 1er Bataillon, 10e Régiment, 5e US ID, attaquèrent Putscheid. Malgré le tir intensif d'artillerie, de mortier et de chars, les troupes américaines ne réussirent pas à déloger les ennemis. Le 27.01.1945, le 3e Bataillon, 11e Régiment, 5e US ID, occupa le village de Weiler après un combat de dix heures. Une batterie d'artillerie s'installa au lieu-dit « Braetfeld ». Le 28.01.1945, les Américains attaquèrent de nouveau. Toutefois les Allemands ripostèrent avec une contre-offensive avec des fantassins, des chars et des canons d'assaut automoteurs qui arrêtèrent net les unités US. Une nouvelle attaque des compagnies A, B, C et E du 1er et 2e Bataillon était couronnée de succès. Putscheid était définitivement libéré le 28 janvier 1945 à 13.40 heures. Le village était en ruines.

Chapel of Putscheid

School of Putscheid

German „Panther" tank, House Nosbusch Dominique

House Nosbusch Dominique

House Boewer

House Tibesz: 03.04.1946

House Miller

DER KAMPF UM PÜTSCHEID

Am 16. Dezember 1944 griffen drei deutsche Armeen, 250.000 Mann, 1000 Panzer, die 83.000 amerikanischen Truppen im Westen an, die auf einer Linie von Echternach (L) bis nach Monschau (D) stationiert waren. Dank des tapferen Widerstandes der unterbesetzten amerikanischen Truppen, konnten die Deutschen ihr Ziel Antwerpen (B) nicht erreichen.

Die Dritte US Armee unter dem Kommando von General George S. Patton Jr. griff am 22. Dezember 1944 die deutschen Stellungen von Süden her an. In harten, sechs Wochen andauernden Kämpfen gelang es den Amerikanern und Briten die Angreifer auf ihre Ausgangspositionen zurückzudrängen.

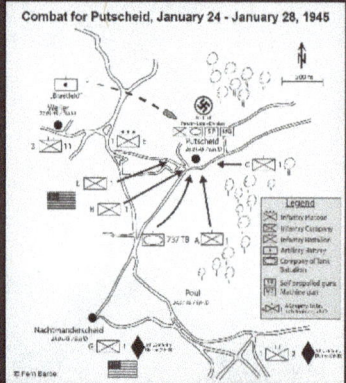

Combat for Putscheid, January 24 - January 28, 1945

Die 5. und 4. US Infanteriedivisionen (ID) überquerten am 18. Januar 1945 den Fluss Sauer und begannen am 19. Januar 1945 mit der Befreiung der nördlich von Diekirch gelegenen Dörfer.

Pütscheid wurde von Einheiten der Panzer Lehr Division gehalten und stark verteidigt. Am 24.01.1945 griffen A und B Kompanie, 10. Regiment, 5. US ID, Pütscheid an. Trotz heftigen Artillerie-, Mörser- und Panzerbeschusses, gelang es den US Truppen nicht, das Dorf zu befreien. Am 27.01.45 besetzte das 3. Bataillon, 11. Regiment, 5. US ID, nach zehnstündigem Kampf das Dorf Weiler. Eine Batterie Artillerie nahm Position in „Braetfeld". Am 28.01.45, griffen die Amis wieder an. Doch die Deutschen starteten einen Gegenangriff mit Infanterie, Panzern und Sturmgeschützen, der die US Einheiten festnagelte. Ein neuer Angriff der Kompanien A, B, C & E mit Unterstützung der Panzer des 737. US Tank Battalions, brachte den Erfolg. Am 28. Januar 1945, um 13.40 Uhr war Pütscheid endgültig befreit. Das Dorf lag in Schutt und Asche.

Sources; Quellen:
The Fifth Infantry Division in the ETO, 1945, prepared by the Fifth Division Historical Section Headquarters Fifth Infantry Division.
Jean Milmeister, Die Ardennenschlacht 1944-1945 in Luxembourg, Editions Saint-Paul, Luxembourg, 1994.
Photos: © Robert LEER, architecte diplômé

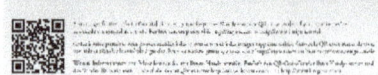

Putscheid (Church - continued)

Just inside of Church entrance is a plaque to the liberation of Putscheid by the 5th US Infantry Division, 10th Infantry Regiment, 7th Engineer Combat Battalion, 46th and 50th Field Artillery Battalions. It lists the 45 US soldiers Killed in Action for the liberation of Putscheid between 18 – 28 January 1945.

Liberation of PUTSCHEID

5th US Infantry Division, 10th Infantry Regiment
7th Engineer Combat Battalion, 46th & 50th Field Artillery Battalions

358 Casualties, January 18 – 28, 1945
235 Wounded in Action, 78 Missing in Action, 45 Killed in Action

January 18
Pfc Arnold R. Brown, 50th Field Artillery Battalion, 5th ID
Private Clifford L. Beavers
Sgt Donald L. Ickes, 7th Engineer Combat Battalion, 5th ID
Private Sigismund W. Ligas, 7th ECB, 5th ID
Private James H. Lyon
First Lieutenant Robert N. Massonet, 50th FA Bn, 5th ID
Private First Class James W. Mitchell
Pfc Leroy R. Thomas, 7th ECB, 5th ID

January 19
Corporal Charles E. Amerson, 46th FA Bn, 5th ID
Pfc John J. Gaddis Jr.
Private George S. Gorham
Private Winfried C. Hannah
Pfc Generoso Martinez
Private Otis Mather
Staff Sergeant Charles D. Mellendorf
Sergeant Andrew S. Nixon
Pfc Rudolph G. Rico
Pfc Frank A. Weddle

January 20
Private Frederick J. Gardner
Pfc Joseph Martin

January 22
Pfc Attilio V. Bisesi
Pfc Wendell S. Canoy
Pfc James W. De Bello
Second Lieutenant Virgil N. Hawthorne
Private Charles L. Hoskins
Technician Fourth Grade Dominick H. Melillo, 46th FA Bn, 5th ID

January 23
Private Frank J. Caldiero
Corporal Warren R. Hastings
Private Calvin R. Marklein
Pfc John Silvia

January 24
Pfc Edward E. Anderson
Private Francis W. Blanchard
Pfc Enario Boccagni
Pfc Jarold W. Johnson
Pfc Lawrence Mouret

January 25
Private Joseph A. Davidson
Private Edward J. Yosick

January 28
Staff Sergeant John R. Ahearn
Pfc Arthur Beck
Pfc Casper Cassano
Pfc John Chichilla
Pfc Ell B. Crowe
Corporal Carl L. Crowther
Pfc David J. Davidson
Pfc Melvin W. Dunn

The exact number of the German losses remains unknown.
We include the military victims of both sides and the civilian casualties in our prayers in the hope
that no young person will ever have to go to war again.

Putscheid (The Pavillon)

Two information panels, one side about the facts of the war in the commune of Putscheid, and the other side "Remember US Info Points" which list key battle locations in Luxembourg.

See Panels Section for details.

The Pavillon

Intersection of Haaptstross and Highway 322,

Putscheid

Latitude: 49.962196

Longitude: 6.130807

Information Panel at Putscheid (The Pavillon)

REMEMBER US
PUTSCHEID

FACTS OF WAR IN THE COMMUNE OF PUTSCHEID – INFO POINTS

1. **Stolzembourg:**
First crossing into the Third Reich

On September 11, 1944, a small unit of the 5th US Armored Division accompanied by a French liaison officer crossed the River Our and entered for the first time mainland Germany. This news was radioed to the entire world.

2. **Weiler: Beginning of the Battle of the Bulge on December 16, 1944**

In Weiler, I-Company, 3rd Batallion, 110th Regiment, 28th US Infantry Division, defended the village all day long against the German aggressors. During the night, part of the unit retreated to 3rd Batallion's Headquarters in Consthum, the remainder of the company was taken prisoner the next day.

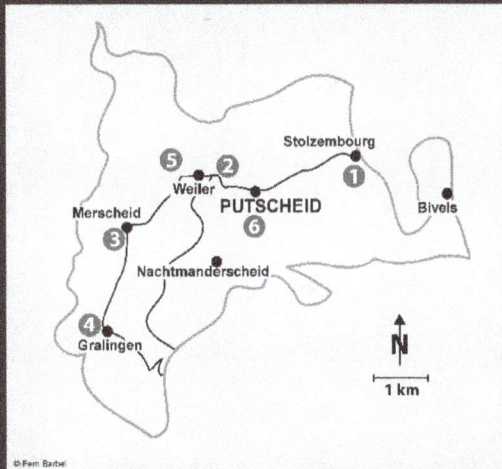

3. **Merscheid: December 16, 1944**

US soldiers and members of the Luxembourg armed resistance (militiamen) defended the village. The small garrison was defeated, and the inhabitants were taken hostages. The militiamen not belonging to a regular army, were considered by the Germans as terrorists.

4. **Gralingen: Second Liberation, January 18 to 28, 1945**

The Third US Army under the command of General George S. PATTON, jr. attacked the Germans at the southern flank already on December 22, 1944. It took until January 28, 1945, before the region north of Diekirch was re-liberated.

5. **Weiler/Putscheid "Eng Plaz fir eis all": US Field Artillery Position**

After the liberation of Weiler on January 27, 1945, an artillery battery of six 155-mm-howitzers took position at "Braerfeld". The artillery shelling helped liberate Putscheid.

6. **Putscheid: Second Liberation on January 28, 1945**

The Germans defended the village of Putscheid with all means to allow their troops to retreat on the road Putscheid-Vianden. With the support of artillery and attached tanks, 1st Battalion, 10th Regiment, 5th US Infantry Division, succeeded in liberating Putscheid on January 28, 1945.

ÉVÉNEMENTS DE GUERRE DANS LA COMMUNE DE PUTSCHEID – POINTS D'INFORMATION

1. **Stolzembourg: La première entrée dans le Troisième Reich**

Le 11 septembre 1944, une petite unité de la 5e Division Blindée américaine accompagnée d'un officier de liaison français traversa la rivière Our et entra pour la première fois en Allemagne. Cette nouvelle fut diffusée dans le monde entier.

2. **Weiler : Commencement de la Bataille des Ardennes le 16 décembre 1944**

A Weiler, Company I, 3e Bataillon, 110e Régiment, 28e Division d'Infanterie américaine, défendait le village contre les agresseurs allemands. Pendant la nuit une partie de l'unité se retira à Consthum, quartier général du 3e Bataillon. Le reste de la compagnie fut fait prisonnier le jour suivant.

3. **Merscheid : 16 décembre 1944**

Des soldats américains et des miliciens luxembourgeois défendaient le village. La petite garnison fut écrasée et les habitants furent pris comme otages. Les miliciens n'appartenant pas à une armée régulière, les Allemands les considéraient comme des terroristes.

4. **Gralingen : Seconde libération du 18 au 28 janvier 1945**

La 3e Armée américaine sous le commandement du Général George S. PATTON Jr. attaqua les troupes allemandes par le flanc sud déjà le 22 décembre 1944. Les combats durèrent jusqu'au 28 janvier 1945 avant que les villages au nord de Diekirch fussent libérés.

5. **Weiler/Putscheid "Eng Plaz fir eis all": Positions de l'artillerie américaine**

Après la libération de Weiler le 27 janvier 1945, une batterie de six 155-mm-Howitzer prit position dans le lieu-dit « Braerfeld ». Le bombardement de l'artillerie aida à libérer Putscheid.

6. **Putscheid: Seconde libération le 28 janvier 1945**

Les Allemands défendaient le village de Putscheid de toutes leurs forces pour permettre à leurs troupes de se retirer par la route Putscheid-Vianden. Grâce au support de l'artillerie et des chars, le 1er Bataillon du 10e Régiment de la 5e Division d'Infanterie américaine, réussit à libérer Putscheid le 28 janvier 1945.

KRIEGSEREIGNISSE IN DER GEMEINDE PÜTSCHEID – INFO-PUNKTE

1. **Stolzemburg: Die erste Überquerung der deutschen Grenze**

Am 11. September 1944 überquerte eine kleine Einheit der 5. US-Panzerdivision, von einem französischen Verbindungsoffizier begleitet, den Fluss Our und betrat als erste alliierte Einheit Boden des Dritten Reiches. Diese Nachricht wurde in der ganzen Welt verbreitet.

2. **Weiler: Beginn der Ardennenoffensive am 16. Dezember 1944**

In Weiler verteidigte die I-Kompanie, 3. Bataillon, 110. Regiment der 28. US-Infanteriedivision das Dorf gegen die deutschen Aggressoren. Während der Nacht zog sich ein Teil der Einheit ins Hauptquartier des 3. Bataillons nach Consthum zurück. Der Rest der Kompanie wurde am folgenden Tag gefangen genommen.

3. **Merscheid: 16. Dezember 1944**

Amerikanische Soldaten und luxemburgische Widerstandskämpfer (Miliz) verteidigten das Dorf. Die kleine Einheit wurde überrannt und die Dorfbewohner als Geiseln genommen. Da die Widerstandskämpfer keiner regulären Armee angehörten, betrachteten die Deutschen sie als Terroristen.

4. **Gralingen: Zweite Befreiung, 18. - 28. Januar 1945**

Die Dritte US-Armee unter dem Befehl von General George S. Patton Jr. griff am 22. Dezember 1944 die deutschen Stellungen von Süden her an. Es dauerte bis zum 28. Januar 1945, bis die nördlich von Diekirch gelegenen Dörfer endgültig befreit waren.

5. **Weiler/Putscheid "Eng Plaz fir eis all": Position der US-Feldartillerie**

Nach der Befreiung von Weiler am 27. Januar 1945 bezog eine Batterie von sechs 155-mm-Haubitzen in „Braerfeld" Stellung. Der Artilleriebeschuss half bei der Befreiung von Pütscheid.

6. **Pütscheid: Zweite Befreiung am 28. Januar 1945**

Die Deutschen verteidigten Pütscheid mit allen Mitteln, um es ihren Truppen zu erlauben, sich auf der Pütscheid-Viandener Straße zurückzuziehen. Mit Artillerie- und Panzer-Unterstützung gelang es dem 1. Bataillon, 10. Regiment der 5. US-Infanteriedivision, Pütscheid am 28. Januar 1945 zu befreien.

For more information
Pour en savoir plus
Für weitere Informationen

Information Panel at Putscheid (The Pavillon)

REMEMBER US INFO POINTS

1. **STOLZEMBOURG, Bridge, Pont, Brücke:**
The First Crossing into the Third Reich on Sept 11, 1944
La première entrée dans le Troisième Reich, le 11 sept. 1944
Die erste Überquerung der deutschen Grenze am 11.09.1944

2. **PUTSCHEID, Chapel, Chapelle, Kapelle:**
Liberation of Putscheid on Jan 28, 1945
La libération de Putscheid, le 28 janvier 1945
Die Befreiung von Putscheid am 28. Januar 1945

PUTSCHEID, A Place for us all / Une place pour nous tous / Ein Platz für uns alle:
A 155mm Battery / REMEMBER US Itineraries / Facts of War in the Commune of Putscheid, Info Points
Une batterie 155-mm-Howitzer / Itinéraires des panneaux REMEMBER US / Evénements de guerre dans la commune de Putscheid, Points d'information · Eine 155-mm-Haubitze-Batterie / Reiserouten zum Entdecken der REMEMBER US Tafeln / Kriegsereignisse in der Gemeinde Putscheid, Informationspunkte

3. **WEILER, Square Tom Myers, Place Tom Myers, Tom-Myers-Platz:**
Beginning of the Battle of the Bulge on Dec 16, 1944
Commencement de la Bataille des Ardennes, le 16 décembre 1944 · Beginn der Ardennenoffensive am 16. Dezember 1944

4. **MERSCHEID, Communal Square, Place communale, Gemeindeplatz:**
War Victims of Merscheid · *Victimes de guerre de Merscheid · Kriegsopfer von Merscheid*

5. **GRALINGEN, Church, Église, Kirche:**
Liberation of the Commune of Putscheid, January 1945
La libération de la commune de Putscheid au mois de janvier 1945 · Die Befreiung der Gemeinde Putscheid im Monat Januar 1945

6. **MOERSDORFF, Bridge, Pont, Brücke:**
The Vincent J. Festa Bridge · *Le pont Vincent J. Festa · Die Vincent-J.-Festa-Brücke*

7. **DIEKIRCH, Municipal Parc, Parc municipal:**
Sauer River Crossing Jan 18, 1945 / Evacuation of the civilian population Dec 19/20, 1944 · *La traversée de la Sûre le 18 janvier 1945 / L'évacuation de la population civile de Diekirch le 19/20 décembre 1944 · Die Sauer-Überquerung am 18. Januar 1945 / Die Evakuierung der Zivilbevölkerung am 19/20. Dezember 1944*

8. **ETTELBRUCK, Square Patton, Place Patton, Patton Platz:** Sherman Tank M4A1 · *Char Sherman M4A1 · Sherman M4A1 Panzer*

ETTELBRUCK, Private School Sainte Anne, Ecole privée Sainte Anne, Privatschule Sainte Anne:
Lt. Col. James Rudder, Commanding Officer 109th Regt. 28th US Infantry Division · *Lt. Col. James Rudder,*

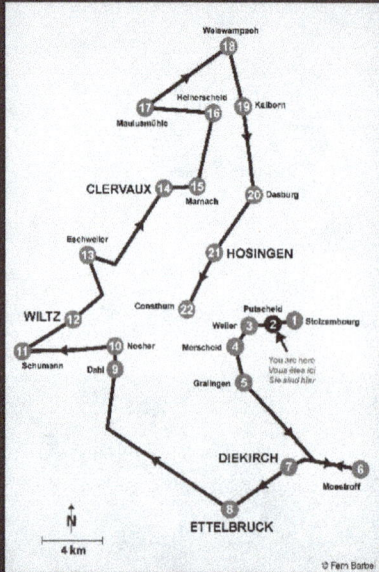

commandant du 109e Régiment, 28e Division d'Infanterie US · Oberst James Rudder, Kommandant des 109. Regimentes, 28. US-Infanteriedivision

9. **DAHL "An Aastert":**
Congressional Medal of Honor for Sergeant Day G. Turner · *Médaille d'Honneur du Congrès pour Sergent Day G. Turner · Congressional Medal of Honor für Feldwebel Day G. Turner*

10. **Nocher, "Um Knupp":**
The long Battle for Nocher · *La longue Bataille pour Nocher · Der lange Kampf um Nocher*

11. **SCHUMANNS CORNER, National Liberation Memorial:**
The Battle around Schumann / Timetable 1839 – 1945
La Bataille autour du carrefour Schumann / Calendrier 1839 – 1945 · Der Kampf um die Kreuzung Schumanns Ecke / Zeitleiste 1839 – 1945

SCHUMANNS CORNER, Mass Grave, Tombeau commun, Gemeinschaftsgrab:
Reconciliation · *Réconciliation · Versöhnung*

12. **WILTZ, 28th US Infantry Division Square, Place de la 28e Division d'Infanterie américaine, 28. US Infanteriedivision-Platz:**
The American St Nick · *Le Saint Nicolas américain · Der amerikanische Heilige Nikolaus*

13. **ESCHWEILER, Church, Église, Kirche / CR 328:**
G.I. George Mergenthaler, Ambush at Café Halt
G.I. George Mergenthaler, Embuscade au Café Halt · G.I. George Mergenthaler, Hinterhalt am Café Halt

14. **CLERVAUX, Square G.I.:**
The American Soldier · *Le soldat américain · Der amerikanische Soldat*

CLERVAUX, Monument for the Victims, Monument A Nos Morts, Totendenkmal:
Clervaux in the Battle of the Bulge · *Clervaux dans la Bataille des Ardennes · Clerf in der Ardennenoffensive*

CLERVAUX, Exterior courtyard of the castle, Cour extérieure du château, Außenhof des Schlosses:
US Sherman Tank M4A3 (76)

15. **MARNACH, Square Cube 521:**
Battle around Marnach, The 707th Tank Battalion in Support of the 110th Regt · *La Bataille autour de Marnach, Le 707e Bataillon de chars en support du 110e Régiment · Der Kampf von Marnach, Das 707. Panzerbataillon in Unterstützung des 110. Regimentes*

16. **HEINERSCHEID, „Am Pesch", Bicycle Trail, Piste cyclable, Radfahrweg:**
The Heinerscheid "Bunker" Disaster · *Le désastre du bunker de Heinerscheid · Die Tragödie im Bunker von Heinerscheid*

17. **MAULUSMÜHLE Wood between Boxhorn and Maulusmühle, Forêt entre Boxhorn et Maulusmühle, Wald zwischen Boxhorn und Maulusmühle:**
The Crash of the Hudson FK-308 · *Le crash du Hudson FK-308 · Der Absturz der Hudson FK-308*

18. **WEISWAMPACH, Lancaster Memorial:**
The 112th Infantry Regiment defending the Northern Tip of Luxembourg · *Le 112e Régiment défend la partie nord du Luxembourg · Das 112. Regiment verteidigt die Nordspitze Luxemburgs*

19. **KALBORN, South facade of the church, Façade sud de l'église, Südfassade der Kirche:**
The Events of September 22, 1944, in Kalborn · *Les événements tragiques du 22 septembre 1944 à Kalborn · Die tragischen Ereignisse des 22. September 1944 in Kalborn*

20. **DASBURG, Bridge, German Territory, Pont, territoire allemand, Brücke, deutsches Hoheitsgebiet:**
Siegfried Line Breakthrough, Feb 20-23, 1945 · *Percée de la Ligne Siegfried, février 20-23, 1945 · Durchbruch in die Siegfriedlinie, Februar 20-23, 1945*

21. **HOSINGEN, Town South Entry, Entrée sud de la localité, Südlicher Ortseingang:**
Defense on Dec 16-18, 1944, Water Tower, Liberation Jan 27, 1945 · *Défense du 16 au 18 décembre 1944, Château d'eau, Libération, le 27 janvier 1945 · Verteidigung 16. - 18. Dezember 1944, Wasserturm, Befreiung am 27. Januar 1945*

22. **CONSTHUM, Church, Église, Kirche:**
Lt. Col Daniel B. Strickler

For more information
Pour en savoir plus
Für weitere Informationen

Rambrouch

Dedicated to our liberators, the 26th Infantry Division in December 1944 and in remembrance of 2nd Lt. Charles E. Parmellee, 78th Fighter Group, 84th Fighter Squadron, 8th US Air Force who died in Rambrouch on September 10, 1944.

Rue Principale 21A

8805 Rambrouch

Latitude: 49.82976

Longitude: 5.84724

A NOS LIBÉRATEURS
DE LA 26e DIV. D'INF. U.S.
DÉCEMBRE 1944

8th Air Force

C.A.G.G-D

We remember in deep gratitude
Mir erenneren ons an déiwer Dankbarkeet un den

2nd Lt. Charles E. PARMELEE   N 0700211
78th F.G. 84th F.Sq. 8th USAF
A.C. P47 N :4325593
† Rambrouch 10.09.1944

Reckange-les-Mersch

Dedicated to pilot First Lieutenant Charles Goffin who was killed in action on September 8, 1944. While Goffin was from Belgium, however, at the time of his death he was flying for the US Army Air Force. Goffin was assigned to the 14th Photo Squadron, 7th Reconnaissance Group and was flying a Spitfire PR MK XI.

Enelter-Kapell

7598 Mersch

Latitude: 49.759658

Longitude: 6.078489

1940 — 1945
FIR EIS KRICHSAFFER
+ KREMER CAMILLE   19 JOER
+ GANSEN EMILE   20 JOER
+ ARENDT CATHERINE   87 JOER
+ REUTER NORBERT   22 JOER
+ CONRAD NICOLAS   43 JOER
+ GANSEN JEAN   20 JOER
+ MAJERUS JOSEPH   22 JOER
+ KUGENER JOSEPH   24 JOER
  AVIATEUR BELGE   31 ANS
+ GOFFIN CHARLES J.J.
  GRAIDE / NEUFCHATEAU
1798 — 1815
+ 15 NAPOLEONSDÉNGER

Redange

Offered by veterans of the 511th Engineer Light Ponton Company and 814th Engineer Company (FB) to the people of Redange.

Rue du Lavoir 2

8506 Redange

Latitude: 49.763898

Longitude: 5.888310

Offert par 511th Engineer Light Ponton Company/814th Engineer Company (FB) à la population de Redange en remerciement de son patriotisme ses sacrifices et son hospitalité 1944/45.

Inscription reads: From the 511th Engineer Light Ponton Company/814th Engineer Company (FB) to the people of Redange in gratitude for their patriotism, sacrifices and hospitality 1944/45.

Reisdorf-Beaufort Road

Dedicated to Company "C" 60th Armored Infantry Battalion, 9th Armored Division who defended this position from December 16-19, 1944 and to the last soldier to leave this area T/SGT Robert N. Hebert.

CR128

6311 Reisdorf

Latitude: 49.85656

Longitude: 6.27491

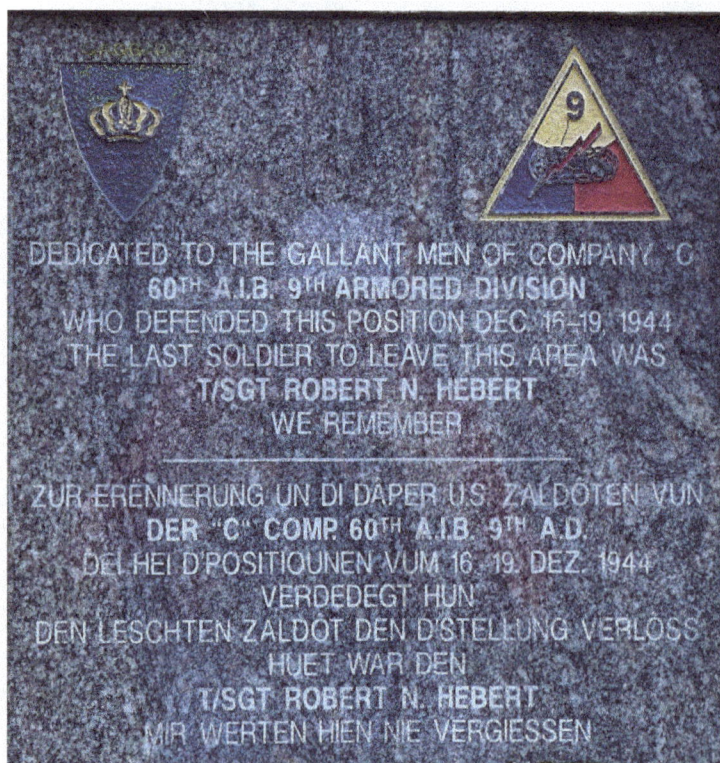

Rodenbourg

Monument to Ernest Hemingway, who stayed in this mill (now a house) during the Battle of the Ardennes in December 1944 as a war correspondent.

Rue de Wormeldange 55

6955 Junglinster

Latitude: 49.6854
Longitude: 6.2937

Sandweiler

In honor of the 5th US Armored Division and in memory of Sgt. Joseph L. Passet and Sgt Jesse Mittiga, who were killed in Sandweiler on September 10, 1944.

Rue de la Chapelle 1

5213 Sandweiler

Latitude: 49.617318

Longitude: 6.213553

Information Panel Sandweiler

LIBERATION 1944

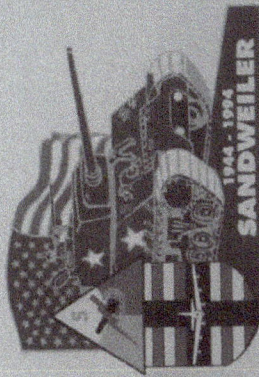

On Sunday, September 10th, 1944 the American troops liberated the City of Luxembourg. In the afternoon, the 5th Armored Division nicknamed the "Victory Division" started advancing towards Echternach. When the column approached the airport German tanks ambushed the advance guard and hit two American tanks. Their occupants fled from the destroyed vehicles and took shelter in a nearby farm. Sgt. Joseph L. PASSET who had been mortally wounded was brought back to "Kalchesbréck" and laid alongside the road. People came and covered the body of the first American liberator killed on the soil of Sandweiler with red, white and blue asters.

A further victim of the ambush, Sgt. Jesse MITTIGA, had been severely burnt. He died soon afterwards.

For the town of Sandweiler, September 10th, 1944 proved to be a black day. As the American forces suspected German tanks and infantry to hide in and around Sandweiler, they bombed the town and set it under heavy artillery and mortar fire. Three people were killed and dozens of houses were completely destroyed or burnt out.

On the following day, Monday September 11th, 1944, two courageos men, René Massard and Nik Molling, decided that it was time for them to mount their bicycle to go to meet the American liberators and to bring them to the village. The town of Sandweiler was liberated that same day. When the first Jeeps passed "Kapellebierg", the villagers noticed at the head of the convoy the two brave Luxemburgish men, their bicycles tied on to the Jeeps. The population of Sandweiler was exuberant in cheering their American liberators.

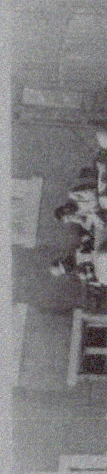

Savelborn

Dedicated to Combat Command A (CAA) of the 9th US Armored Division who held the Ermsdorf-Savelborn-Waldbillig-Christnach line.

Savelborn 6

6380 Vallée de l'Ernz

Latitude: 49.810558

Longitude: 6.252539

Inscription reads: On December 18, 1944 the German attack through the Ardennes was stopped on the Ermsdorf-Savelborn-Waldbillig-Christnach line by soldiers of the following units of Combat Command A of the 9th US Armored Division.

Headquarters and Headquarters Company, Combat Command A; 3rd Armored Field Artillery Battalion; 60th Armored Infantry Battalion; Company A, 9th Armored Engineer Battalion; 19th Tank Battalion; Battery A, 482nd Anti-Aircraft Artillery Automatic Weapons Battalion; Troops A^x, B^x, C, and E^x 89th Cavalry Reconnaissance Squadron; Headquarters Company^x Reconnaissance Company (-) Company B 811th Tank Destroyer Battalion; Company A 131st Ordnance Maintenance Battalion; Company A, 2nd Medical Battalion (Armored).

^X Elements of these units fought here

Offered in memory to all those who died in combat by the Diekirch Historical Society and the veterans of CCA of the 9th US Armored Division.

Schoos (Fischbach)

Monument to 1st Lt. Donald H. Huff, 395th Fighter Squadron, 368th Fighter Group killed in a crash landing of his P-47 aircraft after a bombing mission in Germany on 16 February 1945. Site is located on a farming road just to the west of Schoos.

Natur-and Geopark Mëllerdall

7475 Fischbach

Latitude: 49.75219

Longitude 6.16252

Commune de Fischbach
24.02.2023

Donald Hartley HUFF
* Maine, February 10, 1920
† Schoos, February 16, 1945

Pilot Donald H. HUFF
with his Crew Chief Sgt. Charles MATTHEW

1st Lt Donald H. HUFF

On 16th February 1945 at 5.50 PM
1st Lt Donald H. HUFF from the 368th Fighter Group / 395th Fighter Squadron lost his life tragically on this spot after returning from a bombing mission in Germany.
The railway facilities near St. Thomas and Densborn were the target on that day. While on return he informed his Squadron by radio that he has some smoke in his cockpit and had to do immediately an emergency landing. During the belly landing near the "Ollergronn" at Schoos, he struck a tree and his P-47 caught fire. 1st Lt HUFF didn't manage to leave the plane and burned dreadfully in his cockpit.
We would like to commemorate him and all the pilots and soldiers who gave their lives in WWII for our freedom.

Le 16 février 1945 à 17h50
1er lieutenant Donald H. HUFF du 368th Fighter Group, 395th Fighter Squadron a perdu sa vie tragiquement à cet endroit après son retour d'une mission de bombardement des installations ferroviaires près de St. Thomas et Densborn en Allemagne.
En retournant il a informé par radio son escadrille qu'il y avait de la fumée dans son cockpit et devait faire immédiatement un atterrissage d'urgence. Lors de cet atterrissage forcé près de l'« Ollergronn » à Schoos, il a frôlé un arbre et son P-47 a pris feu.
Le 1er Lt HUFF n'arrivait plus à quitter l'avion et brûlait dans son cockpit.
Nous tenons à le commémorer ainsi que tous les pilotes et soldats qui ont donné leur vie pendant la Seconde Guerre mondiale pour notre paix et pour notre liberté.

Am 16. Februar 1945 um 17:50 Uhr
kam 1st Lt. Donald H. HUFF vom 368th Fighter Group, 395th Fighter Squadron auf dem Rückweg von einem Bombenangriff in Deutschland an dieser Stelle auf tragische Weise ums Leben.
Ziel an diesem Tag waren die Bahnanlagen bei St. Thomas und Densborn. Bei der Rückkehr teilte er seiner Staffel per Funk mit, dass er Rauch im Cockpit habe und sofort notlanden müsse. Bei der Bauchlandung im „Ollergronn" bei Schoos, streifte er einen Baum woraufhin seine P-47 Feuer fing. 1st Lt HUFF schaffte es nicht, das Flugzeug zu verlassen und verbrannte in seinem Cockpit. Wir möchten ihm, sowie allen Piloten und Soldaten gedenken, welche im 2. Weltkrieg ihr Leben für unseren Frieden und unsere Freiheit ließen.

Schumannseck Café (Nothum)

Located on the side of the Schumanns Eck (corner) Café the plaque is dedicated to Private Ted B. Long of Malakoff, Texas, Company L, 359th Infantry Regiment, 90th Infantry Division, who was killed in action in this area.

Béiwenerstrooss 19

9678 Wilz

Latitude: 49.945975

Longitude: 5.87936

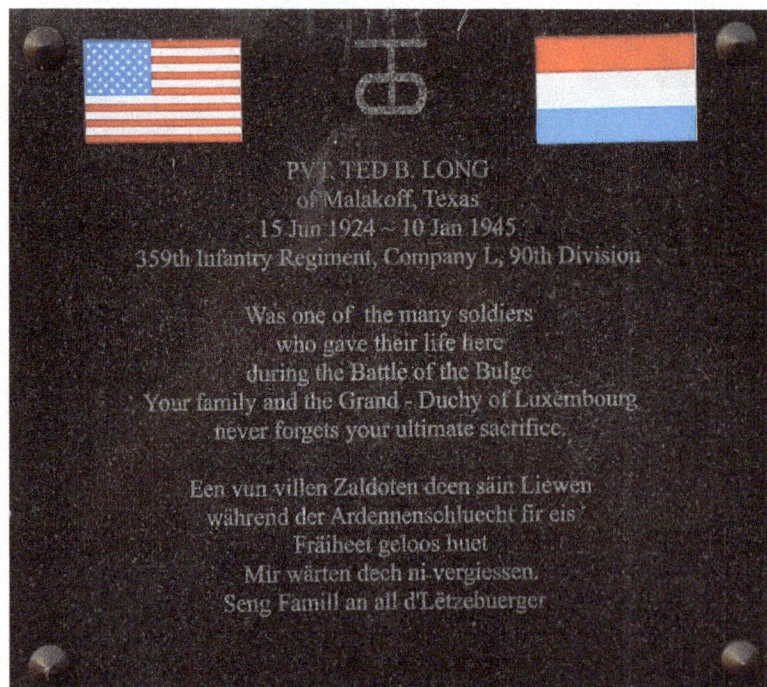

PVT. TED B. LONG
of Malakoff, Texas
15 Jun 1924 ~ 10 Jan 1945
359th Infantry Regiment, Company L, 90th Division

Was one of the many soldiers
who gave their life here
during the Battle of the Bulge
Your family and the Grand - Duchy of Luxembourg
never forgets your ultimate sacrifice.

Een vun villen Zaldoten deen säin Liewen
während der Ardennenschluecht fir eis
Fräiheet geloos huet
Mir wärten dech ni vergiessen.
Seng Famill an all d'Lëtzebuerger

Luxembourg Inscription is the same as the English paragraph above it.

Schumannseck (National Liberation Memorial)

National Liberation Memorial dedicated to all Allied Forces engaged in Luxembourg 1944-1945 and specifically to the units that fought in the area in Dec 1944 and January 1945: 28th US Infantry Division, 26th US Infantry Division, 90th US Infantry Division, and the 6th US Cavalry Group.

Intersection of N 26 and N15

Near Nothum

4546 Wilz

Latitude: 49.9469

Longitude: 5.8812

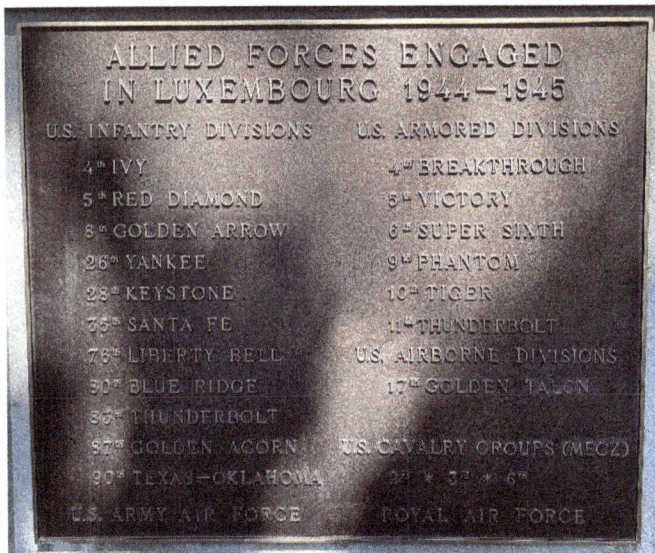

From this monument you can take a walk through a series of US and German foxholes used in the defense/attack of this area (see trails section).

Also at the site is a German 7.5 cm PAK 40 anti-tank gun with the muzzle brake missing.

Information Panel Schumannseck

HELL AT SCHUMANNS ECK

The BATTLE of the BULGE

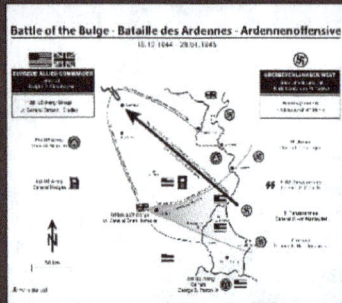

On **December 16, 1944**, Hitler launched a surprise attack against the US troops holding the front from Echternach (L) to Monschau (Germ.). This became known as the Battle of the Bulge. It was intended to reach the port of Antwerp (B) within three days, to split the British and US forces and to urge the Allies to an armistice. The stiff resistance of the US defenders thwarted that plan and allowed General George S. Patton's (Jr.) Third Army to liberate the encircled town of Bastogne (B) on 26th of December 1944 and to attack the Germans on the southern flank.

Le **16 décembre 1944**, Hitler lança une attaque surprise contre les soldats américains défendant le front d'Echternach (L) jusqu'à Monsjoie (Allem.). C'était la Bataille des Ardennes qui avait pour but de s'emparer du port d'Anvers (B) dans trois jours, de séparer les troupes britanniques et américaines et de forcer les Alliés à signer un armistice. L'esprit de combat des défenseurs américains prévenait ce plan et permettait à la Troisième Armée du Général George S. Patton Jr. de libérer la ville de Bastogne (B) encerclée le 26 décembre 1944 et d'attaquer les Allemands par le flanc sud.

Am **16. Dezember 1944** startete Hitler einen Überraschungsangriff gegen die amerikanischen Soldaten, die von Echternach (L) bis Monschau (D) die Front hielten. Es war die Ardennen-Offensive, die zum Ziel hatte, den Hafen von Antwerpen (B) binnen drei Tagen zu erobern, die amerikanischen und britischen Truppen zu spalten und die Alliierten zu einem Waffenstillstand zu zwingen. Die Entschlossenheit der amerikanischen Verteidiger machte den Plan zunichte und gab der Dritten US Armee von General George S. Patton Jr. die Möglichkeit, die eingekesselte Stadt Bastnach (B) am 26. Dezember 1944 zu befreien und die Deutschen an der Südflanke anzugreifen.

The DEFENSE of WILTZ
December 19, 1944

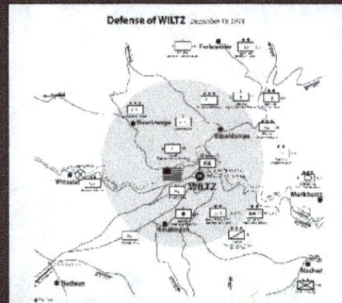

At Wiltz, on **December 17, 1944**, at about 4:00 p.m., General Norman D. Cota ordered the 28th US Infantry Division's headquarters to be moved to Sibret (Belgium). The town of Wiltz was defended by a "Provisional Battalion" (including all the personnel such as the bandsmen, the cooks and the military police), the 600 men of the 44th Engineer Combat Battalion, the remainders of the 707th Tank Battalion, the 447th Anti Aircraft Battalion and the 687th Field Artillery Battalion. On Dec 19, Colonel Daniel B. Strickler retreated with the remains of the 3rd Battalion, 110th Regiment, 28th US Infantry Division, totalling 250 men, to Wiltz. He assumed responsibility for the defense of Wiltz.

A Wiltz, le **17 décembre 1944**, vers 16:00 heures, le Général Norman D. Cota ordonna que le quartier général de la 28e Division d'Infanterie américaine fût transféré à Sibret (B). La ville de Wiltz était défendue par un « Provisional Battalion » (Bataillon Provisoire) qui comprenait le personnel non combattant de la division tel que les musiciens, les cuisiniers et la police militaire, les 600 hommes du 44e Bataillon de Combat du Génie, les restes du 707e Bataillon de Chars, le 447e Bataillon Anti Aérien et le 687e Bataillon d'Artillerie de Campagne. Le 19 décembre, le Colonel Daniel B. Strickler se retira avec les restes du 3e Bataillon, 110e Régiment, 28e Division d'Infanterie Américaine comptant 250 hommes sur Wiltz. La défense de Wiltz lui fut confiée.

In Wiltz, am **17. Dezember 1944** gegen 16:00 Uhr, befahl General Norman D. Cota, dass das Hauptquartier der 28. US Infanteriedivision nach Sibret (B) verlegt werden sollte. Die Stadt Wiltz wurde von einem „Provisional Battalion" (Provisorisches Bataillon), das aus dem nicht kämpfenden Personal der Division (Musiker, Köche, Militärpolizei) zusammengesetzt wurde, von den 600 Mann des 44. Pionierbataillons, den Resten des 707. Panzerbataillons, des 447. Flugabwehrbataillons und des 687. Feldartilleriebataillons, verteidigt. Am 19. Dezember zog sich Colonel Daniel B. Strickler mit den Resten des 3. Bataillons, 110. Regiment, 28. US Infanteriedivision, das 250 Mann zählte, nach Wiltz zurück. Die Verteidigung der Stadt Wiltz wurde unter sein Kommando gestellt.

US RETREAT from WILTZ
December 19, 1944

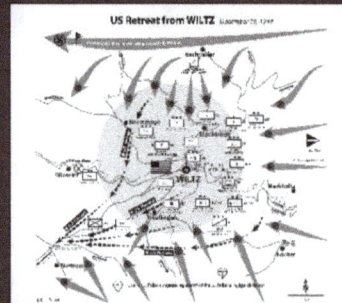

On **December 19**, at 2:00 p.m., the Germans attacked Wiltz from three sides and succeeded in penetrating the center of the town. The Americans defended their positions until they ran out of ammunition. In the evening, Col. Strickler gave the order to retreat in small groups and try to reach Sibret and Bastogne, Belgium. As the town of Wiltz had meanwhile been encircled, many American soldiers were killed, hurt or taken prisoners by using the road via Café Schumann and the Poteau of Harlange or via the Poteau of Doncols to reach Belgium.
110th Regiment:
On December 15, 1944: 150 officers and 2.823 soldiers
On December 23, 1944:   55 officers and   730 soldiers

Le **19 décembre** à 14:00 heures, les Allemands attaquèrent Wiltz de trois côtés et réussirent à pénétrer dans le centre de la ville. Les Américains défendaient leurs positions jusqu'à la dernière cartouche. Le soir, le Colonel Strickler donna l'ordre de se retirer en petits groupes et de rejoindre Sibret et Bastogne (Belgique). La ville de Wiltz ayant été encerclée entretemps, beaucoup de soldats américains furent tués, blessés ou faits prisonniers en prenant le chemin pour la Belgique en passant par le carrefour Schumann et le Poteau de Harlange ou par le Poteau de Doncols.
110e Régiment:
Le 15 décembre 1944 : 150 officiers et 2.823 soldats
Le 23 décembre 1944 :   55 officiers et   730 soldats

Am **19. Dezember**, um 14:00 Uhr, griffen die Deutschen Wiltz von drei Seiten an und es gelang ihnen, bis ins Zentrum der Stadt vorzudringen. Die Amerikaner verteidigten ihre Stellungen bis zur letzten Patrone. Am Abend gab Colonel Strickler den Befehl, sich in kleinen Gruppen nach Sibret und Bastnach, Belgien, zurückzuziehen. Da Wiltz inzwischen eingekesselt war, wurden viele Amerikaner getötet, verletzt oder gefangen genommen, als sie versuchten, Belgien über die Kreuzung Café Schumann und Poteau von Harlingen oder über den Poteau von Doncols zu erreichen.
110. Regiment:
Am 15. Dez. 1944: 150 Offiziere und 2.823 Soldaten
Am 23. Dez. 1944:   55 Offiziere und   730 Soldaten

© National Liberation Memorial a.s.b.l.

Information Panel Schumannseck

② REMEMBER SCHUMANNS ECK

3ʳᴰ ARMY's COUNTERATTACK

December 22 - 25, 1944   **December 24 - 27, 1944**

2nd Liberation of Luxembourg by US Troops

2nd Liberation of Luxembourg by US Troops

On December 22, 1944, at 06:00 a.m., General George S. Patton's (Jr.) 3ʳᵈ Army attacked the German southern flank with three divisions abreast. The 4ᵗʰ US Armored Division was to advance up the Arlon - Bastogne highway with the 26ᵗʰ US Infantry Division ("Yankee Division") in the center and the 80ᵗʰ US Inf Div on the right. The first contacts with the enemy took place at Martelange, Rindschleiden, Rambrouch, Grosbous and Ettelbruck. – On the following days, the enemy resisted vigorously. The 4ᵗʰ US Armored Division captured Martelange and Bigonville. The 26ᵗʰ US Infantry Division pushed the enemy out of Koetschette, Rambrouch, Wahl, Grosbous, Dellen and Buschrodt. The 80ᵗʰ US Infantry Division succeeded in advancing up to Merzig, Merscheid, Heiderscheid, Heiderscheidergrund and to liberate Ettelbruck on Dec 25. – In order to have at hand a rapid and powerful armor-infantry team, "Task Force Hamilton" was organized. It consisted of elements of the 328ᵗʰ Regiment, 26ᵗʰ Division, Company C of 735ᵗʰ Tank Battalion, one Platoon of 818ᵗʰ Tank Destroyer Battalion, one Section of 390ᵗʰ Anti Aircraft Artillery Battalion, one Section of 101ˢᵗ Engineer Combat Battalion.

Through the 24ᵗʰ and 25ᵗʰ of December the progress of the attacking 3ʳᵈ US Army was delayed by the recurrent counter-attacks and the stiff resistance by the enemy. Heavy fighting took place in the sector of the 26ᵗʰ US Infantry Division in Arsdorf and especially in Eschdorf. The battle raged for two days and nights, and the latter town was reduced to rubble. These strategic positions enabled the 26ᵗʰ US Infantry Division to cross the Süre River on the following day near Lultzhausen in spite of a strong enemy resistance. – On Dec 26, the 4ᵗʰ US Armored Div made contact with the defenders of Bastogne. – The battlefield between the 4ᵗʰ US Armored Division and the 26ᵗʰ US Infantry Division was assigned to the 6ᵗʰ US Cavalry Group. On Dec 27, the 35ᵗʰ US Infantry Division attacked through the 6ᵗʰ US Cavalry Group and seized Surré, Boulaide and Baschleiden. The 26ᵗʰ US Infantry Division consolidated the bridgehead on the other side of the Süre River, and, after the capture of Liefrange, advanced to the high grounds to conquer Mecher and Kaundorf. The 80ᵗʰ US Infantry Division still held up by the enemy, was not able to establish contact with the right wing of the 104ᵗʰ Regiment of the 26ᵗʰ US Infantry Division.

Le 22 décembre 1944, à 06:00 heures, la 3ᵉ Armée du Général George S. Patton Jr. attaqua le flanc sud des Allemands avec trois divisions de front. La 4ᵉ Division Blindée US devait avancer sur l'axe Arlon – Bastogne avec la 26ᵉ Division d'Infanterie US ("Yankee Division") au milieu et la 80ᵉ Division d'Infanterie US à sa droite. Les premiers contacts avec l'ennemi se firent à Martelange, Rindschleiden, Rambrouch, Grosbous et Ettelbruck. – Les jours suivants, l'ennemi résista énergiquement. La 4ᵉ Division Blindée US captura Martelange et Bigonville. La 26ᵉ Division d'Infanterie US poussa l'ennemi hors de Koetschette, Rambrouch, Wahl, Grosbous, Dellen et Buschrodt. La 80ᵉ Division d'Infanterie US réussit à avancer jusqu'à Merzig, Merscheid, Heiderscheid, Heiderscheidergrund et à libérer Ettelbruck le 25 décembre. – Afin d'avancer le plus vite possible jusqu'à la rivière Süre, « Task Force Hamilton » fut constituée. Elle était composée d'éléments du 328ᵉ Régiment de la 26ᵉ Division d'Infanterie US, de la Compagnie C du 735ᵉ Bataillon de chars, d'un peloton du 818ᵉ Bataillon blindé antichars, d'une section du 390ᵉ Bataillon anti-aérien, d'une section du 101ᵉ Bataillon de Combat du Génie.

Le 24 et 25 décembre, l'avance de la 3ᵉ Armée américaine fut retardée par les contre-attaques répétées et la résistance farouche de l'ennemi. De rudes combats eurent lieu dans le secteur de la 26ᵉ Division d'Infanterie US à Arsdorf et à Eschdorf. La bataille sévissait pendant deux jours et nuits et réduisit le dernier village en un tas de pierres. Grâce à la conquête de ces positions stratégiques, la 26ᵉ Division d'Infanterie US eut la possibilité de traverser la rivière Süre à Lultzhausen le jour suivant, malgré une résistance farouche de l'ennemi. – Le 26 décembre, la 4ᵉ Division Blindée US brisa l'encerclement de Bastogne et prit contact avec les défenseurs de la ville. – Le champ de bataille entre la 4ᵉ Division Blindée US et la 26ᵉ Division d'Infanterie US fut assigné au 6ᵉ Groupement de Cavalerie US. Le 27 décembre, la 35ᵉ Division d'Infanterie US attaqua à travers les lignes du 6ᵉ Groupement de Cavalerie US et saisit Surré, Boulaide et Baschleiden. La 26ᵉ Division d'Infanterie US consolida la tête de pont de l'autre côté de la Süre, et, après la prise de Liefrange, avança vers les hauteurs pour occuper les villages de Mecher et de Kaundorf. La 80ᵉ Division d'Infanterie US, encore retenue par l'ennemi, ne pouvait établir le contact avec le flanc droit du 104ᵉ Régiment de la 26ᵉ Division d'Infanterie US.

Am 22. Dezember 1944, um 06:00 Uhr, griff General George S. Patton (Jr) mit seiner Dritten Armee die Südflanke der Deutschen mit drei Divisionen an. Die 4. US Panzerdivision folgte der Achse Arlon – Bastnach mit der 26. US Infanteriedivision ("Yankee Division") in der Mitte und der 80. US Infanteriedivision an der rechten Seite. Die ersten Berührungen mit dem Feind waren in Martelingen, Rindschleiden, Rambrouch, Grosbous und Ettelbrück. – An den folgenden Tagen, wehrte sich die Deutschen mit allen Kräften. Die 4. US Panzerdivision besetzte Martelingen und Bigonville. Die 26. US Infanteriedivision warf den Feind aus Koetschette, Rambrouch, Wahl, Grosbous, Dellen und Buschrodt. Die 80. US Infanteriedivision stieß nach Merzig, Merscheid, Heiderscheid, Heiderscheidergrund vor und befreite Ettelbrück am 25. Dezember. – Um nun möglichst schnell einen Angriffskeil bis zur Sauer vorantreiben zu können, wurde „Task Force Hamilton", eine schnelle, schlagkräftige Panzer-Infanterie-Einheit, zusammengestellt. Sie bestand aus Einheiten des 328. Regimentes der 26. US Infanteriedivision, der Kompanie C des 735. Tankbataillons, einem Zuge des 818. Panzerzerstörerbataillons, einer Sektion des 390. Flugabwehrbataillons, einer Sektion des 101. Pionierkampfbataillons.

Am 24. und 25. Dezember wurde der Vorstoß der 3. US-Armee durch die wiederholten Gegenangriffe und den entschlossenen Widerstand des Feindes gebremst. Im Bereich der 26. US Infanteriedivision fanden schwere Kämpfe in Arsdorf und besonders in Eschdorf statt. Die Schlacht dauerte zwei Tage und Nächte und legte das Dorf in Schutt und Asche. Die eroberten strategischen Stellungen gaben der 26. US Infanteriedivision die Möglichkeit, am folgenden Tag den Fluss Sauer, trotz des massiven Widerstandes des Feindes, bei Lultzhausen zu überqueren. – Am 26. Dezember brach die 4. US Panzerdivision den Ring um Bastnach und nahm Kontakt mit den Verteidigern der Stadt auf. – Das Kampfgebiet zwischen der 4. US Panzerdivision und der 26. US Infanteriedivision wurde der 6. US Kavalleriegruppe zugeteilt. Am 27. Dezember griff die 35. US Infanteriedivision durch die Linien der 6. US Kavalleriegruppe an und besetzte Syr, Bauschleiden und Baschleiden. Die 26. US Infanteriedivision festigte den Brückenkopf am anderen Sauerufer und drang nach der Einnahme von Liefringen auf den Höhenrücken vor, um anschließend die Orte Mecher und Kaundorf zu erobern. Die 80. US Infanteriedivision, die immer noch vom Feind aufgehalten wurde, konnte den Anschluss zum rechten Flügel des 104. Regimentes der 26. US Infanteriedivision nicht herstellen.

Information Panel Schumannseck

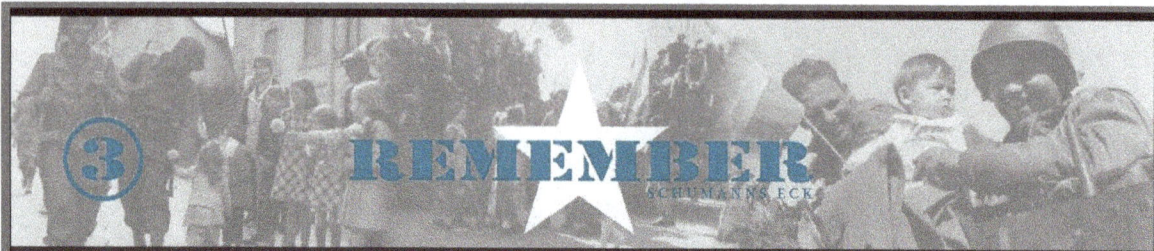

BATTLE of SCHUMANNS ECK

„Night Attack" on Nothum
December 27, 1944

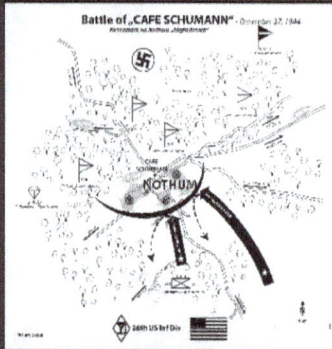

During the night of December 27, 1944, units of the 101st Regiment, 26th US Infantry Division attacked without any success the village of Nothum. The troops had to retreat to their initial jump-off positions after 11:00 p.m.

Dans la nuit du 27 décembre 1944, des unités du 101e Régiment de la 26e Division d'Infanterie US attaquèrent le village de Nothum sans succès. Elles durent se replier après 23 heures sur leurs points de départ.

In der Nacht vom 27. Dezember 1944, griffen Einheiten des 101. Regiments der 26. US Infanteriedivision das Dorf Nothum ohne Erfolg an. Sie mussten sich nach 23 Uhr in ihre Ausgangspositionen zurückziehen.

Second attack on Nothum
December 28, 1944, 07:00 a.m.

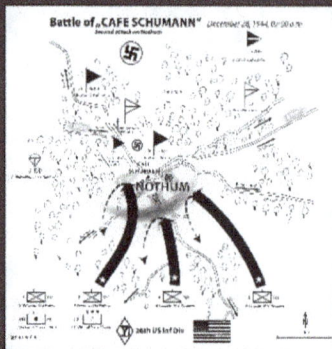

On Dec 28, at 7:00 a.m., the U.S. attack was resumed under heavy supporting artillery fire. But the German soldiers of the 1st Battalion of the 36th Grenadier Regiment of the 9th Volks-Grenadier-Division (VGD Colonel Kolb) with the support of the Panzerjägerabteilung 9 (Tank Destroyer Unit 9) and parts of the Sturmartillerie-Brigade 911 (brigade of assault guns) of the Führer-Grenadier-Brigade (FGB) prevented the American advance. Due to the artificial fog screen laid by the 4.2 Chemical Mortar Platoon, F Company retreated with many casualties. So did E and G Company. The junction with the 1st Battalion operating on the west side of the road to Bavigne had not been successful.

Le 28 décembre, à 7:00 heures, l'attaque américaine fut reprise sous le couvert d'un feu d'artillerie nourri. Mais les soldats du 1er Bataillon du 36e Grenadier Regiment de la 9e Volks-Grenadier-Division (VGD – Oberst Kolb) avec le support de la Panzerjägerabteilung 9 (unité de chasseurs de chars) et des unités de la Sturmartillerie-Brigade 911 (brigade de chars d'assaut) de la Führer-Grenadier-Brigade (FGB) repoussèrent l'avance américaine. Grâce au brouillard artificiel produit par le 4.2 Chemical Mortar Platoon, la Compagnie F se retira avec beaucoup de pertes. Les compagnies E et G firent de même. La jonction avec le 1er Bataillon opérant à l'ouest de la route de Bavigne n'a pas été couronnée de succès.

Am 28. Dezember, um 7:00 Uhr, wurde der amerikanische Angriff mit schwerer Artillerieunterstützung wiederholt. Die Soldaten des 1. Bataillons des 36. Grenadier Regimentes der 9. Volks-Grenadier-Division (VGD – Oberst Kolb) wehrten mit Feuerunterstützung durch die Panzerjägerabteilung 9 und Teile der Sturmartillerie-Brigade 911 der Führer-Grenadier-Brigade (FGB), den amerikanischen Angriff ab. Erst als das 4.2 Chemical Mortar Platoon die feindlichen Stellungen eingenebelt hatte, konnte sich die Kompanie F unter hohen Verlusten zurückziehen. Die Kompanien E und G taten dasselbe. Der Anschluss an das 1. Bataillon, das westlich der Straße nach Bavigne kämpfte, hatte nicht geklappt.

Third attack on Nothum
December 28, 1944, 04:30 p.m.

On Dec 28, at 4:30 p.m., after an intense artillery and mortar fire, US tanks and tank destroyers supported the infantry's attack on Nothum. Unfortunately a tank killed seven of its own troops unintentionally. Finally, after heavy fighting, F Company reached the cemetery, E Company the village of Nothum, G Company was stopped by heavy German fire coming from Café Schumann and retreated to Nothum. 3rd Battalion, delayed by a German counter-attack at Kaundorf, advanced to Nothum and ordered its engineer platoon to demine the road up to Café Schumann. However, strong German fire from Café Schumann stopped this action. – The losses of both sides were considerable. At its first engagement, the 1st Battalion of the 36th Grenadier Regiment had about 40 % of losses. – About midnight a German counter-attack was announced for the next day.

Le 28 décembre, à 16:30 heures, après un feu intense d'artillerie et de mortier et supportées par des unités de chars, les troupes américaines reprirent leur attaque sur Nothum. Malheureusement, un char ami tua sept de ses propres hommes. Après de lourds combats, Compagnie F atteignit le cimetière, Compagnie E entra dans le village de Nothum, Compagnie G fut arrêtée devant le Café Schumann et dut se replier sur Nothum. Le 3e Bataillon, retardé par une contre-attaque allemande sur Kaundorf, avança sur Nothum et ordonna à son peloton de pionniers de déminer la route menant au Café Schumann. Toutefois, un feu bien nourri venant du Café Schumann arrêta cette action. – Les pertes des deux côtés furent considérables. Par ce premier engagement, le 1er Bataillon du 36e Régiment de Grenadiers déployait à peu près 40 % de pertes. – Vers minuit, une contre-attaque allemande fut annoncée pour le lendemain.

Am 28. Dezember, um 16:30 Uhr, nach intensivem Artillerie- und Granatwerferbeschuss und von Panzern und Panzerzerstörern unterstützt, nahmen die amerikanischen Truppen den Angriff auf Nothum wieder auf. Unglücklicherweise tötete ein US Panzer sieben seiner eigenen Leute. Nach heftigen Kämpfen erreichte Kompanie F den Friedhof nördlich von Nothum, Kompanie E drang in das Dorf ein, Kompanie G wurde durch heftiges deutsches Feuer aus dem Café Schumann gestoppt und zog sich nach Nothum zurück. – Das 3. Bataillon, das durch einen deutschen Gegenangriff auf Kaundorf aufgehalten worden war, erreichte ebenfalls Nothum und befahl seinem Pionierzug, die Straße bis zum Café Schumann zu entminen. Heftiges Infanteriefeuer vom Café Schumann kommend, stoppte diese Aktion. Die Verluste waren auf beiden Seiten erheblich. So hatte das 1. Bataillon des 36. Grenadier Regimentes bereits durch diesen ersten Einsatz etwa 40 % an Verlusten zu beklagen. – Gegen Mitternacht wurde ein deutscher Gegenangriff für den folgenden Tag angesagt.

Information Panel Schumannseck

BATTLE of SCHUMANNS ECK

December 29, 1944, 01:30 p.m.	December 30, at 10:45 a.m.	December 31, 1944, 04:00 a.m.

December 29: In order to forestall the German attack, the 2nd Battalion retreated from Nothum to its jump-off positions of the previous day at 4:00 a.m. The 3rd Battalion was ordered back to the defense of Kaundorf. The 9th Volks-Grenadier-Brigade had nearly completely arrived, and the worn-out Führer-Grenadier-Brigade, except the Grenadier Battalion 829 at Berlé, could be retreated from the front. – However, the announced German attack did not take place. At 1:00 p.m., the US troops re-conquered Nothum without any losses. The 3rd Battalion advanced again starting from Kaundorf. K and I. Companies tried to turn the enemy's flank east of Nothum by following the "Schlirbach" brook valley and were to meet I Company at Café Schumann. I Company fell under enemy fire and swerved east. K Company succeeded in occupying the hill 490 on the way to Roullingen. Company L secured the right flank, advanced in the direction of Roullingen where it established the contact with L Company of the 104th US Infantry Regiment, 26th US Infantry Division. In spite of being far away from its initial objective, the Schumann crossroads, the 3rd Battalion held this important strategic position near Roullingen in the direction of Wiltz.

29 décembre : Les troupes américaines du 2e Bataillon, réoccupèrent Nothum sans pertes. – Les compagnies K et L du 3e Bataillon contournèrent les positions ennemies à l'est de Nothum. Compagnie I tomba sous le feu ennemi et s'écarta vers l'est. Compagnie K occupa la colline 490 sur la route de Roullingen. Compagnie L garantit le flanc droit. Le 3e Bataillon gardait cette importante position-clef vers Wiltz.  *

29. Dezember: Die amerikanischen Truppen des 2. Bataillons nahmen Nothum ohne Verluste wieder ein. – Die Kompanien K und L des 3. Bataillons umgingen die deutschen Stellungen östlich von Nothum. Kompanie I geriet unter feindliches Feuer und wich nach Osten aus. Kompanie K besetzte den Hügel 490 an der Roullinger Straße. Kompanie L sicherte die rechte Flanke. Das 3. Bataillon hielt diese wichtige Schlüsselstellung nach Wiltz. *

On December 30, at 10:45 a.m., the 3rd Battalion, close to the bifurcation to Roullingen, advanced into a gap between the 57th Grenadier Regiment and the 36th Grenadier Regiment of the 9th Volks-Grenadier-Division. The road to Wiltz was blocked; the German positions towards the Schumann crossroads and the road towards the village of Roullingen were overrun. The commander of the 1st Battalion, 116th Grenadier Regiment was taken prisoner with his staff. – At 11.15 a.m., after a violent artillery barrage and with the support of heavy machine gun fire, F Company of the 2nd Battalion took the group of houses around Café Schumann in a surprise attack. E Company followed and many German soldiers surrendered. As G Company had missed the objective and had not linked up with the 3rd Battalion, the two companies could not continue their attack to hill 506 ahead of Berlé. They dug their foxholes along the road and covered them with wooden beams they found in the houses. – During the night, many German patrols tried to spy out the American strength.

Le 30 décembre, le 3e Bataillon, avança dans une brèche entre le 57e Grenadier Regiment et le 36e Grenadier Regiment de la 9e Volks-Grenadier-Division. La route vers Wiltz fut bloquée, les positions allemandes vers le carrefour Schumann et vers le village de Roullingen furent conquises. Le commandant du 1er Bataillon, 116e Grenadier Regiment, fut fait prisonnier avec son état-major. – Compagnie E du 2e Bataillon conquit les maisons autour du Café Schumann. Compagnie G, n'ayant pu contacter le 3e Bataillon, celui-ci ne pouvait continuer son attaque. Les soldats se retranchèrent dans leurs trous de fusiliers creusés le long de la route.  *

Am 30. Dez. stieß das 3. Bataillon in eine Frontlücke zwischen dem 57. GrenRgt. und dem 36. GrenRgt. der 9. VGD vor. Die Straße nach Wiltz wurde abgeschnitten, die deutschen Positionen zur Kreuzung Schumann und nach Roullingen wurden aufgerollt. Der Kommandeur des 1. Bataillons, 116. GrenRgt, geriet mit seinem Stab in Gefangenschaft. Die F Kompanie, 2. Bataillon, eroberte die Häuserreihe um das Café Schumann. Das 3. Bataillon konnte seinen Angriff nicht fortsetzen, da keine Verbindung zur G Kompanie bestand. *

On December 31, the German armed forces high command had not yet abandoned the objective of capturing Bastogne and attached the 1st SS-Panzerkorps (with the 9th and the 12th SS-Panzerdivision) to the 5th Panzerarmee. To prevent that the German troops at Bastogne could be cut off, it was a high strategic necessity that the front around Nothum had to be defended at all costs. – At 4:00 a.m., after heavy artillery and mortar shelling, the Germans attacked and re-conquered the American positions near Roullingen, hill 490. All day long, the American artillery fired over 15,000 shells on the German front as well as on the 9th Volks-Grenadier-Division's headquarters at Noertrange. A second defense line was established north of Mecher. The US mortar- and MG-platoons had many casualties. The Germans attacked three times during the night and broke through the first defense line. Only US artillery fire on the own lines stopped them.

Le 31 décembre, les Allemands reconquirent les positions américaines. Pendant toute la journée, l'artillerie américaine bombarda le front allemand ainsi que le quartier général de la 9e Volks-Grenadier-Division à Noertrange avec plus de 15.000 obus. Une deuxième ligne de défense fut établie au nord de Mecher. Les Allemands attaquèrent trois fois pendant la nuit et envahirent la première ligne de défense. Les Américains ne purent les arrêter que par un feu d'artillerie sur leurs propres lignes.  *

Am 31. Dezember eroberten die Deutschen die amerikanischen Stellungen wieder. Während des ganzen Tages feuerte die amerikanische Artillerie über 15.000 Granaten auf die deutsche Front sowie auf das Hauptquartier der 9. Volks-Grenadier-Division in Noertringen. Eine zweite Verteidigungslinie wurde nördlich von Mecher eingerichtet. Während der Nacht griffen die Deutschen dreimal an und durchbrachen die erste Verteidigungslinie. Nur amerikanisches Artilleriefeuer auf die eigenen Linien konnte sie stoppen. *

Information Panel Schumannseck

⑤ REMEMBER SCHUMANNS ECK

BATTLE OF SCHUMANNS ECK

January 1st, 1945

Battle of „CAFE SCHUMANN" January 1, 1945

🇺🇸 At 02:00 a.m., on **January 1st, 1945**, the Americans progressively began replacing the 101st Infantry Regiment by units of the 328th Infantry Regiment. – At 08:56, the Germans again attacked the crossroads with the support of Hetzer-tanks of the Tank Destroyer Unit of the 9th Volks-Grenadier-Division. The attack was however stopped by US artillery fire. After three days of fighting under the most adverse conditions, the units on both sides were worn out. Nevertheless the American command wasn't discouraged and planned a new attack on hill 490 near Berlé for the next day.

🇫🇷 À 02:00 heures du matin du **1er janvier 1945**, les Américains commencèrent à remplacer progressivement le 101e Régiment d'Infanterie par des unités du 328e Régiment d'Infanterie. À 8:56 heures les Allemands attaquèrent de nouveau le carrefour avec le support de chars Hetzer de l'unité des chasseurs de chars de la 9e Volks-Grenadier-Division. Toutefois, un feu d'artillerie nourri américain arrêta l'attaque. Les soldats des unités des deux camps étaient à bout de forces après trois jours de combat sous des conditions inhumaines. Néanmoins, le commandement américain ne se découragea pas et planifiait une nouvelle attaque sur la hauteur 490 près de Berlé pour le lendemain.

🇩🇪 Am Morgen des **1. Januar 1945**, um 02:00 Uhr, begannen die Amerikaner damit, das 101. Infanterie Regiment progressiv durch Einheiten des 328. Infanterie Regiments abzulösen. – Um 8:56 Uhr griffen die Deutschen mit Unterstützung von „Hetzer"-Panzern der Panzerjägerabteilung der 9. Volks-Grenadier-Division das Straßenkreuz erneut an. Der Angriff kam jedoch unter amerikanischem Artilleriefeuer zum Erliegen. Nach drei Tagen Kampf unter unmenschlichen Bedingungen, waren die Männer beider Seiten erschöpft. Trotzdem sollte sich die amerikanische Führung nicht entmutigen und ein neuer Angriff auf die Höhe 490 bei Berl wurde für den folgenden Tag geplant.

January 2, 1945

Battle of „CAFE SCHUMANN" January 2, 1945, 06:00 a.m.

🇺🇸 On **January 2, 1945, at 6:00 a.m.**, the remains of E, F and G Companies should have conquered hill 490 (Berlé). 200 meters from their starting point, heavy German artillery, assault gun and a quadruple 2 cm anti-aircraft gun forced the Americans to retreat to their starting point and pinned them down. – The following day, the 65 remaining soldiers of the initial 871 men of the 2nd Battalion, 101st US Infantry Regiment, were replaced by the 328th Regiment of the 26th US Infantry Division. – The following days, "Task Force Scott" was formed by the soldiers still fit for combat of the 101st Infantry Regiment who, together with "Task Force Ficket" of the 6th Cavalry Group, should roll up the Pocket of Harlange by the south attacking from the Poteau of Harlange.

🇫🇷 Le **2 janvier 1945, à 6:00 heures**, les restes des compagnies E, F et G auraient dû prendre la colline 490 (Berlé). À 200 mètres de leurs positions de départ, le feu nourri de l'artillerie, de char d'assaut et de canon anti-aérien-quadruplé allemand força les Américains à se retirer à leur point de départ et les y cloua au sol. – Le lendemain, les 65 soldats restants des 871 hommes qu'avait comptés le 2e Bataillon, 101e Régiment d'Infanterie, au début des hostilités, furent remplacés par le 328e Régiment de la 26e Division d'Infanterie US. – Les jours suivants, la Task Force Scott fut mise en place avec les soldats encore aptes au combat du 101e Régiment d'Infanterie qui, ensemble avec la Task Force Ficket du 6e Groupe de Cavalerie, devaient éliminer la poche de Harlange par le sud, à partir du Poteau de Harlange.

🇩🇪 Am **2. Januar 1945, um 6:00 Uhr**, sollten die Reste der Kompanien E, F und G die Anhöhe 490 (Berl) einnehmen. 200 Meter von ihren Ausgangsstellungen entfernt, zwang schwere deutsche Artillerie-, Sturmgeschütz- und Vierlingsflakfeuer die Amerikaner sich zu ihrem Anfangspunkt zurückzuziehen und hielten sie nieder. – Am nächsten Tag wurden die restlichen 65 Soldaten des anfangs 871 Mann zählenden 2. Bataillon, 101. Infanterie Regiment, durch das 328. Regiment der 26. US Infanterie Division ersetzt. – Aus den noch kampffähigen Soldaten des 101. Infanterie Regimentes wurde in den folgenden Tagen die Task Force Scott aufgestellt, die zusammen mit der Task Force Ficker der 6. Kavalleriegruppe, vom Poteau von Harlingen aus, den Kessel von Harlingen von Süden her aufrollen sollten.

January 3 to 6, 1945

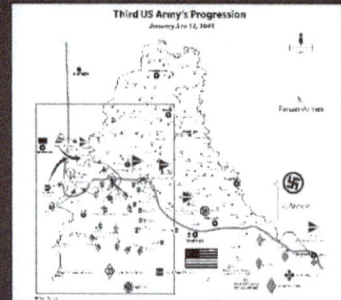

Third US Army's Progression January 3 to 12, 1945

🇺🇸 **January 3 to 6, 1945**. The American high command intended to cut off the German elite troops fighting in the "Pocket of Harlange" by attacking simultaneously from the west and the south-east in order to unify the troops at the important crossroad of Doncols. On January 4, the commander of the 1st Battalion, 36th Grenadier Regiment, was taken prisoner with his staff. Up to the 6th of January, the 328th US Infantry Regiment tried in vain to force the breakthrough to the crossroad via the hill 490 (Berlé). The situation was comparable to a trench-war with terrible losses on both sides. – To weaken the German defense at Nothum, the 80th US Infantry Division crossed the Sûre River in the east and attacked Goesdorf and Dahl on **January 6**. The capture of Dahl was a new threat in the back of the German troops fighting around Wiltz. In order to clear up the danger at Dahl, Fieldmarshal Model ordered the Führer-Grenadier-Brigade from the area of Doncols-Grummelscheid to Nocher.

🇫🇷 **Du 3 au 6 janvier 1945**, la situation ressemblait à une guerre de tranchées. Le 328e Régiment d'Infanterie US ne réussit pas à percer jusqu'au carrefour par la hauteur 490 (Berlé). Le commandement américain voulut couper la retraite aux troupes allemandes qui combattaient dans la « Poche de Harlange » en les attaquant simultanément de l'ouest et du sud. – Pour affaiblir la défense allemande à Nothum, la 80e Division d'Infanterie US traversa la Sûre et attaqua Goesdorf et Dahl, le **6 janvier**. Pour contrecarrer cette nouvelle menace dans le dos des unités allemandes, la Führer-Grenadier-Brigade fut envoyée à Nocher. 🔲

🇩🇪 **Vom 3. – 6. Januar 1945** sah die Lage einem Stellungskrieg ähnlich. Dem 328. US Infanterie Regiment gelang es nicht, über die Höhe 490 (Berlé) bis zum Straßenkreuz vorzustoßen. Durch gleichzeitige Angriffe von Westen und Südosten strebte die US-Führung die Vereinigung im Bereich des wichtigen Straßenkreuzes von Doncols an, um hierdurch die deutschen Truppen im „Kessel von Harlingen" abzuschneiden. Um die deutsche Verteidigung bei Nothum zu schwächen, setzte die 80. US Infanteriedivision über die Sauer und nahm Goesdorf und Dahl am **6. Januar 1945** ein. Aufgrund dieser neuen Bedrohung im Rücken der deutschen Verbände, wurde die Führer-Grenadier-Brigade nach Nocher verlegt. 🔲

Information Panel Schumannseck

6 REMEMBER SCHUMANNSECK

THE LIBERATION

January 7 - 12, 1945

Winding up of the „Pocket of Harlange"
January 7- 12, 1945

January / February, 1945

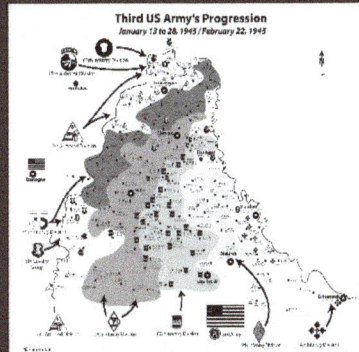

Third US Army's Progression
January 13 to 28, 1945 / February 22, 1945

During the period of January 5 to 9, the 26th US Infantry Division maintained defensive positions, while the 90th US Infantry Division prepared to attack through the 26th US Infantry Division's positions. Despite of the German 9th Volks-Grenadier-Division's and the 5th Parachute Division's fierce resistance, the 357th US Infantry Regiment of the 90th US Infantry Division succeeded in capturing Berlé on January 9 and Sonlez on January 10. – The main burden of the attack on Doncols lay in the sector of the 359th US Infantry Regiment that, indeed, succeeded in conquering hill 490 on January 9, but any further advance was annihilated by the German defense fire on Pommerloch. It was only after a risky night attack that the German defenses of the Poteau of Doncols could be by-passed and that the crossroad as well as the village of Doncols were conquered on January 11 by daylight. – The 6th US Armored Division occupied Wardin and Bras. "Task Force Fickett" of the 6th US Cavalry Group attacked Harlange, while "Task Force Scott" of the 26th US Infantry Division kicked the enemy out of the region of Harlange. – Even though the Doncols crossroad changed several times the occupiers due to the fierce counter-attacks of the 9th Volks-Grenadier-Division and the combat teams of the 1st SS-Panzerdivision "LAH" (Leibstandarte Adolf Hitler) and the Führer-Grenadier-Brigade, the "Pocket of Harlange" could definitely be eliminated on January 12 owing to the reunion of the 90th US Infantry Division with the 6th US Armored Division and the 35th US Infantry Division. A large part of the 5th Fallschirmjägerdivision was broken up and many of its soldiers were taken prisoners. In this way the fighting around the heights of Nothum - Doncols came to an end. – Never in History, so many horrors and sufferings had ever taken place in such a short period on the territory of Luxembourg.

Pendant la période du **5 au 9 janvier**, la 26e Division d'Infanterie US maintenait ses positions de défense. Le 357e Régiment d'Infanterie de la 90e Division d'Infanterie US attaqua à travers les lignes de la 26e Division d'Infanterie US et captura Berlé le **9 janvier** et Sonlez le **10 janvier**. – Le carrefour ainsi que le village de Doncols ont été conquis le **11 janvier** après des combats acharnés.- La 6e Division Blindée occupa Wardin et Bras. La « Task Force Fickett » du 6e Groupe de Cavalerie US attaqua Harlange tandis que la « Task Force Scott » de la 26e Division d'Infanterie US chassa l'ennemi de la région de Harlange. Malgré la résistance farouche des troupes allemandes, la « Poche de Harlange » a pu définitivement être éliminée le **12 janvier** grâce à la réunion de la 90e Division d'Infanterie US avec la 6e Division Blindée US et la 35e Division d'Infanterie US. [EN] *

Im Zeitraum vom **5. zum 9. Januar**, hielt die 26. US Infanteriedivision ihre Verteidigungspositionen. Das 357. Infanterie Regiment der 90. US Infanteriedivision griff durch die Linien der 26. US Infanteriedivision an und nahm Berlé am **9. Januar** und Sonlez am **10. Januar** ein. – Das Straßenkreuz sowie Doncols wurden am **11. Januar** nach heftigen Kämpfen eingenommen. – Die 6. US Panzerdivision besetzte Wardin und Bras. „Task Force Fickett" der 6. Kavalleriegruppe griff Harlingen an- während „Task Force Scott" der 26. US Infanteriedivision den Feind aus der Region von Harlingen warf. – Trotz des erbitterten Widerstandes der deutschen Truppen, konnte der Kessel von Harlingen durch die Vereinigung der 90. US Infanteriedivision mit der 6. US Panzerdivision und der 35. US Infanteriedivision am **12. Januar** endgültig abgeschnürt werden. [EN] *

During the week of January **13 to 20, 1945**, the front in the 26th US Infantry Division area remained stable, the 90th US Infantry Division made small gains, while the enemy, reinforced by the 1st SS-Panzerdivision, resisted vigorously. Meanwhile, by **January 20, 1945**, First Army Troops had pushed down from the north and had linked up with Third Army forces. They squeezed the German advance and drove the enemy back to the East. On **January 21**, two regiments of the 26th US Infantry Division crossed the Wiltz River, fought their way to Noertrange, penetrated into Wiltz and continued liberating the villages north of this town. Furthermore, the 5th and 4th US Infantry Divisions started their attacks north and east of Diekirch on **January 18** and threatened to cut the German supply and retreat routes. – Third Army's Infantry and Armored Divisions succeeded in liberating the north of Luxembourg up to **January 28, 1945**. The upper town of Vianden was liberated by the 1255th Engineer Combat Battalion on **February 12** and downtown by elements of the 6th Cavalry Group on **February 22**. The Battle of the Bulge came to an end and the American troops continued their way to the Rhine. **German losses : 26.000 dead soldiers, American losses : 19.000 dead soldiers, British losses: 200 dead soldiers.** (Source: Charles B. MacDonald)

Durant la semaine du **13 au 20 janvier**, le front dans le secteur de la 26e Division d'Infanterie US restait stable, la 90e Division d'Infanterie fit de légers gains, tandis que l'ennemi, renforcé par la 1ère SS-Panzerdivision, résista farouchement. Entretemps, vers le **20 janvier 1945**, les troupes de la 1ère Armée US avançant du nord, s'étaient jointes aux forces de la 3e Armée. Elles éranglèrent l'avance allemande et poussèrent l'ennemi vers l'est. Le **21 janvier**, deux régiments de la 26e Division d'Infanterie US traversèrent la rivière Wiltz, se frayèrent un chemin vers Noertrange, entrèrent dans la ville de Wiltz et continuèrent à libérer les villages au nord de cette ville. En plus, le **18 janvier**, les 5e et 4e Divisions d'Infanterie US commencèrent leurs attaques au nord et à l'est de Diekirch et menacèrent de couper les routes d'approvisionnement et de retrait des Allemands. – Les divisions d'infanterie et de blindés de la 3e Armée US réussirent à libérer les localités au nord du Luxembourg jusqu'au **28 janvier 1945**. – La Bataille des Ardennes était terminée et les troupes américaines continuèrent leur chemin vers le Rhin. **Pertes allemandes : 26.000 morts, pertes américaines : 19.000 morts, pertes britanniques : 200 morts.** (Source: Charles B. MacDonald) [EN] *

Während der Woche vom **13. bis zum 20. Januar** blieb die Front im Sektor der 26. US Infanteriedivision stabil, die 90. Infanteriedivision machte leichte Geländegewinne, während der Feind, verstärkt durch die 1. SS-Panzerdivision, sich hartnäckig wehrte. Inzwischen, um den **20. Januar 1945**, hatten sich die 1. und 3. US Armee vereint. Sie würgten den deutschen Vorstoß ab und drückten den Feind gegen Osten. Am **21. Januar** überquerten zwei Regimenter der 26. US Infanteriedivision den Fluss Wiltz, bahnten sich einen Weg nach Noertringen, nahmen Wiltz und fuhren fort, weitere Ortschaften nördlich dieser Stadt zu befreien. Des Weiteren begannen die 5. und 4. US Infanteriedivisionen am **18. Januar 1945** ihre Angriffe nördlich und östlich der Stadt Diekirch und drohten, die deutschen Verpflegungs- und Rückzugswege abzuschneiden. – Bis zum **28. Januar 1945**, befreiten die Infanterie- und Panzerdivisionen der 3. US Armee die Ortschaften im Norden Luxemburgs. Die Ardennenschlacht war beendet und die amerikanischen Truppen setzten ihren Weg zum Rhein fort. **Deutsche Verluste: 26.000 Tote, amerikanische Verluste: 19.000 Tote, britische Verluste: 200 Tote.** (Quelle: Charles B. MacDonald) [EN] *

Information Panel Schumannseck

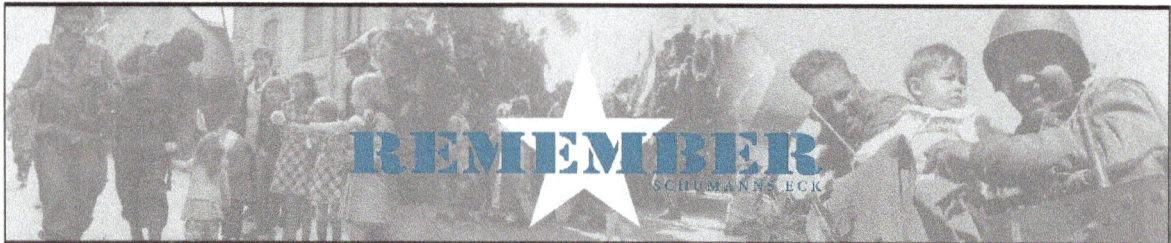

REMEMBER
SCHUMANNS ECK

DATES	TIME TABLE	CHRONOLOGIE	ZEITLEISTE
1839	Independency of the Grand Duchy of Luxembourg.	Indépendance du Grand-Duché de Luxembourg.	Unabhängigkeit des Großherzogtums Luxemburg.
1867	Treaty of London: Eternal neutrality of Luxembourg.	Traité de Londres: Neutralité éternelle du Luxembourg.	Londoner Vertrag : Ewige Neutralität Luxemburgs.
1870 – 1871	Franco-Prussian War. Luxembourg's neutrality is respected.	Guerre franco-allemande : la neutralité du Luxembourg est respectée.	Deutsch-Französischer Krieg : Luxemburgs Neutralität wird respektiert.
1914 – 1918	World War I. Imperial troops occupy Luxembourg.	1re Guerre Mondiale: des troupes allemandes occupent le Luxembourg.	Erster Weltkrieg : Deutsche Truppen besetzen Luxemburg.
Sept 01, 1939	Nazi Germany attacks Poland.	L'Allemagne nazie attaque la Pologne.	Nazi-Deutschland greift Polen an.
Sept 03, 1939	The United Kingdom and France declare war on Germany.	La Grande-Bretagne et la France déclarent la guerre à l'Allemagne.	Großbritannien und Frankreich erklären Deutschland den Krieg.
Sept 17, 1939	The Soviet Union attacks Poland. The two aggressors divide the country.	L'Union Soviétique attaque la Pologne. Les deux agresseurs partagent le pays.	Die Sowjetunion greift Polen an. Die beiden Aggressoren teilen das Land.
Apr 04, 1940	Germany attacks Denmark and Norway.	L'Allemagne attaque le Danemark et la Norvège.	Deutschland greift Dänemark und Norwegen an.
May 10, 1940	German troops invade the three neutral countries - the Netherlands, Belgium as well as Luxembourg - and attack France.	Des troupes allemandes agressent les trois pays neutres - les Pays-Bas, la Belgique ainsi que le Luxembourg - et attaquent la France.	Deutsche Truppen überfallen die drei neutralen Länder – die Niederlande, Belgien sowie Luxemburg – und greifen Frankreich an.
June 10, 1940	Italy declares war on the United Kingdom and France.	L'Italie déclare la guerre à la Grande-Bretagne et à la France.	Italien erklärt Großbritannien und Frankreich den Krieg.
June 22, 1940	France surrenders and is divided into an occupied and a free zone.	La France capitule et est divisée en une zone occupée et une zone libre.	Frankreich kapituliert und wird in eine besetzte und eine freie Zone eingeteilt.
July – Oct 1940	Germany massively bombs England. The Royal Air Force resists to the aggressor. Hitler abandons his plan to invade England.	L'Allemagne bombarde massivement l'Angleterre. Le Royal Air Force résiste à l'agresseur. Hitler abandonne son plan d'envahir l'Angleterre.	Deutschland bombardiert massiv England. Die Royal Air Force bietet dem Aggressor die Stirn. Hitler lässt den Plan, England einzunehmen, fallen.
Apr 06, 1941	Germany attacks Yugoslavia.	L'Allemagne attaque la Yougoslavie.	Deutschland greift Jugoslawien an.
Apr 09, 1941	Germany attacks Greece.	L'Allemagne attaque la Grèce.	Deutschland greift Griechenland an.
June 22, 1941	Germany attacks the Soviet Union.	L'Allemagne attaque l'Union Soviétique.	Deutschland greift die Sowjetunion an.
Oct 10, 1941	As the population is submitted under a census, most of the Luxembourg people state that they are Luxembourgers, that their language is Luxembourgish and that they belong to the people of Luxembourg.	La population luxembourgeoise étant soumise à un recensement de la population, la majorité des Luxembourgeois déclare qu'ils sont des Luxembourgeois, qu'ils parlent le luxembourgeois et qu'ils appartiennent au peuple luxembourgeois.	Bei einer Volkszählung, antwortet die überwiegende Mehrheit der Luxemburger auf die Fragen der Staatsangehörigkeit, der Muttersprache und der Volkszugehörigkeit, dass sie Luxemburger sind, luxemburgisch sprechen und zum Volk der Luxemburger gehören.
Dec 07, 1941	Japan destroys most of the U.S. Pacific fleet in Pearl Harbor, Hawaii.	Le Japon détruit la plus grande partie de la flotte américaine du Pacifique à Pearl Harbor, Hawaï.	Japan zerstört den größten Teil der amerikanischen Pazifikflotte in Pearl Harbor, Hawaii.
Dec 08, 1941	The USA and the United Kingdom declare war on Japan.	Les USA et la Grande-Bretagne déclarent la guerre au Japon.	Die USA und Großbritannien erklären Japan den Krieg.
Dec 11, 1941	Germany and Italy declare war on the USA.	L'Allemagne et l'Italie déclarent la guerre aux USA.	Deutschland und Italien erklären der USA den Krieg.
Aug 30, 1942	The Chief of the Civil Affairs ("Gauleiter") Gustav Simon proclaims that all the young men born in Luxembourg between 1920 –1924 (risen later to 1927) have to serve in the German Army, the "Wehrmacht".	Le Chef des Affaires Civiles ("Gauleiter") Gustav Simon proclame que tous les jeunes hommes, nés entre 1920 et 1924 (plus tard jusqu'à 1927) doivent joindre l'Armée allemande, la « Wehrmacht ».	Der Chef der Zivilverwaltung Gustav Simon proklamiert, dass alle in Luxemburg zwischen 1920 und 1924 (später bis 1927) geborenen jungä Männer in die Wehrmacht einberufen werden.
Aug 31, 1942	A general strike breaks out in the whole country of Luxembourg.	Une grève générale éclate dans le pays entier.	Ein Generalstreik bricht in ganz Luxemburg aus.
From Sep 1942	Deportation of 1,140 Luxembourg families to East Germany.	Déportation de 1.140 familles luxembourgeoises en Allemagne de l'Est.	Umsiedlung von 1.140 luxemburgischen Familien nach Ostdeutschland.
Nov 11, 1942	US Forces land in North Africa. Thereupon the Germans occupy Free France.	Des forces américaines débarquent en Afrique du Nord. En revanche, les Allemands occupent la France libre.	Amerikanische Truppen landen in Nordafrika. Daraufhin besetzen die Deutschen die freie Zone Frankreichs.
Nov 1942	The English Field Marshall Bernard Law Montgomery beats the German Afrikakorps near El Alamein, Egypt.	Le Maréchal anglais Bernard Law Montgomery bat l'Afrikakorps allemand près d'El Alamein, Egypte.	Der englische Feldmarschall Bernard Law Montgomery schlägt das deutsche Afrikakorps bei El Alamein, Ägypten.
Feb 2, 1943	The Sixth German Army surrenders at Stalingrad, Soviet Union.	La 6e Armée allemande capitule à Stalingrad, URSS.	Die 6. Deutsche Armee ergibt sich in Stalingrad, UdSSR.
May 1943	The Americans succeed in eliminating the German submarines from the North Atlantic.	Les Américains réussissent à chasser les sous-marins allemands de l'Atlantique Nord.	Es gelingt den Amerikanern, die deutschen U-Boote aus dem Nordatlantik zu vertreiben.
1943	Massive bombings on the industrial areas and the cities of Germany by the Royal Air Force and the allied air forces.	Des bombardements massifs sur les zones industrielles et les villes allemandes par la Royal Air Force et les forces aériennes alliées.	Massives Bombardement auf deutsche Industrieanlagen und Städte durch die Royal Air Force und die alliierten Luftstreitkräfte.
July 9, 1943	The Allies land on Sicily.	Les troupes alliées débarquent en Sicile.	Die Alliierten landen auf Sizilien.
Sep 3, 1943	Italy surrenders and signs an armistice. German troops disarm the Italian soldiers and continue fighting.	L'Italie capitule et signe un armistice. Les troupes allemandes désarment les soldats italiens et continuent à combattre.	Italien ergibt sich und unterzeichnet einen Waffenstillstand. Deutsche Truppen entwaffnen die italienischen Soldaten und kämpfen weiter.
June 5, 1944	US soldiers enter into Rome.	Des troupes américaines entrent à Rome.	Amerikanische Truppen rücken in Rom ein.
June 6, 1944	The Allied Forces (U.S. British and Canadian troops with French, Belgian, Netherland, Polish, Czech, Luxembourg ... volunteers) land in Normandy.	Les troupes alliées (Américains, Anglais et Canadiens, avec des volontaires français, belges, néerlandais, polonais, tchèques, luxembourgeois ...) débarquent en Normandie.	Alliierte Truppen (Amerikaner, Engländer und Kanadier mit französischen, belgischen, holländischen, polnischen, tschechischen, luxemburgischen ... Freiwilligen) landen in der Normandie.
Aug 15, 1944	US and French troops land in the "Provence", South France.	Des troupes américaines et françaises débarquent en Provence, au sud de la France.	Amerikanische und französische Truppen landen in der "Provence", Südfrankreich.
Aug 25, 1944	Liberation of Paris.	Libération de Paris.	Befreiung von Paris.
Sept 3, 1944	Liberation of Brussels.	Libération de Bruxelles.	Befreiung von Brüssel.
Sept 9, 10, 11, 1944	5th US Armored Division and 28th US Infantry Division liberate the capital and the country of Luxembourg.	La 5e Division Blindée US et la 28e Division d'infanterie libèrent la capitale et le pays du Luxembourg.	Die 5. US Panzerdivision und die 28. US Infanteriedivision befreien die Hauptstadt und das Land Luxemburg.
Sept 11, 1944	First allied patrol crosses the border of the "Third Reich" at Stolzembourg (L).	Une patrouille alliée traverse la frontière du « 3e Reich » pour la première fois à Stolzembourg (L).	Ein alliierter Spähtrupp überquert in Stolzemburg (L) zum ersten Mal die Grenze zum Dritten Reich.
Dec 16, 1944	The beginning of the Battle of the Bulge.	Le commencement de la Bataille des Ardennes.	Der Beginn der Ardennenoffensive.
Dec 22, 1944	Counterattack of General George S. Patton's (Jr.) Third Army.	Contre-attaque de la 3e Armée américaine du Général George S. Patton Jr.	Gegenangriff der Dritten Armee von General George S. Patton Jr.
Dec 26, 1944	4th US Armored Division breaks the ring around the encircled town of Bastogne (B).	La 4e Division Blindée US brise l'encerclement autour de la ville de Bastogne (B).	Die 4. US Panzerdivision sprengt den Belagerungsring um die Stadt Bastnach (B).
Jan 12, 1945	The Russians start their offensive in the East.	Les Russes commencent leur offensive à l'est.	Die Russen beginnen ihre Offensive im Osten.
Jan 21, 1945	Liberation of Wiltz.	Libération de Wiltz.	Befreiung von Wiltz.
Feb 04 - 11, 1945	1945: Yalta Conference: Roosevelt, Churchill and Stalin debate about the post war era.	Conférence de Yalta: Roosevelt, Churchill et Staline débattent au sujet du temps après la guerre.	Konferenz von Jalta: Roosevelt, Churchill und Stalin debattieren über die Zeit nach dem Krieg.
Feb 12 & 22, 1945	Liberation of Vianden (L).	Libération de Vianden (L).	Befreiung von Vianden (L).
April 30, 1945	Hitler commits suicide.	Hitler se suicide.	Hitler nimmt sich das Leben.
May 08, 1945	Germany surrenders unconditionally.	L'Allemagne capitule inconditionnellement.	Deutschland kapituliert bedingungslos.
Aug 6, 1945	First A-bomb on Hiroshima, Japan.	Première bombe atomique sur Hiroshima, Japon.	Erste Atombombe auf Hiroshima, Japan.
Aug 09, 1945	Second A-bomb on Nagasaki, Japan.	Deuxième bombe atomique sur Nagasaki, Japon.	Zweite Atombombe auf Nagasaki, Japan.
Sept 02, 1945	Japan surrenders.	Le Japon capitule.	Japan kapituliert.
	END OF WORLD WAR II	**FIN DE LA DEUXIÈME GUERRE MONDIALE**	**ENDE DES ZWEITEN WELTKRIEGES**

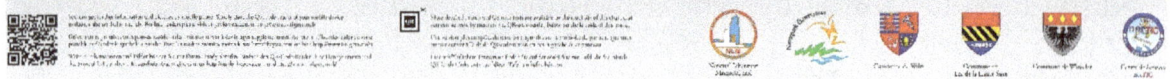

Schumannseck (Mass Grave)

In Memory of all soldiers who died in these woods in the Winter 1944-1945. There are four information panels at this site.

On N26 at Intersection with CR318

4546 Wilz

Latitude: 49.94937

Longitude: 5.88839

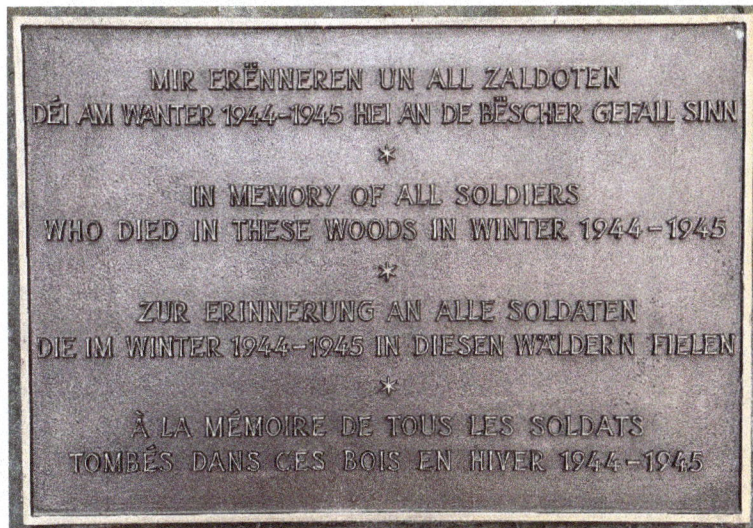

From this monument you can take a walk through a series of US and German foxholes used in the defense/attack of this area (see trails section). A selecti8on of phots of foxholes, shell craters, reconstructed bunkers etc. and their matching life-sized photos can be seen.

Information Panel Schumannseck (Mass Grave)

REMEMBER
SCHUMANNS ECK

HELL AT SCHUMANNS ECK

CASUALTIES – PERTES – VERLUSTE

On June 12, 2004, the "Sentier du Souvenir" ("Trail of Remembrance") and the adjacent monument in remembrance of all soldiers who died in these woods during the Battle of the Bulge in winter 1944-1945, were both inaugurated.

The struggle around the important strategic crossroads at 'Café Schuman' cost the lives of thousands of soldiers on both sides. A bomb crater, in which 157 fallen German soldiers had been temporarily buried after the battle, became known thereafter as the "mass grave". Local eyewitnesses testified that several hundred soldiers of both nations had been buried here. Colonel Peale, Commander of the 1st Battalion, 101st Infantry Regiment, 26th Infantry Division, and Col. Dellert, Commander of the 3rd Battalion, 101st Infantry Regiment, 26th Infantry Division, confirmed that the bodies of the dead soldiers had been assembled during the Battle of the Bulge before being transferred to various military cemeteries.

The same day, the US veteran and former Private First Class Harry W. Harvey, 90th US Infantry Division, and the German veteran Erhard Mitzinnek, 9. Volksgrenadierdivision, representing their killed – in action – and surviving comrades, shook hands as a sign of reconciliation.

The bronze plaque in Luxembourg language reads: "A mass grave was located on this spot". This memorial certainly is unique in Luxembourg as it reminds us of the fallen American – as well as of the dead German soldiers.

In front of the memorial remembering all the fallen soldiers, US Veteran Harry W. Harvey (right side) and German Veteran Erhard Mitzinnek shake hands as a sign of reconciliation between former enemies.

En face du mémorial qui commémore tous les soldats tombés, les deux ennemis d'antan, le vétéran US Harry W. Harvey (à droite) et le vétéran allemand Erhard Mitzinnek se serrent la main en signe de réconciliation.

Vor dem Mahnmal, das aller gefallenen Soldaten gedenkt, reichen sich der US-Veteran Harry W. Harvey (rechts) und der deutsche Veteran Erhard Mitzinnek die Hand als Zeichen der Versöhnung ehemaliger Feinde.

Sources/Quelle: *Archiv Luxemburger Wort.*
Foto: *Ady Richard*

German soldiers' graves in Wiltz
Des tombes allemandes à Wiltz
Deutsche Gräber in Wiltz

Le 12 juin 2004, le "Sentier du Souvenir" et le monument à la mémoire de **tous** les soldats tombés dans ces bois en hiver 1944 - 1945 durant la Bataille des Ardennes, furent inaugurés.

Les combats autour du carrefour 'Schumann', important point stratégique, coûtèrent la vie à des milliers de soldats des deux côtés. Un cratère de bombe dans lequel 157 soldats allemands furent temporairement enterrés après les combats, fut connu sous le nom de 'tombe commune'. Des témoins oculaires locaux attestèrent que des centaines de soldats des deux nations étaient enterrés ici. Le Colonel Peale, Commandant du 1er Battalion, 101st Infantry Regiment, 26th Infantry Division, et le Colonel Dellert, Commandant du 3e Battalion, 101st Infantry Regiment, 26th Infantry Division, confirmèrent que les corps des soldats morts ont été rassemblés pendant la Bataille des Ardennes avant d'être transférés vers différents cimetières militaires.

Ce même jour, le vétéran américain, Private First Class Harry W. Harvey, 90th US Infantry Division, et le vétéran allemand, Gefreiter Erhard Mitzinnek, 9. Volksgrenadierdivision, représentant leurs camarades survivants et morts au combat, se serraient la main en signe de réconciliation.

Le texte de la plaque en bronze est libellé en luxembourgeois: «A cet endroit, il y avait une tombe commune».
Ce mémorial est certainement unique au Luxembourg, vu qu'il commémore les soldats américains aussi bien que les soldats allemands.

German graves
Des tombes allemandes
Deutsche Gräber

Am 12. Juni 2004 wurden am „Schumanns Eck" der „Sentier du Souvenir" (Erinnerungspfad) und der Gedenkstein zur Erinnerung an alle Soldaten, die im Winter 1944 - 1945 während der Ardennenoffensive in diesen Wäldern fielen, eingeweiht.

Die Kämpfe um die strategisch wichtige Straßenkreuzung 'Café Schumann' kosteten vielen Tausenden Soldaten beider Seiten das Leben. Ein Bombentrichter, in dem 157 gefallene deutsche Soldaten nach der Schlacht provisorisch beigesetzt wurden, wurde zum Massengrab. Lokale Augenzeugen bezeugten, dass mehrere hundert Soldaten beider Nationen hier begraben waren. Oberst Peale, Kommandeur des 1st Battalion, 101st Infantry Regiment, 26th Infantry Division, und Oberst Dellert, Kommandeur des 3. Battalion, 101st Infantry Regiment, 26th Infantry Division, bestätigten, dass die Körper der toten Soldaten während

der Ardennenschlacht zusammen getragen wurden, bevor sie nach verschiedenen Militärfriedhöfen überführt wurden. An diesem Tag reichten sich der amerikanische Veteran, Private First Class Harry W. Harvey, 90th US Infantry Division, und der deutsche Veteran, Gefreiter Erhard Mitzinnek, 9. Volksgrenadierdivision, stellvertretend für ihre überlebenden und im Kampf gefallenen Kameraden die Hände zum Zeichen der Versöhnung.
In Luxemburger Sprache steht über der bronzenen Platte geschrieben: „An dieser Stelle befand sich ein Massengrab". Dieses Denkmal ist wohl einzigartig in Luxemburg, da es sowohl an die getöteten Amerikaner wie auch an die gefallenen Deutschen erinnert.

Lt. Col. George Randolph
commander of the 712th US Tank battalion
commandant du 712e US Tank Battalion
Kommandeur des 712e US Tank Battalion

Lt. Col. G. Randolph
dead beside a tank destroyer of the 773rd Tank Destroyer Battalion
mort à côté d'un chasseur de char du 773e TD Battalion
tot neben einem Panzerzerstörer des 773rd TD Battalion

Fotos: Lëtzebuerg am 2. Weltkrich 1944-1945, R. Gaul, F. Karen, F. Rockenbrod, Editions St. Paul, 1994

The Company Commander informs Heinz Draber's wife of her husband's death, killed in action at Doncols (L).

Le Commandant de Compagnie informe l'épouse du décès de son mari Heinz Draber, tué à Doncols (L).

Mitteilung des Kompaniechefs zum Tode des Obergefreiten Heinz Draber, 05.03.1913 - 06.01.1945, in Doncols (L) gefallen.

Information Panel Schumannseck (Mass Grave)

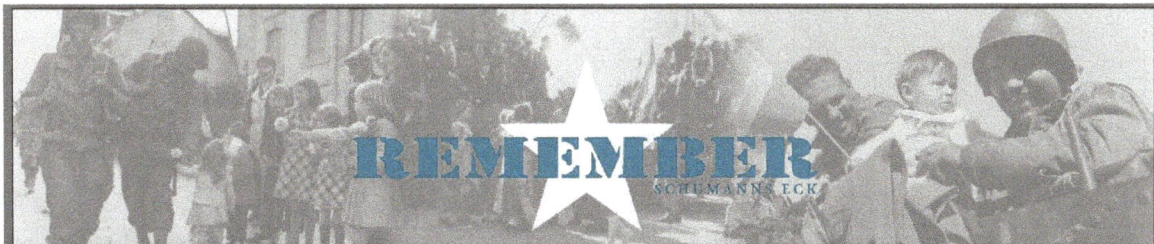

HELL AT SCHUMANNS ECK

CASUALTIES – PERTES – VERLUSTE

The following photos show the circumstances in which the dead soldiers had been recovered.

Les photos suivantes montrent les circonstances dans lesquelles les morts furent récupérés.

Die folgenden Fotos zeigen die Umstände, unter denen die Toten geborgen wurden.

After the fighting, special US graves registration units brought the dead soldiers to collecting points.

Après les combats, des unités spéciales américaines transportèrent les soldats morts sur des points collectifs.

Nach den Kämpfen brachten amerikanische Spezialeinheiten die toten Soldaten an Sammelstellen.

German paratrooper's grave near Harlange

Tombeau d'un para allemand près de Harlange

Grab eines deutschen Fallschirmjägers bei Harlingen

US soldiers seeking cover from German artillery fire between a ruined building and a tank destroyer at Nothum. A dead soldier is lying on the stretcher.

Des soldats américains cherchant abri devant une attaque d'artillerie allemande entre un bâtiment en ruines et un char à Nothum. Un soldat tombé est couché sur la civière.

US Soldaten suchen Deckung vor einem deutschen Artillerieüberfall zwischen einem zerstörten Gebäude und einem Panzerzerstörer in Nothum. Ein gefallener Soldat liegt auf der Tragbahre.

US soldier collecting the frozen bodies of German soldiers.

Soldat américain rassemblant des corps gelés de soldats allemands.

Ein amerikanischer Soldat legt gefrorene Körper deutscher Soldaten beisammen.

Dead US soldiers in their foxhole

Soldats américains morts dans leur trou de fusilier

Tote amerikanische Soldaten in ihrem Schützenloch

The frozen and mutilated body of a German soldier.

Le corps déchiqueté d'un soldat allemand.

Die gefrorene und verstümmelte Leiche eines deutschen Soldaten.

Dead American machinegunner in his position

US tireur de mitrailleuse tué dans son poste

Toter US MG-Schütze in seiner Stellung

Dead US soldier

Soldat américain mort

Toter amerikanischer Soldat

The losses on both sides were extremely high. The Battle of the Bulge cost the lives of 19.000 US soldiers, 200 British soldiers and 26.000 German soldiers. The total losses of dead, wounded, missing and prisoners amount to more than 160.000 soldiers. At the end, more than a million soldiers were engaged in this battle. The fighting was terrible. Entire towns and villages were destroyed; approximately 3.800 civilians lost their lives in Belgium and in Luxembourg during the 6 weeks of the Battle of the Bulge.

Les pertes des deux côtés étaient énormes. La Bataille des Ardennes a coûté la vie à 19.000 soldats américains, 200 soldats britanniques et 26.000 soldats allemands. Les pertes totales en morts, blessés, disparus et prisonniers se chiffraient à plus de 160.000 soldats. A la fin, plus d'un million de soldats étaient engagés dans cette bataille.
Les combats furent terribles. Des villes et des villages entiers furent détruits. En Belgique et au Luxembourg environ 3.800 personnes civiles perdirent leur vie pendant les 6 semaines de l'Offensive des Ardennes.

Die Verluste auf beiden Seiten waren sehr hoch. Die Ardennenoffensive kostete 19.000 amerikanischen, 200 britischen und 26.000 deutschen Soldaten das Leben. Im Ganzen beliefen sich die Verluste an Toten, Verletzten, Vermissten und Gefangenen auf über 160.000 Soldaten. Am Ende war mehr als eine Million Soldaten in dieser Schlacht eingesetzt.
Die Kämpfe waren schrecklich. Ganze Städte und Dörfer wurden zerstört. In Belgien und in Luxemburg verloren ungefähr 3.800 Zivilisten ihr Leben während den 6 Wochen der Ardennenoffensive.

© National Liberation Memorial a.s.b.l.

Information Panel Schumannseck (Mass Grave)

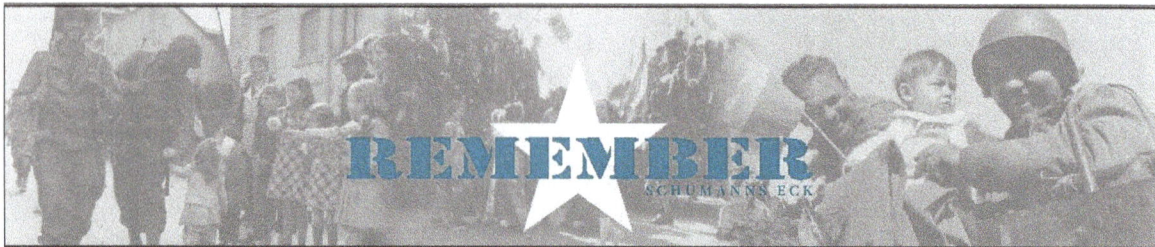

HELL at SCHUMANNS ECK

WOUNDED SOLDIERS – LES BLESSÉS – VERWUNDETE

🇺🇸 „Purple Heart" certificate attesting that Private John A. Dewire, K Company, 101st Regiment, 26th US Infantry Division, had been wounded at Nothum on January 3, 1945. In fact, John was severely wounded at Melchior's house. – The idea to set up a Liberation Memorial at Schumann's Corner is due to John A. Dewire's initiative.

🇫🇷 Certificat attestant que le soldat John A. Dewire, Compagnie K, 101e Régiment, 26e Division d'Infanterie Américaine, a été blessé à Nothum le 3 janvier 1945. En réalité, John a été grièvement blessé à la maison Melchior. L'idée de réaliser un Monument de la Libération au « Schumanns Eck » est due à l'initiative de John A. Dewire.

🇩🇪 Urkunde, die John A. Dewires Verletzung in Nothum am 3. Januar 1945 bescheinigt. In Wirklichkeit wurde John A. Dewire, K Kompanie, 101. Regiment, 26. US Infanteriedivision, beim Haus Melchior verletzt. Die Idee ein Denkmal zur Befreiung Luxemburgs am „Schumanns Eck" zu errichten, geht auf eine Initiative von John A. Dewires zurück.

German wound certificate
Certificat de blessé au combat allemand
Deutsches Verwundetenzertifikat

🇺🇸 Shortly prior to his unit's engagement in the Battle of the Bulge, Private First Class George Fisher, 328th Regiment, 26th US Infantry Division, sent this Christmas card with his picture to his family members from the town of Metz, France. It should be the last news for a long time. The telegram informing his parents of his injury he got at the "Mass Grave" reached his parents two weeks later. George had survived hell at "Café Schumann". – It must have been terrible for the survivors when receiving such "signs of life" in case of death a long time after the electronic confirmation.

🇫🇷 Peu avant l'engagement de son unité dans la Bataille des Ardennes, le soldat première classe George Fisher, 328e Régiment, 26e Division d'Infanterie Américaine,

envoya cette carte de Noël avec sa photo rapidement à sa famille aux États Unis d'Amérique à partir de Metz (F). Cette nouvelle allait être la dernière pour un long terme. Le télégramme au sujet de sa blessure atteignit ses parents deux semaines plus tard. George Fisher avait survécu à l'enfer du « Café Schumann ». – Il devait avoir été particulièrement terrible pour les survivants de recevoir de pareils « signes de vie » en cas de mort longtemps après la confirmation électronique.

🇩🇪 Kurz vor dem Einsatz seiner Einheit in der Ardennenoffensive schickte Private First Class George Fisher, 328. Regiment, 26. US Infanteriedivision, noch schnell aus Metz (F) diese Weihnachtsgrußkarte mit dieser Aufnahme an seine Familie in den USA. Es sollte seine letzte Nachricht für lange Zeit werden. Das Telegramm über seine Verwundung beim „Massengrab" erreichte seine Eltern zwei Wochen später George Fisher hatte die Hölle am Café Schumann überlebt. – Besonders furchtbar muss der Erhalt von solchen „Lebenszeichen" lange nach der elektronischen Bestätigung im Todesfalle für die Hinterbliebenen gewesen sein.

Patton's prayer for favorable weather
La prière de Patton pour un temps favorable
Pattons Gebet für besseres Wetter

To prevent the dreaded "trenchfoot" (that often resulted in the amputation of the frozen toes) it was important to keep the feet dry and to stimulate the blood circulation by massaging them regularly.

Pour prévenir le redoutable "trenchfoot" qui aboutissait à une amputation des orteils gelés - il était important de bien sécher les pieds et de les masser régulièrement afin de stimuler la circulation sanguine.

Um dem gefürchteten "trenchfoot" vorzubeugen, bei dem die gefrorenen Zehen amputiert wurden, war es wichtig, die Füße trocken zu behalten und sie regelmäßig zu massieren.

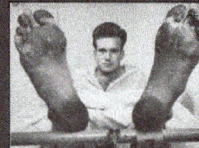

A case of "trenchfoot"
Un cas de « trenchfoot »
Ein Fall von „trenchfoot"

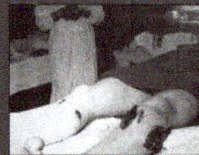

A German machine gun bullet that penetrated both legs could have evil consequences: A "simple wound" that often was considered by the soldiers as a ticket to leave the "white hell", would provoke gangrene and death after an exposition of more than 24 hours to the severe cold or would be followed by an amputation of the wounded limbs.

Une balle d'une mitrailleuse allemande qui a traversé les deux jambes, pouvait avoir de lourdes conséquences: Une « simple blessure » qui était souvent considérée par les soldats comme un ticket pour quitter « l'enfer blanc », pouvait provoquer la gangrène, la mort ou l'amputation après une exposition de la blessure au froid extrême de plus de 24 heures.

Eine deutsche Maschinengewehrkugel, die beide Beine durchschoss, konnte üble Folgen haben: Eine „einfache" Wunde, die oft von den Soldaten als Freifahrtschein aus der „weißen Hölle" angesehen wurde, konnte Wundbrand, den Tod oder eine Amputation nach sich ziehen, wenn die Wunde mehr als 24 Stunden der extremen Kälte ausgesetzt war.

US soldiers rescuing a wounded buddy
Des soldats US secourtant un camarade blessé
Amerikanische Soldaten helfen einem verwundeten Kameraden

Even civilians were treated
Même des personnes civiles furent traitées
Auch Zivilisten wurden behandelt

Information Panel Schumannseck (Mass Grave)

HELL AT SCHUMANNS ECK

The sufferings of the civilians
Les souffrances de la population civile
Das Leid der Zivilbevölkerung

Many civilians fled the battle fields to the south of the country taking along as much as possible of their property.

Beaucoup de personnes civiles s'enfuyaient des champs de bataille vers le sud du pays en emportant le plus possible de leurs possessions.

Viele Zivilisten flohen aus den Kampfgebieten und brachten sich mit ihren Habseligkeiten im Süden des Landes in Sicherheit.

Older people lost their possessions they had worked for their entire life in a few hours. Often they couldn't manage their lives anymore and fell into deep depressions.

Les habitants plus âgés perdirent leurs possessions pour lesquelles ils avaient travaillé toute leur vie en quelques heures. Souvent ils ne pouvaient plus gérer leur vie et tombèrent dans une profonde dépression.

Ältere Menschen verloren ihr Hab und Gut, für das sie ihr Leben lang gearbeitet hatten. In ein paar Stunden. Sie fanden sich oft nicht mehr zurecht und fielen in eine tiefe Depression.

The civilians coming from the north of the country were transported by buses or US trucks to Luxembourg-City where they were accommodated with families or in public buildings.

Les réfugiés venant de l'Oesling furent transportés à partir de Mersch au moyen de bus et de camions US à Luxembourg-Ville où ils étaient logés chez des familles ou dans des bâtiments publics.

Die Flüchtlinge aus dem Oesling wurden ab Mersch mit Bussen oder US-Lastwagen nach Luxemburg-Stadt gebracht, wo sie bei Familien oder in öffentlichen Gebäuden beherbergt wurden.

Refugees' misery – Misère
des réfugiés – Flüchtlingselend

Civilians seeking shelter in the cellars of a brewery

Des civilistes se réfugiant dans les caves d'une brasserie

Zivilisten, die in den Kellern einer Brauerei Schutz suchen

Many villagers had stayed in the cellars of their houses for weeks and lived off preserved food, cans, apples and potatoes. They had been exposed to artillery fire and feared for their lives up to the liberation.

Beaucoup de villageois restaient dans les caves de leurs maisons pendant des semaines et vivaient de conserves, de pommes et de pommes de terre. Ils étaient exposés au feu d'artillerie et craignaient pour leur vie jusqu'à la libération.

Viele Dorfbewohner blieben während Wochen in den Hauskellern und ernährten sich von Eingemachtem, Äpfeln und Kartoffeln. Sie waren dem Artilleriebeschuss ausgesetzt und bangten bis zur Befreiung um ihr Leben.

To prevent epidemics, dead cattle had to be removed as quickly as possible. – Many farmers had lost their livestock and horses.

Pour éviter des épidémies, il fallait enlever les animaux morts le plus vite possible. – Beaucoup de fermiers avaient perdu tout leur bétail et leurs chevaux.

Um den Ausbruch von Seuchen zu verhindern, mussten die verendeten Tiere weggeschafft werden. – Viele Bauern hatten ihren ganzen Viehbestand und ihre Pferde verloren.

Wiltz:
the "battered" street "Scheergaass"

la rue « Scheergaass » en ruines

die verwüstete „Scheergaass"

Ruins of "Café Schumann" at the "Schumann" crossroads

Le „Café Schumann" au carrefour „Schumann", en ruines

Das Gasthaus „Schumann", an der Straßenkreuzung „Schumann", in Ruinen

The ruins of the Melchior House served as a model for the "National Liberation Memorial"

Les ruines de la Maison Melchior servaient de modèle pour le « National Liberation Memorial »

Die Ruinen des Hauses Melchior dienten als Vorlage zum "National Liberation "Memorial

Ruins of the Melchior House
Ruines de la Maison Melchior
Ruinen des Hauses Melchior

We shall not forget our Belgian friends and neighbors who hosted not only many refugees, but who, at the risk of their lives, hid hundreds of Luxembourg deserters who refused to join the German Army, during the Nazi occupation. Their solidarity and their engagement proved of their courage and their friendship. We owe them a great THANK YOU. We express our gratitude to all Luxembourgers who, during the war, have hidden one or more deserters, have resisted to the oppressors or have hosted some refugees.

Nous ne voulons pas oublier nos amis et voisins belges qui accueillirent non seulement beaucoup de réfugiés, mais qui, au péril de leur vie, cachèrent des centaines de réfractaires luxembourgeois pendant le temps de l'occupation nazie. Leur solidarité et leur engagement ont fait preuve de leur courage et de leur amitié. Nous leur devons un très grand MERCI. Nous exprimons notre gratitude à tous les Luxembourgeois qui, pendant la guerre, ont caché des réfractaires, se sont activement opposés aux occupants ou ont accueilli des réfugiés.

Wir vergessen unsere belgischen Freunde und Nachbarn nicht, die nicht nur viele Flüchtlinge aufnahmen, aber die Hunderte Luxemburger Fahnenflüchtige, die nicht in der Wehrmacht dienen wollten, während der Nazi-Besetzung unter Lebensgefahr versteckten. Ihre Solidarität und ihr Einsatz sind Zeugen ihres Mutes und ihrer Freundschaft. Wir schulden ihnen unseren aufrichtigen DANK. Allen Luxemburgern, die während des Krieges Fahnenflüchtige versteckt haben, aktiven Widerstand gegen die Besatzungsmacht geleistet haben oder die Flüchtlinge aufgenommen haben, drücken wir unsere Dankbarkeit aus.

Photos of the 4 panel – Photos des 4 panneaux – Fotos der 4 Tafeln: Archives du Musée National d'Histoire Militaire de Diekirch

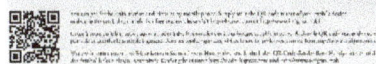

Schuttrange

To those who fought and died for our liberty 1940-1945.

Rue Principale 96

5367 Schuttrange

Latitude: 49.6235

Longitude: 6.2682

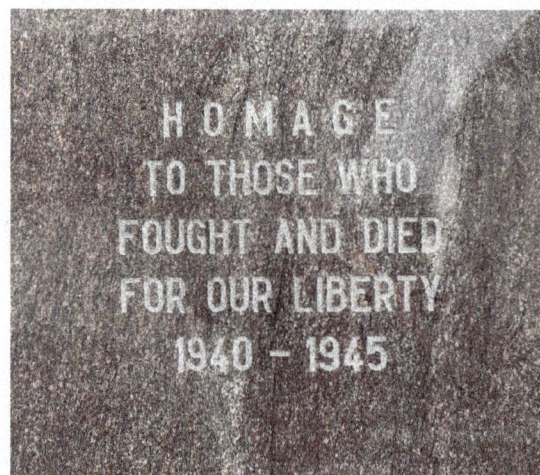

Sonlez

In honor of the 16 men of E and G Companies, 357th Infantry Regiment, 90th US Infantry Division who gave their lives for the liberation of Sonlez on January 11, 1945.

Um Kanal 4

9647 Winseler

Located at the church in Sonlez

Latitude: 49.96401

Longitude: 5.82545

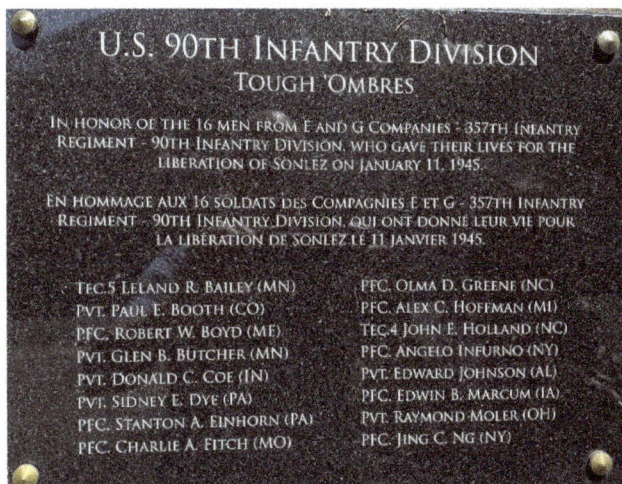

Dedicated to: Tec5 Leland R. Bailey; PVT Paul E. Booth; PFC Robert W. Boyd; PVT Glen B. Butcher; PVT Donald C. Coe; PVT Stanton A. Einhorn; PFC Charles A. Fitch; PFC Olma D. Greene; PFC Alex C. Hoffman; TEC4 John E. Holland; PFC Angelo Infurno; PVT Edward Johnson; PFC Edwin B. Marcum; PVT Raymond Moler; PFC Jing C. Ng

Steinfort

Dedicated to George S. Patton, Jr. 3rd US Army.

Square General Patton 4

8443 Steinfort

Latitude: 49.65978

Longitude: 5.91391

Stolzembourg

5th US Armored Division, crossing point for the first allied soldiers crossing into Germany on September 11, 1944.

Rue Principale

9463 Putscheid

Latitude: 49.96315

Longitude: 6.16990

This monument is due to be rebuilt and will be re-inaugurated in Sept. 2024; below is a plan of the new monument.

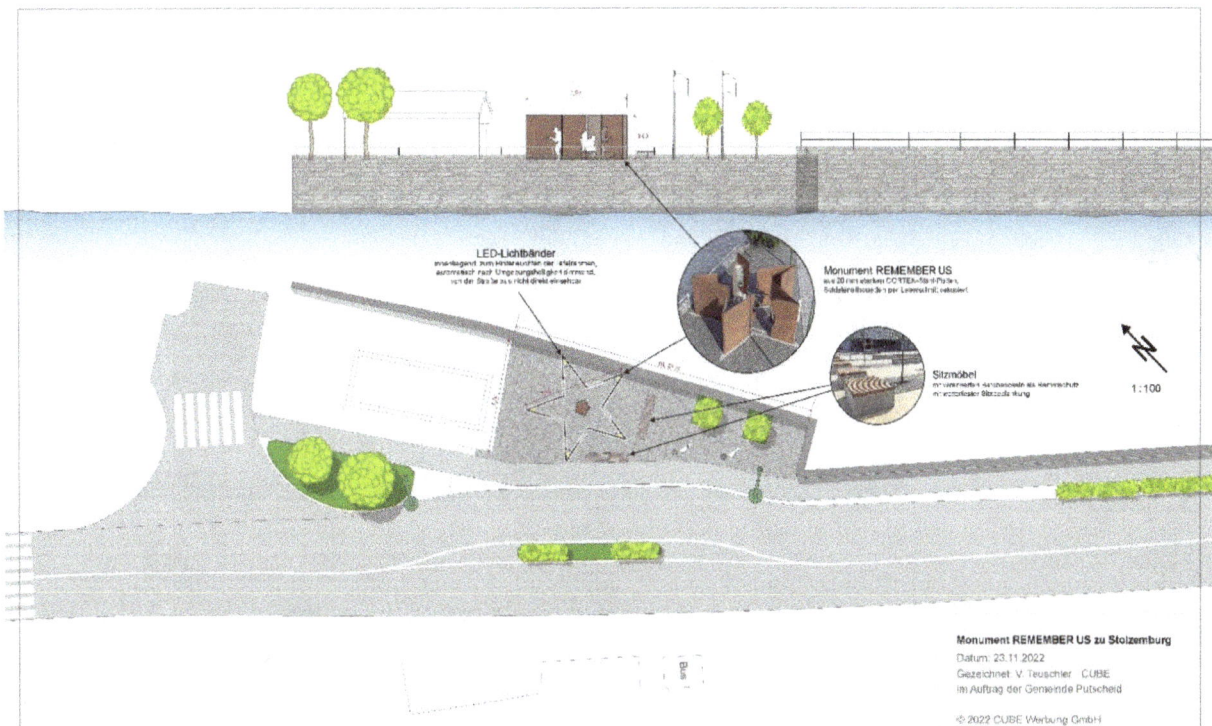

Information Panel Stolzembourg

THE FIRST CROSSING INTO THE THIRD REICH

The first allied soldiers who crossed here into Germany, 96 days after the Landing in Normandy, on September 11, 1944 at 16.30 hours, were Sgt. Warner W. Holzinger, Cpl. Ralph E. Diven, T/5 Coy T. Locke, Pfc. William McColligan, Pfc. George F. McNeal, Pfc. Jesse Stevens and the French Lieutenant Lionel Delille of the 2nd Platoon, Troop B, 85th Cavalry Reconnaissance Squadron, 5th US Armored Division. They made their way to Keppeshausen, Germany, reconnoitered the pillboxes of the Siegfried Line near Waldhof, Germany, and returned at 18.15 hours to report to Lt. Loren L. Vipond. The information was radioed to the Headquarters of the 1st US Army, which flashed that night the news to the world that it had crossed the German border.

LA PREMIÈRE ENTRÉE DANS LE TROISIÈME REICH

Les premiers soldats alliés qui ont traversé ici la frontière allemande, 96 jours après le Débarquement en Normandie, le 11 septembre 1944 à 16.30 heures, étaient Sgt. Warner W. Holzinger, Cpl. Ralph E. Diven, T/5 Coy T. Locke, Pfc. William McColligan, Pfc. George F. McNeal, Pfc. Jesse Stevens et le Lieutenant français Lionel Delille du 2e Peloton, Compagnie de Cavalerie B, 85e Escadron de Cavalerie de Reconnaissance, 5e Division Blindée américaine. Ils avancèrent vers Keppeshausen, observèrent les fortifications de la Ligne Siegfried près de Waldhof, retournèrent vers 18.15 heures à Stolzembourg et firent leur rapport au Lieutenant Loren L. Vipond. La nouvelle fut transmise à l'État Major de la 1re Armée américaine qui informa le même soir le monde entier qu'elle venait de traverser la frontière allemande.

DIE ERSTE ÜBERQUERUNG DER DEUTSCHEN GRENZE

Die ersten alliierten Soldaten, die hier 96 Tage nach der Landung in der Normandie am 11. September 1944 um 16.30 Uhr, die deutsche Grenze überschritten, waren Sgt. Warner W. Holzinger, Cpl. Ralph E. Diven, T/5 Coy T. Locke, Pfc. William McColligan, Pfc. George F. McNeal, Pfc. Jesse Stevens und der französische Leutnant Lionel Delille vom 2. Zug, Kavalleriekompanie B, 85. Kavallerieaufklärungsschwadron, 5. US-Panzerdivision. Sie drangen nach Keppeshausen vor, beobachteten die Bunker des Westwalls bei Waldhof, kehrten gegen 18.15 Uhr nach Stolzemburg zurück und statteten Lt. Loren L. Vipond Bericht. Die Nachricht wurde dem Hauptquartier der 1. US-Armee gefunkt, die noch am selben Abend die Welt benachrichtigte, dass sie die deutsche Grenze überschritten hatte.

STOLZEMBOURG
September 11, 1944

The first crossing into the Third Reich on September 11, 1944
La première entrée dans le Troisième Reich le 11 septembre 1944
Die erste Überquerung der deutschen Grenze am 11. September 1944

Voie de la libération
06.06.1944 - 10.09.1944

The Liberty Road from June 6 to September 10, 1944.
La Voie de la Liberté du 6 juin au 10 septembre 1944.
Der Weg zur Freiheit vom 6. Juni bis zum 10. September 1944.

Source: Histoire de la Seconde Guerre Mondiale de Sir Basil H. Liddell Hart, collection marabout-université, 1973, traduction du livre "HISTORY OF THE SECOND WORLD WAR" by Sir Basil H. Liddell Hart, Cassell et Company L. D. London

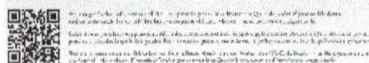

Stuppicht

In memory to our liberators 1944-1945. Dedicated by the Allied Military Vehicle Drivers (AMVD) and all allied military vehicle collectors from Europe.

Stuppicht 1

6155 Fischbach

Latitude: 49.7258

Longitude: 6.18849

Troisvierges

Monument to the 6th US Armored Division. The gun next to the memorial is a German 8.8cm PAK43 anti-tank gun.

Rue de Binsfeld

9912 Troisvierges

Latitude: 50.1182

Longitude: 6.0045

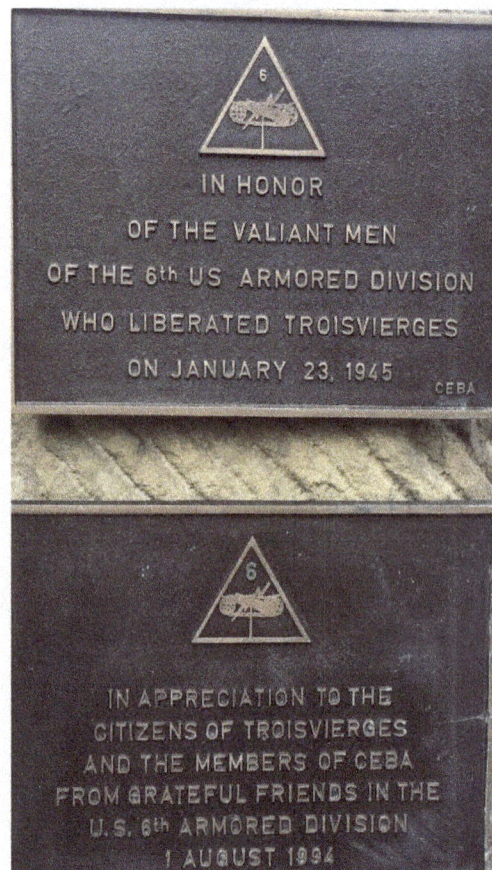

IN HONOR
OF THE VALIANT MEN
OF THE 6th US ARMORED DIVISION
WHO LIBERATED TROISVIERGES
ON JANUARY 23, 1945
CEBA

IN APPRECIATION TO THE
CITIZENS OF TROISVIERGES
AND THE MEMBERS OF CEBA
FROM GRATEFUL FRIENDS IN THE
U.S. 6th ARMORED DIVISION
1 AUGUST 1994

Urspelt (Château)

Monument to 1st Battalion, 110th Infantry Regiment located in front of the Château d'Urspelt hotel.

Planned plaque installation before December 2024.

Am Schlass

9774 Urspelt

Latitude: 50.07508

Longitude: 6.045350

Proposed Text for Plaque

In memory of the men of HQ and HQ Company, 1st Battalion 110th Infantry Regiment.

These valiant men, under the command of Lt.-Col. Donald E. PAUL, had installed 1st Battalion headquarters from December 10th to December 17th, 1944, in the Château d'Urspelt. The brave GIs tried to defend the village against the German onslaught, before being pushed back by overwhelming German forces.

Vianden

In honor of the 6th US Cavalry and 1255th Combat Engineers. Vianden was the last town in Luxembourg to be liberated.

Route de Diekirch

9409 Vianden

Latitude: 49.9324

Longitude: 6.2015

IN WORLD WAR II
THE SOIL OF LUXEMBOURG
WAS FREE AGAIN
AFTER THE LIBERATION OF
VIANDEN   BY THE 6TH U.S.   CAVALRY
FEBRUARY 12,1945

DEN 12. FEBRUAR 1945 GOUF VEIANEN
ALS LESCHT UERTSCHAFT VUM LAND ERËM FRÄI
1984

VETERANS OF THE 1255TH COMBAT ENGINEERS
HONOR THE MEMORY OF
JACK BENDER   DAVID GLATTER
NATHAN CORLEY   EDWARD GRIFFIN
CYREL EVANOW   MARION HANSON
IRA GAMBILL   CHARLES NANCE
VINCENT GAMBINO   HAROLD SMITH
WILLIAM TIFF
WHO GAVE THEIR LIVES ON FEBRUARY 12,1945
AT VIANDEN
SI HUN D'LIEWE GELOOSS FIR D'FRÄIHEET
FEBRUARY 12, 1995

Waldbillig

Dedicated to Combat Command A (CAA) of the 9th US Armored Division who held the Ermsdorf-Savelborn-Waldbillig-Christnach line.

Rue de la Montange 8

7681 Waldbillig

Latitude: 49.7939

Longitude: 6.2807

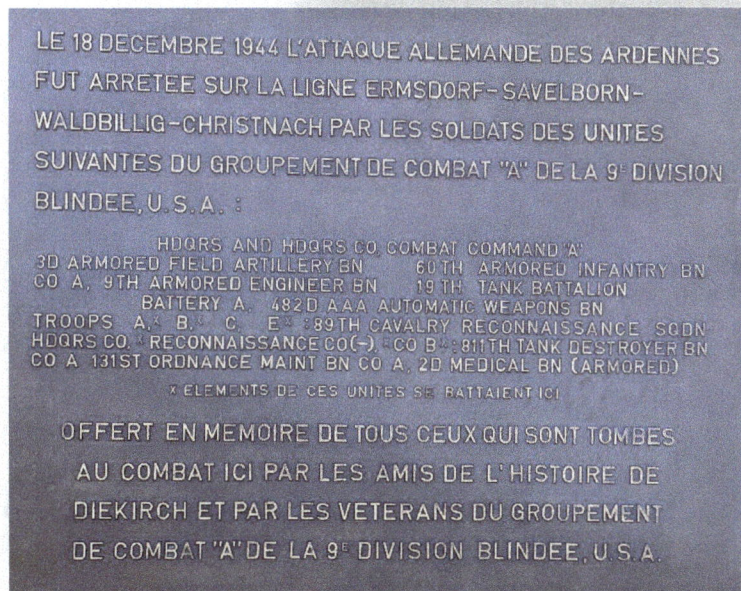

Inscription reads: On December 18, 1944 the German attack in the Ardennes was stopped on the Ermsdorf-Savelborn-Waldbillig-Christnach line by soldiers of the following units of Combat Command A of the 9th US Armored Division.

Headquarters and Headquarters Company, Combat Command A; 3rd Armored Field Artillery Battalion; 60th Armored Infantry Battalion; Company A, 9th Armored Engineer Battalion; 19th Tank Battalion; Battery A, 482nd Anti-Aircraft Artillery Automatic Weapons Battalion; Troops A[x], B[x], C, and E[x] 89th Cavalry Reconnaissance Squadron; Headquarters Company[x] Reconnaissance Company (-) Company B 811th Tank Destroyer Battalion; Company A 131st Ordnance Maintenance Battalion; Company A, 2nd Medical Battalion (Armored).

[x] Elements of these units fought here

Offered in memory of all those who died in combat by the Diekirch Historical Society and the veterans of CCA of the 9th US Armored Division.

Information Panel Waldbillig

Wasserbillig

Monument to the 87th US Infantry Division who liberated Wasserbillig on January 23, 1945 and to three volunteers from Belgium.

Route de Luxembourg 18

6633 Mertert

Latitude: 49.712697

Longitude: 6.49647

Weiler – Tom Myers Square

In memory of the soldiers of I-Company, 110th Regiment, 28th Infantry Division who fought in Weiler in December 1944. Monument is located in Tom Myers Square which was named after a US soldier of I Company, 110th Regiment who was captured the first day of the Battle of the Bulge. Tom survived the war and attended the inauguration of the memorial.

Rue Principale 7A

9466 Putscheid

Beside the fire station

Latitude: 49.961686

Longitude: 6.124756

PLAAZ
TOM MYERS
SQUARE

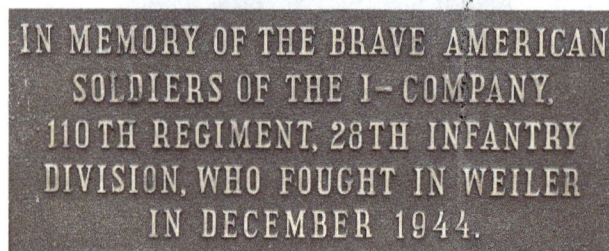

IN MEMORY OF THE BRAVE AMERICAN
SOLDIERS OF THE I-COMPANY,
110TH REGIMENT, 28TH INFANTRY
DIVISION, WHO FOUGHT IN WEILER
IN DECEMBER 1944.

Information Panel Weiler

After the liberation of Luxembourg on September 10, 1944, the front in the North of Luxembourg stabilized along the road leading from Ettelbruck to Wemperhardt, called the Skyline Drive by the GIs. The villages of Weiler and Wahlhausen were outposts defended by I-Company, 3rd Battalion, 110th Regiment, 28th US Infantry Division, reinforced by an anti-tank platoon as well as mortar and heavy machine-gun sections.

On December 16, 1944, Hitler launched a surprise attack with 240.000 German soldiers against the 83.000 US soldiers holding the front from Echternach (L) to Monschau (Germ.). It was the Battle of the Bulge and was intended to reach the port of Antwerp (B) within three days, to split the British and US troops and to force the Allies to an armistice. The stiff resistance of the US defenders thwarted that plan and allowed General George S. Patton Jr. to liberate the encircled town of Bastogne (B) on 26th of December 1944 and to attack the Germans at the southern flank.

The garrison of Weiler resisted the aggressors on December 16, until it ran out of ammunition. The same day, in Wahlhausen, Lt Fisher ordered friendly artillery fire on his own position in order to stop the German advance. In the following night, Company commander Captain Floyd K. McCutchan retreated with his troops to Consthum, 3rd Battalion's headquarters. Many of them were killed or made prisoners of war by the pursuing Germans. Only a few reached Consthum.
During this battle a large part of the North of Luxembourg was destroyed.

Hitler's plan for the Offensive in the Ardennes and actual German advance

Après la libération de la Ville de Luxembourg le 10 septembre 1944, le front se stabilisa le long de la route Ettelbruck – Wemperhardt, appelée par les Américains « Skyline Drive ». Les villages de Weiler et de Wahlhausen étaient des avant-postes défendus par la Compagnie I, 3e Bataillon, 110e Régiment, 28e Division d'Infanterie Américaine, renforcée par un peloton de canons anti-chars ainsi que de sections de mortiers lourds et de mitrailleuses .30.

Le 16 décembre 1944, Hitler lança une attaque surprise avec 240.000 soldats contre les 83.000 soldats américains défendant le front à partir d'Echternach (L) jusque Monschau (Allem.). C'était la Bataille des Ardennes qui avait pour but de s'emparer du port d'Anvers (B) dans trois jours, de séparer les troupes britanniques et américaines et de forcer les Alliés à signer un armistice. L'esprit de combat des défenseurs américains prévenait ce plan et permettait au Général George S. Patton Jr de libérer la ville de Bastogne (B) encerclée le 26 décembre 1944 et d'attaquer les Allemands par le flanc sud.

La garnison de Weiler résistait aux agresseurs pendant toute la journée du 16 décembre jusqu'à épuisement des munitions. Le même jour à Wahlhausen, le Lieutenant Fisher demanda un tir d'artillerie sur sa propre position pour arrêter l'avance des Allemands. Le Capitaine Floyd K. McCutchan, commandant de la compagnie, se retira pendant la nuit avec ses troupes à Consthum, quartier général du 3e Bataillon. Se battant en retraite, beaucoup de ses hommes furent tués ou faits prisonniers ; seulement un nombre restreint atteignit Consthum.
Une grande partie du nord du Luxembourg fut détruite pendant cette bataille.

WEILER, 1945, northern and southern front of the village-hall in ruin (school, teacher's lodgings, dairy, fire-brigade's hangar, assembly room with stage)
© Photo: Rob. LEHR, architecte dipl. Luxembourg

Nach der Befreiung Luxemburgs am 10. September 1944, stabilisierte sich die Front an der Straße Ettelbrück – Wemperhardt, von den Amerikanern „Skyline Drive" genannt. Die Dörfer Weiler und Wahlhausen waren vorgeschobene Posten, die von der I-Kompanie, 3. Bataillon, 110. Regiment, 28. US Infanterie Division, verstärkt durch einen Panzerabwehr-Zug, zwei schwere Granatwerfer-Gruppen und eine .30-MG-Gruppe, verteidigt wurden.

Am 16. Dezember 1944, startete Hitler einen Überraschungsangriff mit 240.000 Soldaten gegen die 83.000 amerikanischen Verteidiger, die von Echternach (L) bis Monschau (D) die Front hielten. Die Ardennen-Offensive hatte zum Ziel, den Hafen von Antwerpen (B) binnen drei Tagen zu erobern, die amerikanischen und britischen Truppen zu spalten und die Alliierten zu einem Waffenstillstand zu zwingen. Die Entschlossenheit der amerikanischen Verteidiger machte den Plan zunichte und erlaubte es General George S. Patton Jr. die eingekesselte Stadt Bastogne (B) am 26. Dezember 1944 zu befreien und die Deutschen an der Südflanke anzugreifen.

Am 16. Dezember, boten die amerikanischen Verteidiger in Weiler den Aggressoren einen ganzen Tag lang die Stirn, bis ihnen die Munition ausging. Am selben Tag, befahl Leutnant Fisher in Wahlhausen Artilleriefeuer auf seine eigene Position, um den Vormarsch der Angreifer aufzuhalten. Kompanie-Chef Kapitän Floyd K. McCutchan zog in der Nacht seine Truppen nach Consthum, Hauptquartier des 3. Bataillons, zurück. Während des Rückzuges, wurden viele seiner Männer getötet oder gefangen genommen. Nur wenige erreichten Consthum.

Ein großer Teil des Öslings wurde während der Ardennen-Offensive zerstört.

Position of I-Company, 3rd Battalion, 110th Regiment, 28th US Infantry Division in WEILER-PUTSCHEID on Dec 16, 1944 and German attack

Weilerbach

In honor of the 5th Infantry Division which crossed the Sauer river here on February 7, 1945 under heavy enemy fire and adverse weather conditions.

Route de Diekirch

6590 Berdorf

Latitude: 49.8338

Longitude: 6.3834

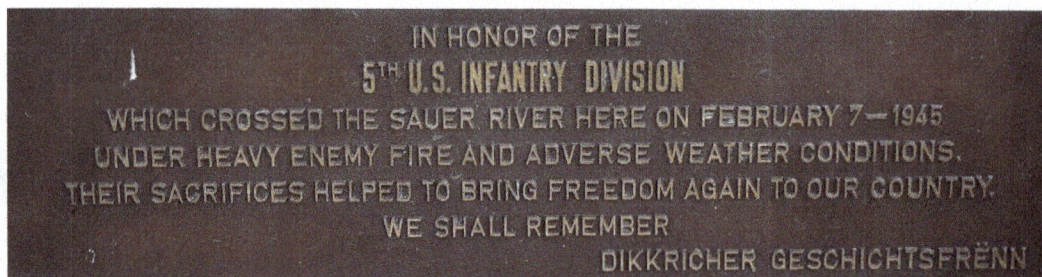

IN HONOR OF THE
5TH U.S. INFANTRY DIVISION
WHICH CROSSED THE SAUER RIVER HERE ON FEBRUARY 7—1945
UNDER HEAVY ENEMY FIRE AND ADVERSE WEATHER CONDITIONS.
THEIR SACRIFICES HELPED TO BRING FREEDOM AGAIN TO OUR COUNTRY.
WE SHALL REMEMBER
DIKKRICHER GESCHICHTSFRËNN

Weiswampach

Dedicated to the 112th Infantry Regiment, 28th Infantry Division which defended the northern tip of Luxembourg. Located on the Luxembourg-Belgium border between Weiswampach and Ouren, Belgium on the CR 363. Located at the site is also a monument to the Royal Airforce crews of two Lancaster bomber that crashed/shot down near Weiswampach during the war.

Latitude: 50.141336

Longitude: 6.113767

Inaugurated December 15, 2019

Information Panels Weiswampach

REMEMBER US
WEISWAMPACH

THE 112ᵀᴴ US INFANTRY REGIMENT DEFENDS THE NORTHERN TIP OF LUXEMBOURG

When the 28ᵗʰ US Infantry Division left the "Hürtgen Forest" south of Aachen (D) in mid-November 1944, the Division was suffering from the heavy casualties from the bitter fights in that area encountered the weeks before. The division was then sent to a so called 'rest area', which it had liberated about 2 months earlier: the northern parts of Luxembourg.

The three main regiments occupied positions along the Luxembourg-German border, marked by the Our and Süre rivers and facing the "Siegfried Line", also called "Westwall" by the Germans. This was a more than 600 km long barrier of concrete bunkers and obstacles, built by the Nazis in the late 1930s and meant to stop any armies trying to invade their "Reich".

The area, which lies in front of you, was held from mid-November 1944 on by the 112ᵗʰ Infantry Regiment of the 28ᵗʰ US Infantry Division. The zone of this regiment extended from Lützkampen (D) in the north to Lieler (L) in the south. Most of the positions of this unit lay on the east side of the Our river, since American units had pierced the German "Westwall" here already in September 1944.

When all hell broke loose early in the morning of Saturday, December 16, 1944, the 112ᵗʰ Infantry Regiment was hit hard by the 560ᵗʰ Volks Grenadier Division and the 116ᵗʰ Panzer Division on their push towards the west.

While initial German attacks could be repulsed by the men of the 112ᵗʰ Infantry Regiment and their supporting units, the enemy pressure grew however stronger and stronger so that all units were ordered to withdraw across the Our river and assemble around Weiswampach (L) on the evening of December 17, 1944.

With enemy ranks being encountered in Troisvierges (L), Colonel Gustin Nelson, Commanding Officer of the 112ᵗʰ Infantry Regiment, ordered his men to pull back towards Huldange (L) early in the morning of December 19, 1944 and to establish defensive positions in the Beiler (L) – Leithum (L) area.

By that time, all communication with Division Headquarters in Wiltz (L) had been severed, so that the 112ᵗʰ Infantry Regiment became attached to the 106ᵗʰ US Infantry Division and was ordered to occupy new positions in the Beho (B) – Salmchâteau (B) area, to cover the southern flank of the 106ᵗʰ US Infantry Division zone.

This troop movement was accomplished by December 22, 1944 and the 112ᵗʰ Infantry Regiment and its supporting units continued covering the withdrawal of the 106ᵗʰ US Infantry and 7ᵗʰ US Armored Division from the St. Vith (B) area.

With their stubborn resistance against overwhelming enemy forces in this area, the outnumbered and outgunned men of the 112ᵗʰ Infantry Regiment and its supporting units definitely slowed down the German advance in this sector and bought precious time for other American units to move in this area and organize new defensive positions that ultimately stopped the enemy advance. The price for this endeavour was however high and the men of the Keystone Division were often neglected the credit they deserve for their efforts and sacrifices.

Dan Reiland 2019

LA DÉFENSE DE LA POINTE NORD DU LUXEMBOURG PAR LE 112ᵉ RÉGIMENT D'INFANTERIE US

Quand la 28ᵉ Division d'Infanterie US quittait la « forêt de Hürtgen », au sud d'Aix-la-Chapelle (D) vers la mi-novembre 1944, elle souffrait des pertes considérables, encourues lors des lourds combats des semaines précédentes dans cette région. La division fut alors transférée dans une zone de repos qu'elle avait libérée 2 mois avant : le nord du Luxembourg.

Les trois régiments principaux occupaient des positions le long de la frontière luxembourgeoise avec l'Allemagne, marquée par les rivières Our et Sûre, face à la « Ligne Siegfried », appelée par les Allemands « Westwall ». Ceci constituait une barrière de casemates et d'obstacles en béton armé, longue de plus de 600 km et construite par les Nazis à la fin des années 1930 pour empêcher une invasion de leur « Reich ».

La région devant vous a été tenue depuis la mi-novembre 1944 par le 112ᵉ Régiment d'Infanterie de la 28ᵉ Division d'Infanterie US. La zone de ce régiment s'étendait de Lützkampen (D) au nord jusqu'à Lieler (L) au sud, avec le quartier général du régiment à Ouren (B). La plupart des positions de cette unité étaient situées du côté est de l'Our, car des unités américaines avaient déjà percé ici le « Westwall » au mois de septembre 1944.

Quand les combats éclataient au petit matin du samedi, le 16 décembre 1944, le 112ᵉ Régiment d'Infanterie subissait de féroces attaques par la 560ᵉ Volksgrenadier Division et la 116ᵉ Panzer Division.

Bien que les hommes du 112ᵉ Régiment et leurs unités de support aient pu repousser les attaques initiales des agresseurs, la pression ennemie augmentait tellement que toutes les unités américaines engagées furent ordonnées de se replier autour de Weiswampach (L) le soir du 17 décembre 1944.

Des chars ennemis localisés à Troisvierges (L) forcèrent le colonel Gustin Nelson, Commandant du 112ᵉ Régiment d'Infanterie, à replier ses hommes vers Huldange (L) au matin du 19 décembre 1944 et à établir des positions défensives autour de Beiler (L) et Leithum (L).

À partir de ce moment, le 112ᵉ Régiment d'Infanterie fut rattaché à la 106ᵉ Division d'Infanterie US, qui se situait au nord de cette unité, car toutes les voies de communication avec le quartier général divisionnaire, situé à Wiltz (L), étaient coupées. Ainsi le régiment allait occuper de nouvelles positions dans la région de Beho (B) – Salmchâteau (B) pour protéger le flanc sud de la 106ᵉ Division d'Infanterie US. Ce changement de position fut accompli le 22 décembre 1944 et ainsi le 112ᵉ Régiment d'Infanterie et ses unités de support continuaient à couvrir la retraite de la 106ᵉ Division d'Infanterie US et de la 7ᵉ Division Blindée US du secteur de St. Vith (B).

Avec leur résistance acharnée contre des forces ennemies en surnombre, les hommes du 112ᵉ Régiment d'Infanterie et leurs unités de support ont réussi à ralentir l'offensive allemande dans ce secteur et ont ainsi permis à d'autres unités américaines de se déployer dans la région et d'organiser la défense qui finalement brisait l'avance ennemie. Le prix de cet engagement fut lourd, mais les efforts et sacrifices des hommes de la « Keystone Division » n'ont trop souvent pas été appréciés à leur juste valeur.

DIE VERTEIDIGUNG DER NORD-SPITZE LUXEMBURGS DURCH DAS 112. US INFANTERIE-REGIMENT

Als die 28. US Infanterie-Division Mitte November 1944 aus dem Hürtgenwald, südlich von Aachen, abgezogen wurde, litt die Division unter den schweren Verlusten, die sie während den heftigen Kämpfen der vergangenen Wochen in dieser Gegend hatte einbüßen müssen. Die Division wurde dann in eine als ruhige Zone eingestufte Gegend entsandt, die sie zwei Monate zuvor befreit hatte: der Norden Luxemburgs.

Die drei Hauptregimente hielten ab dann Stellungen entlang der Flüsse Our und Sauer, die die luxemburgisch-deutsche Grenze bildeten und der „Siegfried Linie", von den Deutschen „Westwall" genannt, gegenüber lagen. Hierbei handelte es sich um einen rund 600 km langen Wall aus Betonbunkern und Hindernissen, der Ende der 1930er Jahre von den Nazis errichtet wurde, um etwaige Feinde daran zu hindern, in ihr „Reich" einzufallen.

Die Gegend, die vor Ihnen liegt, wurde ab Mitte November 1944 vom 112. Infanterie Regiment der 28. US Infanterie-Division gehalten. Der Bereich dieses Regiments erstreckte sich von Lützkampen (D) im Norden, bis Lieler (L) im Süden. Das Regimentshauptquartier befand sich in Ouren (B). Die meisten Stellungen dieser Einheit lagen östlich der Our, da amerikanische Einheiten den Westwall hier bereits im September 1944 durchbrochen hatten.

Als am frühen Samstagmorgen, den 16. Dezember 1944, das 112. Infanterie-Regiment auf einmal schweren Angriffen der 560. Deutschen Volksgrenadier- und der 116. Deutschen Panzer-Division ausgesetzt war, war die Hölle in diesem Bereich los. Während die Männer des 112. Infanterie-Regiments und deren Unterstützungseinheiten anfangs die deutschen Angriffe Richtung Westen noch abwehren konnten, so wurde der gegnerische Druck allerdings immer stärker, sodass alle Einheiten am Abend des 17. Dezember 1944 befehligt wurden, sich über die Our zurückzuziehen und im Raume Weiswampach (L) neue Stellungen zu beziehen.

Als Oberst Gustin Nelson, Befehlshaber des 112. Infanterie-Regiments, am frühen Morgen des 19. Dezember 1944 feindliche Panzer in Troisvierges (L) gemeldet wurden, befahl er seinen Truppen sich nach Huldange (L) zurückzuziehen und Verteidigungsstellungen im Raume Beiler (L) – Leithum (L) zu errichten.

Da zu dieser Zeit die gesamte Kommunikation mit dem Divisionshauptquartier in Wiltz (L) unterbrochen war, wurde das Regiment der 106. US Infanterie-Division angegliedert und sogleich befehligt, neue Stellungen im Raume Beho (B) – Salmchâteau (B) zu beziehen, um hier die südliche Flanke dieser Division abzusichern.

Diese Truppenbewegung war am 22. Dezember 1944 abgeschlossen und das 112. Infanterie Regiment und seine Unterstützungseinheiten sicherten dann weiterhin den Rückzug der 106. US Infanterie und der 7. US Panzer-Division aus dem Großraum St. Vith (B) ab.

Mit ihrem hartnäckigen Widerstand gegen überwältigende feindliche Kräfte gelang es den zahlenmäßig und waffentechnisch unterlegenen Männern des 112. Infanterie-Regiments und seinen Unterstützungseinheiten den deutschen Vorstoß in diesem Sektor deutlich zu verzögern. Diese wertvolle Zeit ermöglichte die Verlegung weiterer amerikanischer Truppen in diesen Frontabschnitt, die schlussendlich den feindlichen Vorstoß stoppen konnten. Der Preis für die Bemühungen war hoch und den Männern der „Keystone Division" blieb die Anerkennung für ihre Leistungen und Opfer leider oftmals verwehrt.

For more information
Pour en savoir plus
Für weitere Informationen

Information Panel Weiswampach

Information Panel Weiswampach

REMEMBER US
WEISWAMPACH

THE 112TH US INFANTRY REGIMENT DEFENDS THE NORTHERN TIP OF LUXEMBOURG

DEC 16 - 22, 1944

To Weiswampach | Bridge | Bridge | To Welchenhausen

To Sevenig

To Lutzkampen

River OUR

Ouren

River OUR

Source: Hugh M. Cole
The Ardennes: Battle of the Bulge

For more information
Pour en savoir plus
Für weitere Informationen

Information Panel Weiswampach

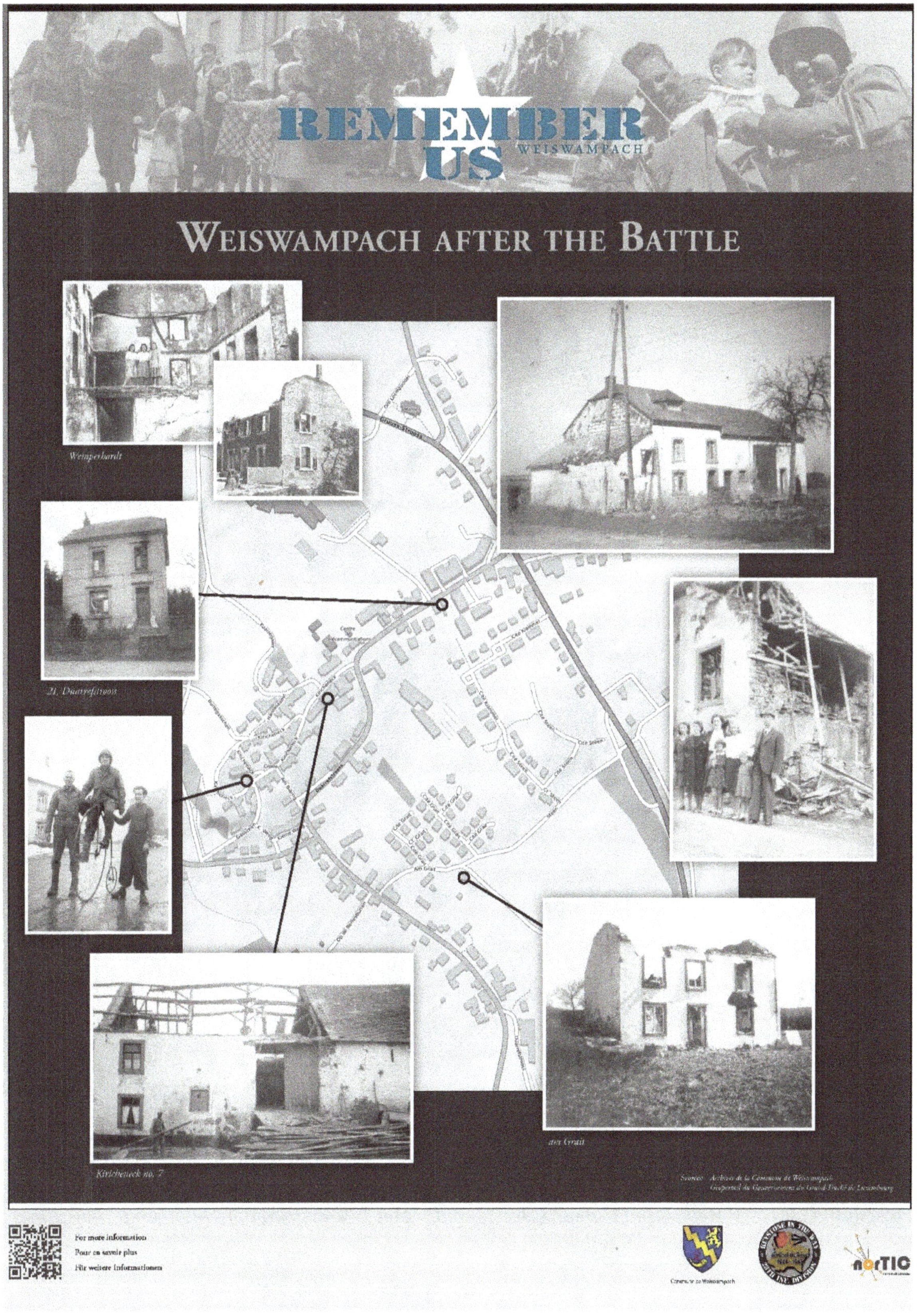

Weiswampach

Monument to C Company, 134th Infantry Regiment, 35th Infantry Division

Route N7

Weiswampach

Latitude: 50.1417

Longitude: 6.0771

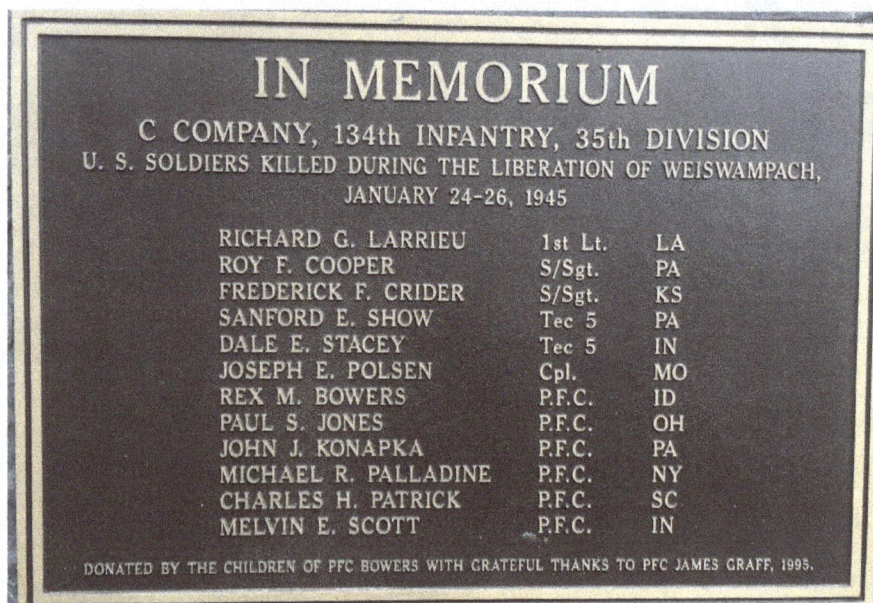

Monument lists the following names as Killed in Action: 1st Lt Richard G. Larrieu; S/Sgt Roy F. Cooper; S/Sgt Frederick F. Crider, Tec5 Sanford E. Show; Tec5 Dale E. Stacey; CPL Joseph E. Polsen; PFC Rex M. Bowers; PFC Paul S. Jones; PFC John J. Konapka; PFC Michael R. Palladine; PFC Charles H. Patrick; PFC Melvin E. Scott.

Wiltz

Memorial to the 28th Infantry Division which liberated Wiltz on September 10, 1944. There are two information panels at the site also.

Grand-Rue 2

9530 Wilz

Latitude: 49.966473

Longitude: 5.937924

Information Panel Wiltz

REMEMBER WILTZ

THEY PLAYED THE BAZOOKA BOOGIE

Those young men, who had brought their musical instruments into the training camps of the US Army, came from all parts of Pennsylvania.

Everywhere they played their tunes, during the maneuvers in the United States, their stay in England or on their way to Paris, where the proud boys of the 28th Division Band marched down the Champs-Élysées on August 29th, 1944. They were unaware that in December, at the beginning of the Battle of the Bulge a few months later, many of those young musicians had to exchange their instruments against carbines, bazookas and hand grenades.

When it became obvious in Wiltz that the Wehrmacht was getting closer and closer and that every available man was urgently needed on the slopes around the city, the musicians of the Division Band were no exception. Whether it was clarinet players, such as T/5 Joseph Gambo, T/4 Francis Telesca, or trumpet players like T/4 Kenneth Myers or Paul McCoy, the latter joining the army straight out of the Jimmy Dorsey Band, they all grabbed weapons to defend the city.

They all showed enormous bravery in trying to defy the approaching enemy. Many of these young musicians lost their lives during those cold December days in the futile attempt to defend Wiltz.

Young men and members of the 28th Division Band like T/5 Jimmy Pitlik, Pfc John Ikey Forman or Joseph Gambo were never to take up their instruments again. Their fate befell them on the hills of Wiltz and the last thing they heard were not melodic sounds, but the sounds of approaching tanks and exploding shells.

In the middle of the bullet hail, brave men like Pfc Leighton Collins and Ben Walker tried in vain to save the band's instruments and musical scores. The loaded trucks came under German fire and burned down a few kilometers from Bastogne.

The trombone player T/5 Elwood N. Dorward, the Drum Major like T/Sgt John Shuart or the second bandleader like Warrant Officer Richard I. Purvis, all of them played the "Bazooka Boogie" as well as they could.

All of them gave their very best and fought bravely.

From left to right, sitting: T/Sgt John F. Noon, PFC Stanley H. Moyer, T/4 Frank Telesca; standing: T/4 Elwood N. Dorward, PFC Leighton Collins

ILS JOUENT LE BAZOOKA BOOGIE

Les jeunes hommes qui amenaient leurs instruments aux camps d'entraînement de l'armée américaine venaient de tous les coins de la Pennsylvanie.

Ils jouaient partout leurs mélodies pendant les manœuvres aux États-Unis, en Angleterre, même en route pour Paris où les jeunes soldats de l'orchestre de la 28ème Division marchaient fièrement le long des Champs Élysées le 29 août 1944. Ils ne s'attendaient pas qu'en décembre, au début de la Bataille des Ardennes, ils devaient échanger leurs instruments contre des carabines, bazookas et des grenades à main.

Lorsqu'il était clair que l'armée allemande s'approchait vite de Wiltz et que chaque soldat était nécessaire pour défendre les alentours de la ville, les musiciens étaient aussi concernés. Des clarinettistes comme T/4 Joseph Gambo, T/4 Francis Telesca, ou des trompettistes comme T/4 Kenneth Myers ou Paul McCoy joignant directement l'armée en venant de la Jimmy Dorsey Band, saisissaient des armes pour défendre la ville. Tous les musiciens essayaient d'être de bons soldats et de tenir tête contre l'ennemi.

Beaucoup de ces jeunes hommes perdaient leur vie en défendant la ville de Wiltz et des musiciens comme T/5 Jimmy Pitlik, PFC John « Ikey » Forman ou Joseph Gambo ne retournaient plus jamais pour jouer leur instrument. Ils étaient frappés par leur sort. Avant de mourir, ils n'entendaient pas le son d'une mélodie, mais le bruit des explosions et des chars qui s'approchaient.

Pendant la bordée de balles, des soldats courageux comme PFC Leighton Collins et Ben Walker essayaient de sauver en vain les instruments et la bibliothèque d'orchestre, car les camions avec les instruments prenaient feu une dizaine de kilomètres avant Bastogne.

Les musiciens ont tout donné. Des hommes courageux comme le joueur de trombone T/5 Elwood N. Dorward, le tambour-major T/Sgt John Shuart ou le dirigent assistant Warrant Officer Richard I. Purvis ont joué le « Bazooka Boogie » le mieux possible.

SIE SPIELTEN DEN BAZOOKA BOOGIE

Aus alle Teilen Pennsylvanias kamen jene jungen Männer, welche ihre Musikinstrumente mit in die Ausbildungslager der US Armee brachten.

Überall spielten sie ihre Melodien, während der Manöver in den Staaten, ihres Aufenthaltes in England bis hin nach Paris, wo die stolzen Jungs der 28. Divisionsmusikkapelle am 29. August 1944 die Champs Élysées entlang marschierten, nichtsahnend, dass im Dezember bei Beginn der Ardennenoffensive viele der jungen Musiker ihre Instrumente gegen Karabiner, Bazookas und Handgranaten tauschen mussten.

Als in Wiltz klar wurde, dass die Wehrmacht immer näher rückte und jeder verfügbare Mann irgendwie und irgendwo an den Hängen um die Stadt gebraucht wurde, traf es auch die Musiker der Divisionsband. Ob Klarinettisten wie T/4 Joseph Gambo, T/4 Francis Telesca, Trompeter wie T/4 Kenneth Myers oder Paul McCoy, welcher geradewegs aus der „Jimmy Dorsey Band" kam, sie alle schnappten sich Waffen um die Stadt zu verteidigen.

Alle versuchten sie gute Soldaten zu sein und dem immer näher kommenden Feind die Stirn zu bieten. Viele der Musiker verloren in jenen Dezembertagen ihr Leben bei dem aussichtslosen Versuch Wiltz zu verteidigen.

Junge Männer und Angehörige der 28. Divisionskapelle wie T/5 Jimmy Pitlik, PFC John „Ikey" Forman oder Joseph Gambo sollten nie wieder zu ihren Instrumenten greifen können. Ihr Schicksal ereilte sie auf den Anhöhen von Wiltz und das Letzte was sie hörten, war nicht der Klang von Melodien, sondern die Geräusche von heranrückenden Panzern und explodierenden Granaten.

Tapfere Männer wie PFC Leighton Collins und Ben Walker versuchten im Hagel der Geschosse die Instrumente und die Bibliothek der Band zu retten, doch ergebnislos. Die beladenen Laster gerieten unter deutschen Beschuss und verbrannten einige Kilometer vor Bastogne.

Ob Posaunisten wie T/5 Elwood N. Dorward, Tambourmajor wie T/Sgt John Shuart oder stellvertretende Bandleader wie Warrant Officer Richard I. Purvis, sie alle spielten den „Bazooka Boogie", so gut wie sie es vermochten.

Alle gaben ihr Bestes und schlugen sich tapfer.

Information Panel Wiltz

REMEMBER WILTZ

THE SURROUNDING OF THE CITY OF WILTZ

While trying to escape the surrounding of the city of Wiltz by the German army, several American soldiers belonging to the 28th Infantry Division were captured. Among those were Sergeant Frank McClelland of the 28th Military Police Platoon and PFC Wendell H. House of the 28th Signal Company.

Frank McCLELLAND

DR. KIMMELMAN

Captain Benedict B. Kimmelman served as a dentist in the 28th Infantry Division. His aid station was located next to the Café Thillens in Niederwiltz.

When on December 18th 1944, the Germans approached the town of Wiltz and forced the American soldiers to retreat, the doctor volunteered to stay with the wounded and continue operating the infirmary. Another volunteer, Sergeant James L. Moffett, also stayed behind to support Dr. Kimmelman.

In the afternoon of December 19th, when it became more and more obvious that the aid station would also have to evacuate its wounded soldiers, Dr. Kimmelman and Sgt. Moffett put together a convoy and tried to escape from the town in order to eventually catch up with the columns which had already left Wiltz.

After a long odyssey and desperate attempts to escape the Germans, Dr. Kimmelman, Sgt. Moffett and the adjutant band-leader of the Division, Richard (Dick) Purvis, together with their wounded, were taken prisoner around 6.30 a.m. at the spot where the road Nothum-Wiltz branches off towards Roullingen.

The infirmary and temporary dental practice of Dr Kimmelman in Wiltz, taken after the offensive in 1945.

L'ENCERCLEMENT DE LA VILLE DE WILTZ

En essayant de s'échapper de l'encerclement de la ville de Wiltz par l'armée allemande, mains soldats américains de la 28ème Division d'Infanterie ont été capturés.

Parmi eux se trouvaient Sergent Frank McClelland de la 28ième Compagnie de la Police Militaire, ainsi que PFC Wendell H. House de la 28ème Compagnie de Transmissions.

DR. KIMMELMAN

Le Capitaine Benedict B. Kimmelman servait comme dentiste à la 28ème Division d'Infanterie.

Sa station de premiers soins était installée à côté du café Thillens à Niederwiltz.

Lorsque les troupes allemandes s'approchaient de Wiltz le 18 décembre 1944, les soldats américains se voyaient forcés de se retirer. Le docteur B. Kimmelman se portait alors volontaire pour rester avec les blessés afin d'assurer le bon fonctionnement de la station de premiers secours.

Sans hésiter, Sergent James L. Moffett donnait volontairement support au Dr. B. Kimmelman et restait avec lui.

Pendant l'après-midi du 19 décembre 1944, il devenait de plus en plus évident que la station de premiers secours devait être évacuée. Le Dr. Kimmelman et Sgt. Moffett ont alors organisé un convoi et ont essayé de s'échapper de la ville pour rattraper les colonnes américaines déjà en route.

Après une longue odyssée et de nombreux essais pour échapper aux Allemands, le Dr. Kimmelman, Sgt. Moffett et le dirigent-assistant Warrant Officer Richard (Dick) Purvis de la Musique Divisionaire de la 28ème, ont été capturés avec les blessés le 20 décembre 1944 vers 6.30 heures à la bifurcation Nothum-Wiltz-Roullingen.

DIE UMSCHLIESSUNG DER STADT WILTZ

Beim Versuch der drohenden Umschliessung der Stadt Wiltz durch die Wehrmacht zu entkommen, gerieten weitere amerikanische Soldaten der 28. Infanterie Division in deutsche Gefangenschaft. Unter ihnen befanden sich auch Sergeant Frank McClelland von der 28. Militärpolizeikompanie, sowie PFC Wendell H. House von der 28. Fernmeldeeinheit.

Wendell H. HOUSE

DR. KIMMELMAN

Captain Benedikt B. Kimmelman diente als Zahnarzt bei der 28. Infanterie Division. Seine Erste Hilfe Station befand sich neben dem Café Thillens in Niederwiltz. Als die Wehrmacht sich am 18. Dezember 1944 der Stadt Wiltz näherte und die amerikanischen Soldaten den Rückzug antreten mussten, erklärte sich der Arzt bereit, bei den Verwundeten zu bleiben, um somit die Krankenstation weiter zu betreiben.

Ein weiterer Freiwilliger, Sergeant James L. Moffett, blieb ebenfalls als Unterstützung bei Dr. Kimmelman zurück.

Am späten Nachmittag des 19. Dezember, als immer offensichtlicher wurde, dass auch die Erste Hilfe Station ihre Verwunderten evakuieren musste, stellten Captain Kimmelman und Sergeant Moffett einen Konvoi zusammen und versuchten aus der Stadt zu fliehen, um jene Kolonnen, welche Wiltz schon verlassen hatten, doch noch einzuholen.

Nach einer langen Odyssee und vielen verzweifelten Versuchen, den anrückenden deutschen Einheiten doch noch zu entkommen, gerieten Dr. Kimmelman, Sergeant Moffett, sowie der stellvertretende Dirigent Warrant Officer Richard (Dick) Purvis am 20. Dezember gegen 6.30 Uhr an der Strassengabelung Nothum-Wiltz-Roullingen mitsamt ihren Verwunderten in deutsche Gefangenschaft.

Rechts: Sgt. James L. MOFFETT, links: Sgt. C. E. KUNEN

Wiltz

To PFC Richard Brookins, 28th Infantry Division who starred as Saint Nicolas to the Children of Wilz in December 1944.

Grand-Rue 2

9530 Wilz

Latitude: 49.966473

Longitude: 5.937924

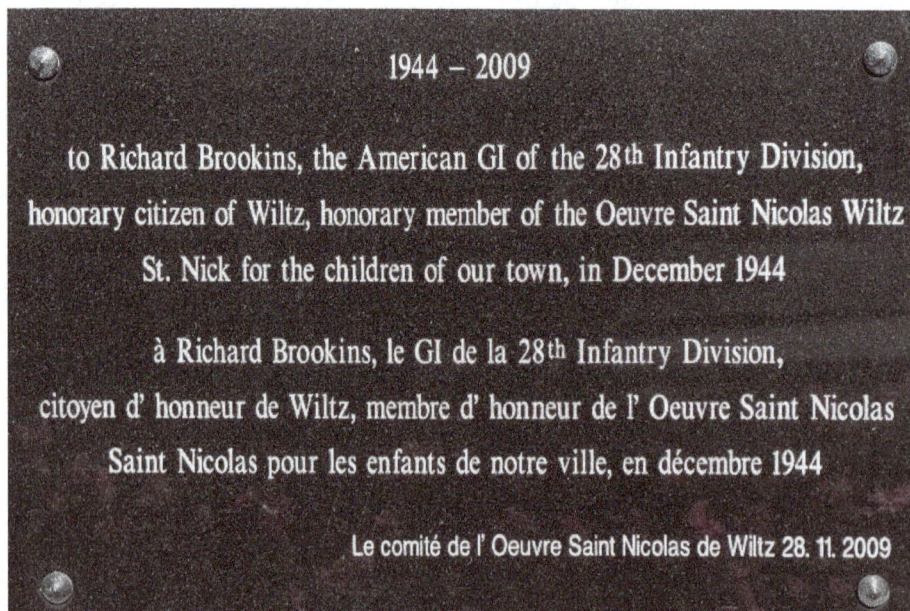

1944 – 2009

to Richard Brookins, the American GI of the 28th Infantry Division,
honorary citizen of Wiltz, honorary member of the Oeuvre Saint Nicolas Wiltz
St. Nick for the children of our town, in December 1944

à Richard Brookins, le GI de la 28th Infantry Division,
citoyen d' honneur de Wiltz, membre d' honneur de l' Oeuvre Saint Nicolas
Saint Nicolas pour les enfants de notre ville, en décembre 1944

Le comité de l' Oeuvre Saint Nicolas de Wiltz 28. 11. 2009

Information Panel Wiltz

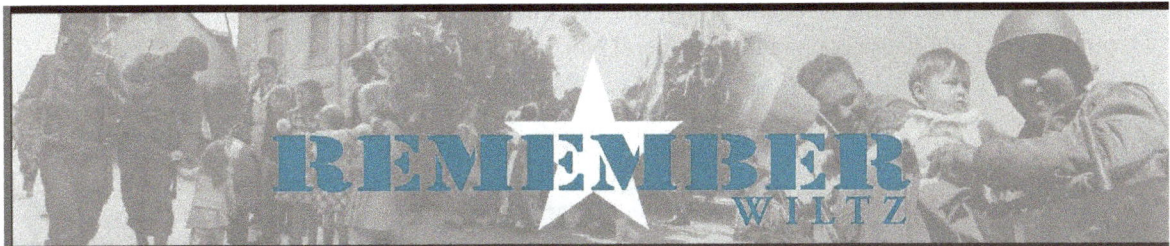

THE AMERICAN SAINT NICK

During November 1944 the soldiers of the 28th Infantry Division were stationed in Wiltz and environs in order to recuperate from the severe pounding which they had suffered in the combats of Hürtgen Forest near Aix-la-Chapelle. Their duties were lighter than in combat and thus it came about that many of the soldiers had conversations with the inhabitants of Wiltz, including the G-2 soldiers Richard Brookins and Harry Stutz.

During these conversations Harry Stutz found out that many local children had never witnessed a St. Nicholas day celebration for lack of candies during the German occupation. Stutz contacted General Norman D. Cota, 28th Infantry division, to ask permission to organize a St. Nick party for the children of Wiltz. After receiving permission, Harry Stutz went to work. The American soldiers gladly donated candies and chocolate while the unit bakery produced donuts and other sweets for the children.

Private First Class Harry Stutz.

Stutz then tried to convince his best friend, Corporal Richard Brookins, who was at first hesitant, to play Saint Nicholas. On December 5th, the American St. Nick appeared in a jeep in the company of two young girls from Wiltz in the attire of angels. Many civilians and soldiers, for whom this was a new experience, were present.

Songs were sung and poems were recited by the local school children in honor of St. Nick, while sweets were handed out by the soldiers to young and old. For many of the children it was the first time in their lives to enjoy chocolate.

This kind and generous gesture of the American soldiers, who had themselves suffered bitterly, has never been forgotten. Richard Brookins returned to Wiltz many times after the end of the war and contributed to keep up and rekindle the spirit of the "American St. Nick".

Saint Nick is driving with his angles Jeanny Schiltbour and Liliane Wampach into the court of the castle in 1944.

LE SAINT NICOLAS AMÉRICAIN

À partir du 18 novembre 1944 les soldats de la 28ième Division d'Infanterie étaient en position de repos à Wiltz et alentours pour régénérer des lourdes pertes subies aux combats au « Hürtgenwald » près d'Aix-la-Chapelle. Comme le service était simple, les soldats ont souvent trouvé le temps de prendre contact avec la population civile. Ainsi faisaient les deux G.Is Richard Brookins et Harry Stutz qui appartenaient à la 28ième Compagnie de Transmissions.

Stutz avait pris connaissance que les enfants de Wiltz n'avaient été comblés de friandises par Saint Nicolas depuis l'occupation du Luxembourg par les Nazis en mai 1940. Il contactait son supérieur le Général Norman D. Cota, commandant de la 28ième Division d'Infanterie, pour demander la permission d'organiser une fête de Saint Nicolas.

L'autorisation lui a été accordée et les soldats ont de suite renoncé à leurs sucreries et chocolats. La boulangerie divisionnaire a aussi préparé des donuts pour les petits.

Alors Harry Stutz a dû convaincre son meilleur ami le corporal Richard Brookins à jouer le rôle de Saint Nicolas. Celui-ci a accepté après de courtes réflexions.

Le 5 décembre 1944 Saint Nicolas faisait son apparition au centre de Wiltz dans une jeep américaine. Il était accompagné de deux anges, deux jeunes filles du pensionnat. C'est tel qu'il se présentait à la population de Wiltz, petits et grands, et aux soldats américains.

Des chansons ont été présentées et des poèmes ont été récités dans la cour du château en honneur du grand Saint. Des enfants de Wiltz recevaient la première fois de leur vie des sucreries, du chocolat et surtout des donuts dans la grande salle du château.

Ce geste des soldats américains, qui avaient beaucoup souffert pendant les durs combats au « Hürtgenwald », n'a jamais été oublié jusqu'à ce jour.

De 1977 à 2011 Richard Brookins est retourné 7 fois à Wiltz pour représenter Saint Nicolas et combler les enfants de Wiltz et de la région de friandises.

Saint Nicolas et t'jemi avec sei angel d'anan en 1994.

DER AMERIKANISCHE «ST. NICK»

Im November 1944 lagen die Soldaten der 28ten U.S. Infanterie Division in und um Wiltz in Ruhestellung, um sich von den Kämpfen im Hürtgenwald nahe Aachen zu erholen. Nebst leichtem Dienst, den die Soldaten verrichteten, kam es auch öfters vor, dass junge Amerikaner sich mit der Zivilbevölkerung von Wiltz unterhielten. So auch die beiden jungen Männer Richard Brookins und Harry Stutz, welche Mitglieder der 28ten Nachrichtenkompanie waren.

Stutz erfuhr bei Gesprächen mit den Einheimischen, dass viele Wiltzer Kinder noch nie ein Nikolausfest erlebt hatten, da dies unter deutscher Besatzung aus Mangel an Süßwaren nicht möglich war.

Harry Stutz kontaktierte den Befehlshaber der Division, General Norman D. Cota, und holte sich dessen Erlaubnis ein, ein Nikolausfest für die Wiltzer Kinder zu organisieren.

Corporal Richard Brookins as the American St. Nick on 5th Dezember 1944 in Wiltz.

Die Soldaten der 28ten Infanterie Division legten ihre Süßigkeiten zusammen und die Divisionsbäckerei stellte Backwaren her.

PFC Stutz überredete seinen besten Freund Richard Brookins den Nikolaus zu spielen. Dieser zögerte erst, willigte jedoch dann ein und erschien am 5. Dezember 1944 als „American St. Nick" in einem Jeep, flankiert von zwei luxemburgischen Mädchen.

Schulkinder trugen dem hohen Himmelsgast ihre Gedichte und Lieder vor, dies im Beisein von zahlreichen US Soldaten, sowie Zivilisten, während Nikolaus seine Geschenke austeilte.

Durch diese schöne Geste seitens der amerikanischen Soldaten, welche den Einwohnern von Wiltz auch nach dem Krieg in Erinnerung blieb, ergab es sich, dass Richard Brookins auch in der Nachkriegszeit, der Tradition folgend, noch oft nach Wiltz reiste, um die Kinder als „American St. Nick" zu beschenken.

Santa Nikolaus feiert 2009 überreich in einem Jeep in der Schloßhof.

Wiltz – Villa Adler

Villa Adler was used at the Headquarters of the 28th Infantry Division from November 20 to December 19, 1944 under command of General Norman Daniel "Dutch" Cota.

Rue du Château 28

9516 Wilz

Latitude: 49.96647

Longitude: 5.93536

REMEMBER WILTZ

«KEYSTONE IN THE WAY» WILTZ

DECEMBER 15, 2019

VILLA ADLER
Headquarters of the 28th US Infantry Division from November 20 to December 19, 1944
Commander:
General Norman Daniel "Dutch" COTA, Sr.
Born on May 30, 1893
Died on October 4, 1971

VILLA ADLER
Quartier général de la 28e Division d'Infanterie US du 20 novembre au 19 décembre 1944
Commandeur:
Général Norman Daniel »Dutch« COTA, senior
Né le 30 mai 1893
Décédé le 4 octobre 1971

VILLA ADLER
Hauptquartier der 28. US-Infanteriedivision vom 20. November bis zum 19. Dezember 1944
Befehlshaber:
General Norman Daniel „Dutch" COTA, Sen.
Geboren am 30. Mai 1893
Gestorben am 4. Oktober 1971

Dwight Eisenhower meeting Norman Cota, 1944

General Marshall outside «Villa Adler», November 1944

Wiltz – Eisenhower Square

Dedicated to General D. Eisenhower who stayed in Wiltz on November 8, 1944.

Rue de Chateau 36

9516 Wilz

Latitude: 49.9665

Longitude: 5.9368

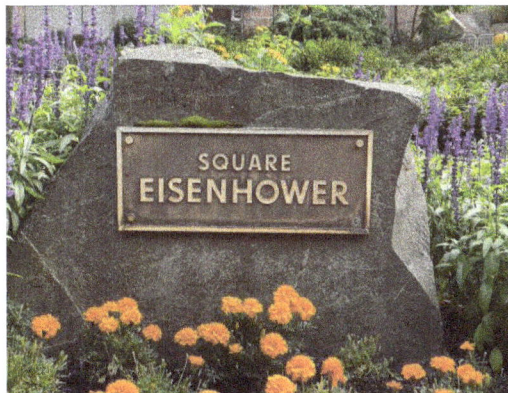

The gun barrel is from a US 3-inch anti-tank gun.

Wiltz

Sherman Tank "Blood and Guts" of the 707th Tank Battalion, used in support of the 110th Regiment, 28th US Infantry Division. Site includes four information panels on the 707th, and their combat operations in the Wiltz area.

Rue d'Ettelbruck

9552 Wiltz

Latitude: 49.9637

Longitude: 5.9383

Information Panel Wiltz

REMEMBER
WILTZ

A Tank of the 707ᵀᴴ Tank Battalion in Wiltz

On December 19, 1944, as the pressure from the German troops on the encircled town of Wiltz became too strong, Lieutenant-Colonel R. W. Ripple's tank had retreated from its position at the Café Halt to Erpeldange/Wiltz. While looking for a good firing position, it crashed backwards into the Clees's house and barn gables. These collapsed and the tank continued towards the Krischler's house and then ran into the dung pit, where it sank. In spite of all efforts, it bogged down in the morass repeatedly and became stuck. The crew had no other choice but to abandon the vehicle. However, minutes later, the Germans captured the colonel with his crew who spent the rest of the war as prisoners of war in a prison camp in Germany.

The tank in the dung pit in Erpeldange/Wiltz
Le char dans la fosse fumière à Erpeldange/Wiltz
Der Panzer in der Mistgrube in Erpeldange/Wiltz
Photo: Archives Vic Weber

In 1946, about 12 months after the war ended, the tank was removed from the dung pit and relocated. A few German prisoners of war, guarded in a camp in Wiltz by Luxembourger soldiers, demolished the enclosure wall of the dung pit. Then with the help of a truck loaded with wood and equipped with a cable winch, they worked for hours to remove the tank from its awkward predicament. They discovered that the engine was damaged, but the steering was operational, so a column of four trucks bound together pulled the tank, while one truck was fixed behind in order to brake the tank when going downhill. Leo Wilmes, who had been a compulsory recruit and a tank driver in the German army, guided the tank together with a German prisoner of war to the Place des Martyrs, where it remained for nearly 70 years.

The tank 3038800 leaves Erpeldange to Wiltz
Le char 3038800 quitte Erpeldange pour Wiltz
Der Panzer 3038800 verlässt Erpeldingen und wird nach Wiltz geschleppt
Photo: Archives Vic Weber

Un char du 707ᴱ Bataillon de chars à Wiltz

Le 19 décembre 1944, la pression des troupes allemandes devenant trop forte, le Lieutenant-Colonel R. W. Ripple décida d'abandonner la position de son char au Café Halt et de le diriger vers Erpeldange/Wiltz. En cherchant une bonne position de tir, le tank recula dans le pignon de la maison et de la grange Clees qui s'écroula. Il s'avança alors vers la maison Krischler. Toutefois, il s'enfonça dans la fosse fumière de cette ferme. En dépit de tous les efforts pour sortir de ce marais, le véhicule s'enlisa de plus en plus et ne parvint pas à se libérer. Il ne resta donc à l'équipage que d'abandonner son char. Quelques minutes plus tard, les Allemands arrêtèrent le colonel et son équipage qui passèrent le reste de la guerre dans un camp de prisonniers en Allemagne.

A few villagers of Erpeldange deliberate how to remove the tank
Quelques habitants d'Erpeldange délibèrent comment ils vont enlever le char
Einige Erpeldinger Bewohner beraten, wie sie den Panzer fortschaffen können
Photo: Archives Vic Weber

Au milieu de l'année 1946, le char fut dégagé de la fosse fumière et remorqué à Wiltz. Des prisonniers de guerre allemands, gardés dans un camp à Wiltz par des soldats luxembourgeois, furent chargés de démolir le mur d'enceinte de la fosse fumière.

Un camion chargé de bois et équipé d'un treuil, réussit, après des heures de travail, à dégager le char de sa situation difficile. Quatre camions furent attelés en file indienne devant le char, tandis qu'un camion fut attaché par derrière pour freiner le char dans une rue en pente. Le moteur était défectueux, mais la direction était toujours opérationnelle. Léon Wilmes, enrôlé de force et conducteur de char dans l'armée allemande, fit manœuvrer le char avec l'aide d'un prisonnier de guerre allemand jusqu'à la Place des Martyrs où il resta près de 70 ans.

Sherman M1 (75) 3038800, 707ᵗʰ Tank Battalion, B-9 "BLOOD & GUTS" in Erpeldange/Wiltz
Photo: Archives Vic Weber

Ein Panzer des 707. Tank-Bataillons in Wiltz

Am 19. Dezember 1944, als der Druck der deutschen Truppen zu stark wurde, entschied Oberstleutnant R. W. Ripple die Stellung seines Panzers am Café Halt aufzugeben und sich nach Erpeldingen/Wiltz zurückzuziehen. Bei der Suche nach einer guten Stellung, fuhr der Panzer rückwärts in den Giebel der Scheune und des Hauses Clees, sodass dieser zusammenkrachte. Er fuhr weiter neben das Haus Krischler. Doch hier geriet er in die Mistgrube, in die er einsackte. Trotz aller Anstrengungen, um aus diesem Morast herauszukommen, versank das Gefährt immer mehr und konnte sich nicht mehr befreien. Der Besatzung blieb nichts anders übrig, als ihr Gefährt zu verlassen. Minuten später, nahmen die Deutschen den Oberstleutnant mit seiner Mannschaft gefangen. Die Kriegsgefangenen verbrachten den Rest des Krieges in einem Kriegsgefangenenlager in Deutschland.

Family Clees, Erpeldange/Wiltz
Photo: Archives Vic Weber

Mitte 1946 wurde der Panzer nach Wiltz abgeschleppt. Deutsche Kriegsgefangene, die in einem Lager in Wiltz von Luxemburger Soldaten bewacht wurden, mussten die Einfriedungsmauer am Mischaufen abreißen. Ein mit einer Seilwinde ausgerüsteter und mit Holz beladener Lastwagen zog den Panzer nach stundenlanger Arbeit aus seiner misslichen Lage. Vier Lastwagen wurden hintereinander vor den Panzer gespannt, während ein LKW hinten angekoppelt wurde, um den Panzer bergabwärts zu bremsen. Der Motor war defekt, die Lenkung war jedoch noch betriebsbereit. Leo Wilmes, Zwangsrekrutierter und Panzerfahrer bei der Wehrmacht, lenkte den Panzer zusammen mit einem deutschen Kriegsgefangenen zur Place des Martyrs, wo er etwa 70 Jahre stand.

Sherman M1 (75) 3038800 in the main street of Wiltz
Dans la Grand-rue de Wiltz
In der Hauptstraße in Wiltz
Photo: Archives Vic Weber

Information Panel Wiltz

Information Panel Wiltz

REMEMBER WILTZ

The 707th Tank Battalion's Support to the 28th U.S. Infantry Division, Dec 44

5. Panzer-Armee (von Manteuffel)

7. Armee (Brandenberger)

The 707th TANK BATTALION in support of the 110th Regiment, 28th US Infantry Division
Le 707e Bataillon de chars en support du 110e Régiment, 28e Division d'Infanterie américaine
Das 707. Panzerbataillon in Unterstützung des 110. Regimentes der 28. US Infanteriedivision

THE 707TH TANK BATTALION IN SUPPORT OF THE 110TH REGIMENT, 28TH US INFANTRY DIVISION

In December 1944, the 707th Tank Battalion was attached to the 28th US Infantry Division and was in the Division's reserve. On December 16, 1944, at the beginning of the Battle of the Bulge, the 707th Tank Battalion was alerted and partitioned as follows:

- Companies A and B were assigned to the 110th Regiment (Clervaux)
- Company C came under the orders of the 109th Regiment (Diekirch)
- Company D (light tanks) was attached to the 112th Regiment (Weiswampach).

For four days, the tanks of Companies A and B, together with the defenders of the 110th Regiment and its attached units, offered stiff resistance to the attacking German units in Marnach, Munshausen, Clervaux, Hosingen, Bockholtz, Consthum, Hoscheid and Wiltz. Gradually, however, the stronger German "Panther" tanks eliminated many of the American tanks. The remaining tanks concentrated in Wiltz and bravely defended the town. When the Germans encircled the town, and the Americans had run out of ammunition and fuel, the crews destroyed their remaining tanks and fled to Belgium. - Total losses: 34 tanks from Company A and B, 17 light tanks of Company D and 19 tanks of Company B, Combat Command R, 9th US Armored Division that had fought in Clervaux, Reuler and Heinerscheid.

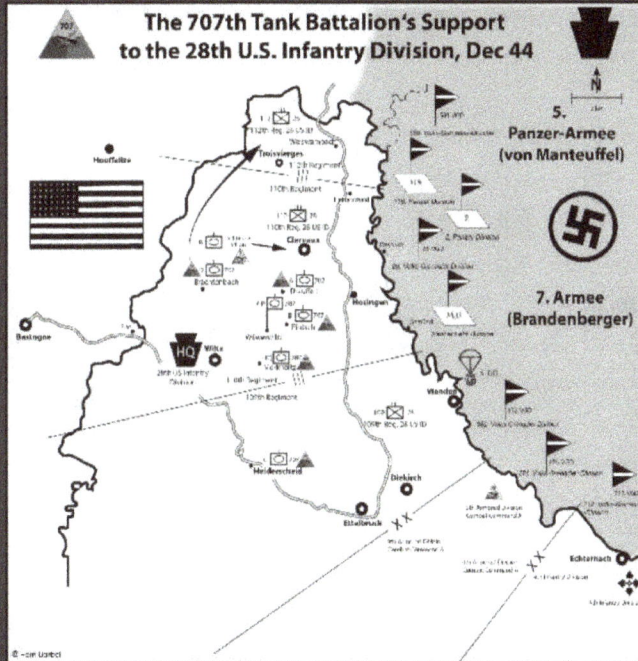

LE 707E BATAILLON DE CHARS EN SUPPORT DU 110E RÉGIMENT, 28E DIVISION D'INFANTERIE AMÉRICAINE

Au mois de décembre 1944, le 707e Bataillon de chars (707th Tank Battalion) était rattaché à la 28e Division d'Infanterie américaine et faisait partie de la réserve de la division. Au matin du 16 décembre 1944, au commencement de la Bataille des Ardennes, le 707th Tank Battalion fut alerté et réparti de la façon suivante :

- Les Compagnies A et B furent attribuées au 110e Régiment (Clervaux),
- la Compagnie C passa sous les ordres du 109 Régiment (Diekirch),
- la Compagnie D (chars légers) rejoignit le 112 Régiment (Weiswampach)

Pendant quatre jours, les chars des compagnies A et B ainsi que les défenseurs du 110e Régiment et les unités qui y étaient rattachées opposèrent une résistance farouche aux attaquants allemands dans les localités de Marnach, Munshausen, Clervaux, Hosingen, Bockholtz, Hoscheid et Wiltz. Toutefois, peu à peu, les chars allemands « Panther » plus forts éliminèrent de nombreux chars américains. Les chars restants se concentrèrent à Wiltz et défendirent la ville avec acharnement. La ville étant encerclée, les Américains manquant de munitions et de carburant, les équipages détruisirent leurs chars et se replièrent sur la Belgique. - Total des pertes: 34 chars des Compagnies A et B, 17 chars légers de la Compagnie D ainsi que 19 chars de la Compagnie B, Combat Command R, de la 9e Division Blindée Américaine qui avaient combattu à Clervaux, Reuler et Heinerscheid.

DAS 707. PANZERBATAILLON IN UNTERSTÜTZUNG DES 110. REGIMENTES DER 28. US INFANTERIEDIVISION

Im Dezember 1944 war das 707. Tank Battalion der 28. US Infanteriedivision angegliedert und befand sich in der Divisionsreserve. Am 16. Dezember 1944, am Beginn der Ardennenoffensive, wurde das 707. Tank Battalion um 07.00 Uhr alarmiert und folgendermaßen aufgeteilt:

- Kompanie A und B wurden dem 110. Regiment (Clerf) zugeordnet,
- Kompanie C wurde dem 109. Regiment (Diekirch) zugeteilt,
- Kompanie D (leichte Panzer) wurde dem 112. Regiment (Weiswampach) unterstellt.

Die Panzer der Kompanien A und B leisteten in Marnach, Munshausen, Clerf, Hosingen, Bockholtz, Consthum, Hoscheid und Wiltz mit den Verteidigern des 110. Regimentes und den angegliederten Einheiten den deutschen Angriffen während vier Tagen heftigen Widerstand. Viele Panzer wurden jedoch nach und nach von den stärkeren deutschen Panther Panzern ausgeschaltet.

Die übriggebliebenen Tanks konzentrierten sich in Wiltz und verteidigten die Stadt hartnäckig. Als die Stadt eingekesselt war, den Amerikanern die Munition und den Treibstoff fehlten, zerstörten die Besatzungen ihre Panzer und flüchteten nach Belgien. - Alle 34 Panzer der Kompanien A und B, die 17 leichten Panzer der Kompanie D sowie die 19 Panzer der Kompanie B, Kampfgruppe R (CCR), der 9. US Panzerdivision, die in Clerf, Reuler und Heinerscheid eingesetzt gewesen waren, waren zerstört.

Sources/Quellen:
- Zeitzeuge Herr Josy Clees
  Témoin de l'époque M. Josy Clees
  Contemporary witness Mr. Josy Clees
- Die Ardennenschlacht 1944-1945 in Luxemburg von Jean Milmeister, Editions Saint-Paul, Luxembourg, 1994, ISBN 2-87963-
- Panzer 1939-1945 von Jim Winchester, Gondrom Verlag GmbH 2009, ISBN 3-8112-1797-6
- U.S.Army (ETO 1944-45) Marquages et organisation von Emile Becker und Jean Milmeister, Imprimerie G. Willems, Dudelange, 1988
- Sherman: A History of the American Medium Tank by R.P. Hunnicutt, page 512, Taurus Enterprises, 1978

Information Panel Wiltz

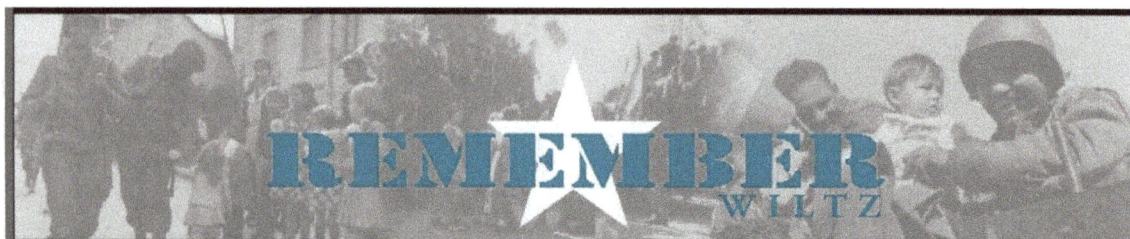

REMEMBER WILTZ

THE 707TH TANK BATTALION IN SUPPORT OF THE 110TH REGIMENT, 28TH US INFANTRY DIVISION

The 707th Tank Battalion was divided into 4 companies and numbered 729 men. It was equipped with 53 medium Sherman M4 (75 gun), 6 Sherman M4 (105mm Howitzer) and 17 light Stuart tanks. The Sherman tank was the most important tank of the Allies.

During the war, 49.234 Shermans of all types were built in the United States of America. The medium tank M4 (75) was equipped with a 75 mm gun, two .30 machine guns and one .50 machinegun. It weighed 30,5 tons, its engine had 350 HP, and its speed limit was 40 km/h. A 5-man crew operated it. The Sherman tank was inferior to the German "Panther" and "Tiger" tanks. It was only when it was equipped with a 76 mm cannon, that it was able to defy its adversaries.

LE 707E BATAILLON DE CHARS EN SUPPORT DU 110E RÉGIMENT, 28E DIVISION D'INFANTERIE AMÉRICAINE

Le 707th Tank Bataillon était divisé en quatre compagnies et comptait 729 hommes. Il était équipé de 53 chars medium Sherman M4 (canon de 75 mm), 6 Sherman M4 (105 mm obusier) et de 17 chars légers Stuart. – Le char Sherman était le char le plus important des Forces Alliées.

Pendant la guerre, les Américains en construisirent 49.234 exemplaires en différentes versions. Le char médium M4 (75) était équipé d'un canon de 75 mm, de 2 mitrailleuses .30 et d'une mitrailleuse .50. Il pesait 30,5 tonnes, son moteur avait une puissance de 350 CV, sa vitesse maximale était de 40 km/h. Un équipage de 5 hommes le manœuvrait. Il était inférieur aux chars allemands « Panther » et « Tiger ». Ce n'est qu'une fois équipé d'un canon de 76 mm qu'il était à même de se mesurer à ses adversaires.

DAS 707. PANZERBATAILLON IN UNTERSTÜTZUNG DES 110. REGIMENTES DER 28. US INFANTERIEDIVISION

Das 707. Tank Bataillon war in 4 Kompanien aufgeteilt und zählte 729 Mann. Es war mit 53 mittelschweren Sherman M4 (75 mm Kanone), 6 Sherman M4 (105 mm Haubitze) und 17 leichten Stuart Panzern ausgerüstet. – Der Sherman Panzer war der wichtigste Panzer der Alliierten.

Während des Krieges, stellten die Amerikaner 49.234 Exemplare in verschiedenen Versionen her. Der Medium Tank M4 (75) war mit einer 75 mm Kanone, 2 MGs .30 und einem MG .50 ausgerüstet. Er wog 30,5 Tonnen, sein Motor hatte eine Pferdestärke von 350 PS, seine Höchstgeschwindigkeit betrug 40 km/h. Eine 5-Mann-Besatzung bediente ihn. Er war den deutschen Panther und Tiger Panzern unterlegen. Erst als er mit einer 76 mm Kanone ausgerüstet wurde, konnte er es mit seinen Gegnern aufnehmen.

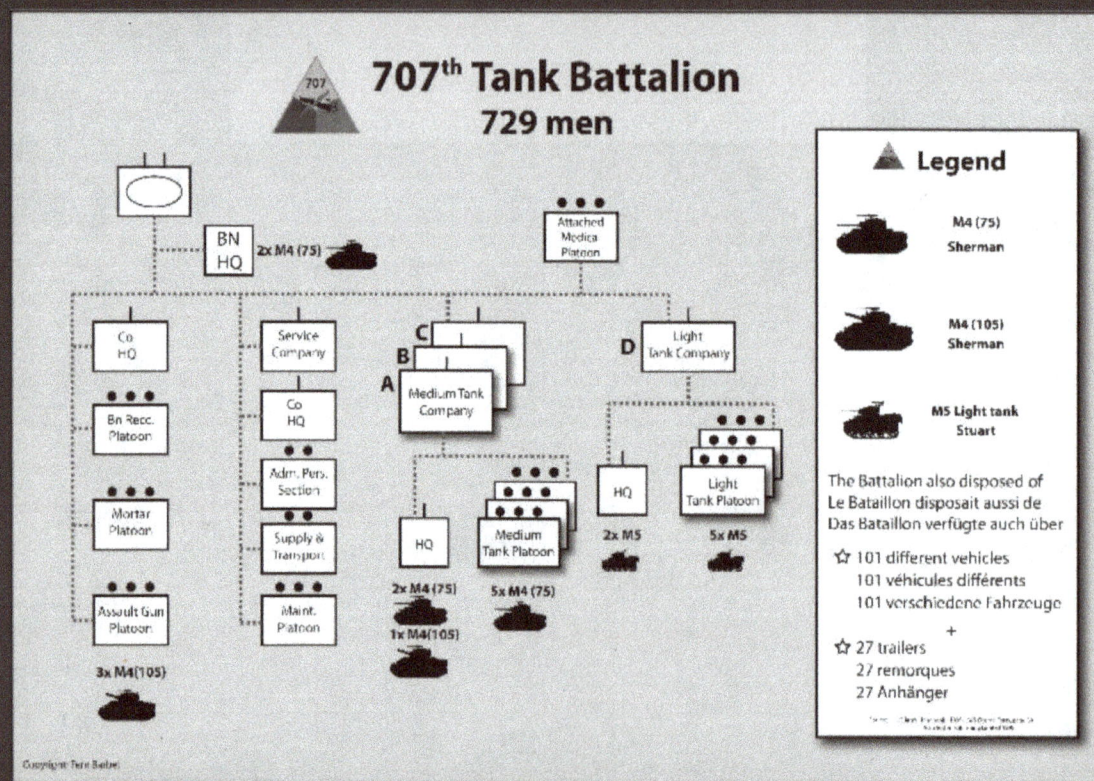

707th Tank Battalion
729 men

BN HQ — 2x M4 (75)

Attached Medical Platoon

Co HQ
Bn Recc. Platoon
Mortar Platoon
Assault Gun Platoon
3x M4 (105)

Service Company
Co HQ
Adm. Pers. Section
Supply & Transport
Maint. Platoon

A B C — Medium Tank Company
HQ — 2x M4 (75)
Medium Tank Platoon — 5x M4 (75)
1x M4(105)

HQ — 2x M5
D — Light Tank Company
HQ
Light Tank Platoon — 5x M5

Legend

M4 (75) Sherman

M4 (105) Sherman

M5 Light tank Stuart

The Battalion also disposed of
Le Bataillon disposait aussi de
Das Bataillon verfügte auch über

☆ 101 different vehicles
101 véhicules différents
101 verschiedene Fahrzeuge
+
☆ 27 trailers
27 remorques
27 Anhänger

707th Tank Battalion's organisation chart · Organigramme du 707th Bataillon de chars · Organigramm des 707. Panzerbataillons

Wiltz (Noertrange)

Dedicated to the 28th Division Band, who helped in the defense of Wiltz by trading their instrument for arms.

Route de Noertrange

9543 Wiltz

Latitude: 49.9705

Longitude: 5.9160

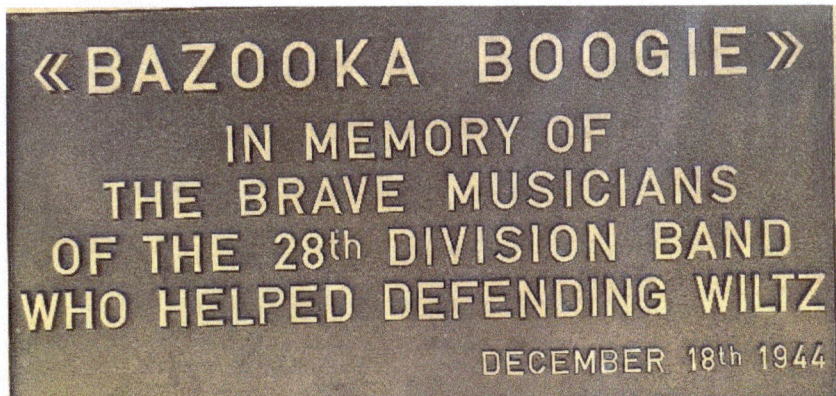

Photo shows German 8.8 cm PAK43/1 anti-tank gun

Wincrange

In memory of the 90th Infantry Division.

Wincrange 85

9780 Wincrange

Latitude: 50.0588

Longitude: 5.9212

Inscription reads: In memory of the valiant soldiers of the 90th US Infantry Division who have fought in this region in January 1945

Walking Trails

Throughout Luxembourg there are several trails that are World War II themed and pass various artifacts or historical markers related to US Forces. While if you walk a lot in the northern part of Luxembourg you are sure to run across Monuments, the following trails are specially related to World War II.

Bavigne-Berlé-Bavigne Trail

Trail starts at a parking lot in the center of Bavigne and takes about 2 hours. The trail passes the US monument in Berlé (see monument page for Berlé). There is also a monument along the way for residents of Berlé who on January 4, 1945 were forced by German occupying forces to leave the village and had to seek shelter in the forest.

Starting point in Bavigne

Trail marker

Monument in Berle (see monument page for more information)

Monument to residents of Berle who had to flee German forces

Fridhaff "Hooldär" Trail

The trail starts at the northbound parking on the N7 in Fridhaff (also known as *Friidhaff* or *Friedhof*) at coordinates (49.89227, 6.13448). Take the trail to the chalets and then the red "Rundweg." The trail is about 1.2 km, plus the walk in from the car park, so about 2km in total. Along the trail there are two information charts (Both in German) that cover the fighting in this area from both a German and an American viewpoint. There are also some German foxholes visible just behind one of the information boards.

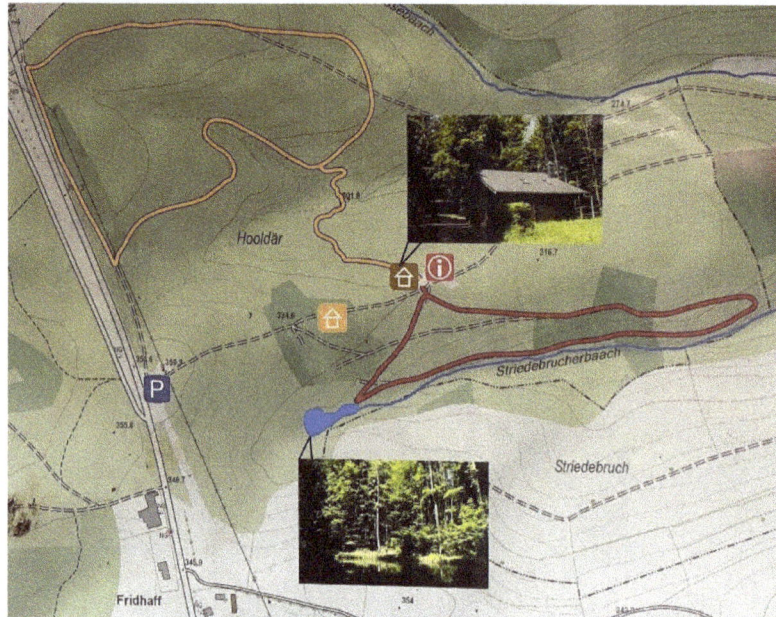

Info board with several German dugouts

Info board from American perspective

Hoesdorf-Bettendorf-Wallendorf Trail

Trail runs along both the German and Luxembourg side of the border marked by the Our River. The Luxembourg side of the trail takes about 3 hours to walk and has 10 information panels. The German side of the border has 6 information charts. Along the walk you pass several dugouts, tank 'footprints', a destroyed farmhouse, and various views across the battlefield and the German Westwall. In this sector, elements of the 109th Infantry Regiment, 28th Infantry Division fought the 352nd Volksgrenadier and 5th Paratroop Division of the German Army at the beginning of Battle of the Bulge.

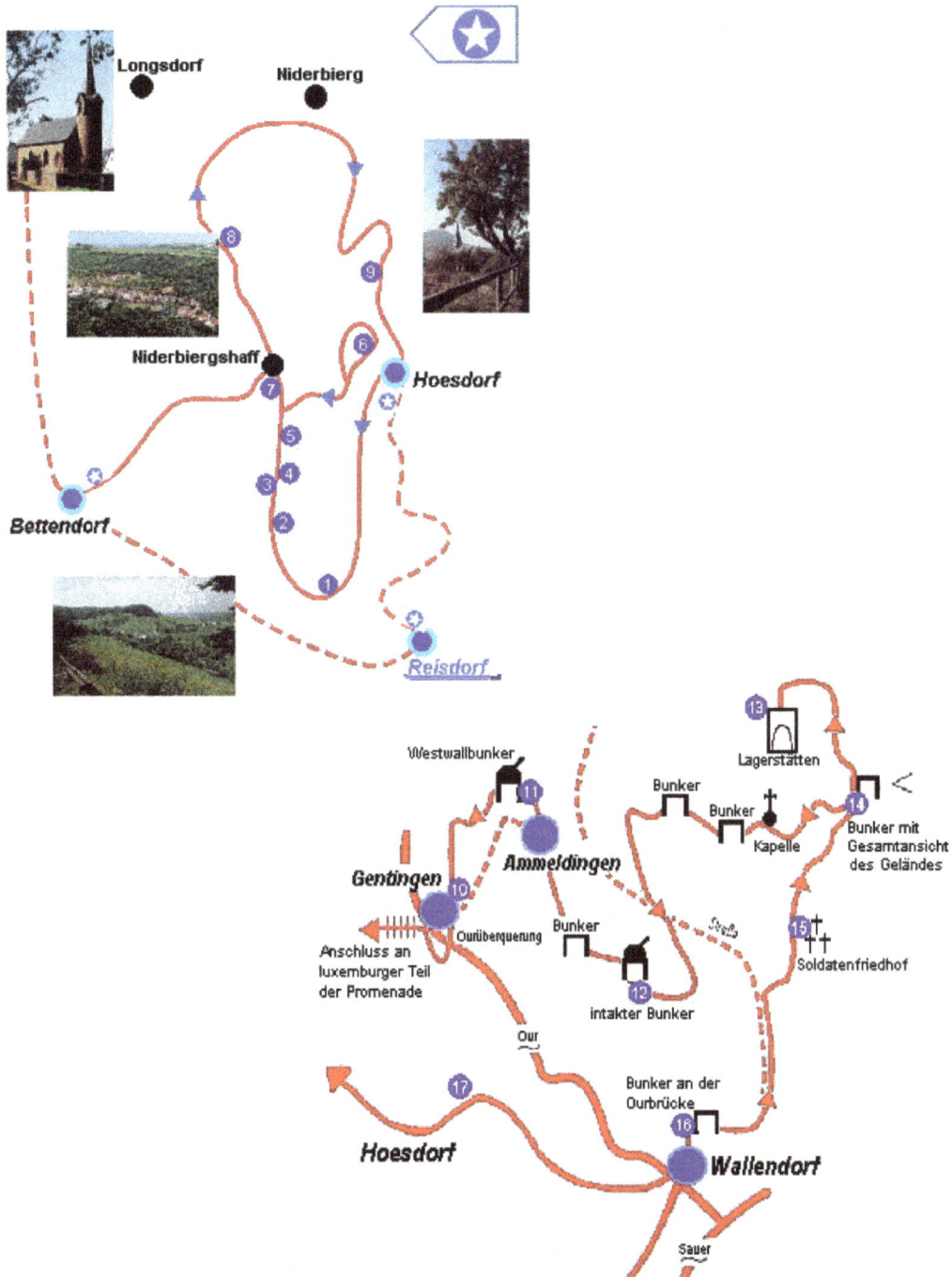

Maps available at the National Museum of Military History in Diekirch, or from www.visitluxembourg.com

Trail Marker

Starting point in Hoesdorf

GI tree carving

GI Dugout

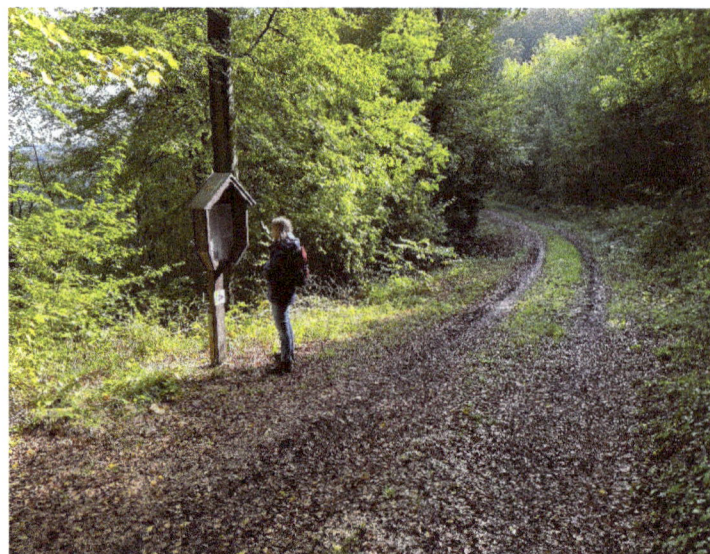

Information panel on trail

Schumannseck Trail

This is a walking trail that covers the "Schumannseck" battle site. The trail starts near the Schumannseck National liberation Memorial (see Monuments page for Nothum Schumannseck). There are two trails, one has a duration of 45 minutes and the other takes about 1:45, The trail passes many fox holes, and bomb craters. The trail has many information panels which tell the story of the close combat in this area during the Battle of the Bulge.

SCHUMANNSECK
MEMORIAL SITE OF THE BATTLE OF THE BULGE
SITE DE MÉMOIRE DE LA BATAILLE DES ARDENNES
GEDENKSTÄTTE DER ARDENNENOFFENSIVE

This is the entrance to the historical strategic site of Schumannseck. The deadliest combat of the Battle of the Bulge took place here, where American soldiers fought against the German Army for several weeks, between 27th December 1944 and 21st January 1945. Thousands of lives were sacrificed.

Vous êtes à l'entrée du site stratégique des combats du Schumannseck. Ici ont eu lieu les combats les plus meurtriers de la Bataille des Ardennes. Ils ont opposé les soldats américains à l'armée allemande pendant plusieurs semaines entre le 27 décembre 1944 et le 21 janvier 1945. Des milliers de vies ont été sacrifiées.

Sie befinden sich am Eingang des strategisch wichtigen Kriegsschauplatzes am Schumannseck. Während der Ardennenoffensive standen sich hier deutsche und amerikanische Truppen wochenlang, vom 27. Dezember 1944 bis zum 21. Januar 1945, in blutigen Kämpfen gegenüber. Mehrere Tausend Menschenleben wurden geopfert.

Discover the fate of those who fought in this forgotten battle, their suffering can still be felt in these woods.

This place is a national memorial site – we kindly ask you to respect it.

Découvrez le destin des personnes enrôlées dans cette bataille oubliée, leurs souffrances hantent encore ces bois...

Ce site est un lieu de mémoire national. Nous vous prions de le respecter.

Entdecken Sie das Schicksal der Menschen, die in diese vergessene Schlacht verwickelt waren. Ihr Vermächtnis geistert bis heute durch die Wälder ...

Diese Gedenkstätte ist ein nationaler Ort der Erinnerung. Wir bitten um den entsprechenden Respekt.

Short path
↔ 1,2 km
🕐 0h45
easy

Full path
↔ 2,8 km
🕐 1h45
easy

Please stay on the marked and signposted trail.

This site is a former battlefield, and although the trail has been secured, duds that remain unrecovered can be dangerous.

Unlawful excavations are strictly prohibited on this site. Archaeological heritage, which testifies to history, is public property protected under the criminal law of 7.12.2016.

Merci de rester sur le sentier balisé et fléché.

Ce site est un ancien champ de bataille. Le sentier a été sécurisé, mais des armes encore enfouies peuvent être dangereuses.

Toute fouille clandestine sur ce site est strictement interdite. Le patrimoine archéologique, témoin de l'histoire, est un bien public protégé par la loi pénale du 7.12.2016.

Bitte bleiben Sie auf dem markierten und ausgeschilderten Weg.

Diese Stätte ist ein ehemaliges Schlachtfeld. Der Weg wurde gesichert, Gefahr durch noch vergrabene Munition kann jedoch nicht ausgeschlossen werden.

Jede nicht genehmigte Grabung auf dem Gelände ist strengstens untersagt. Das archäologische Kulturerbe ist als Zeugnis der Geschichte ein öffentliches Gut, das durch das Gesetz vom 7.12.2016 strafrechtlich geschützt ist.

Start of north end of the trail at Schumannseck

Reconstructed fox holes

Start of south end of the Schumannseck trail

Bomb crater of 500 pound bomb dropped by LT William H. Nellis on 27 December 1944. Nellis was shot down as he made the bombing run and killed. Nellis Air Force Base, Las Vegas Nevada is named after him.

Luxembourg American Cemetery

Luxembourg-Hamm

The Luxembourg American Cemetery Memorial is located near the Luxembourg Airport and holds 5,076 service members, 101 of them unknown, and list another 371 missing in action. There are also 22 sets of brothers buried in the cemetery, and one woman, 2nd Lt. Nancy Leo. General George S. Patton Jr. and Medal of Honor recipients Staff Sergeant Day G. Turner and PVT William D. McGee are buried in the cemetery. Of the 101 unknown graves, some of the remains has been identified and returned to the U.S. at the request of the family.

Val du Scheid 50

2517 Luxembourg

Latitude: 49.6129

Longitude: 6.1862

MOH recipient Day G. Turner
grave marker

MOH recipient William D. McGee
grave marker

Medal of Honor Recipients

During the liberation of Luxembourg two US soldiers were awarded the Medal of Honor for action in Luxembourg (Day G. Turner and Francis J. Clark). The Medal of Honor is the highest US military decoration awarded for acts of valor in combat. A third Medal of Honor recipient (Wiliam D. McGee) was killed in action in Germany but buried in the Luxembourg American Cemetery.

Day G. Turner

Staff Sergeant Day G. Turner is buried in the Luxembourg American Cemetery (Plot E Row 10, Grave 72). Turner, originally from Berwick Pennsylvania was killed in action on February 8, 1945. He was assigned to the 319th Infantry Regiment, 80th Infantry Division. Along with the Medal of Honor, Turner also was awarded the Purple Heart with 2 Oak Leaf Clusters (American Battle Monuments Commission). Turner was originally from Nescopeck, Pennsylvania

Turner received the Medal of Honor for action on 8 January 1945 at Dahl, Luxembourg. His Medal of Honor citation reads:

He commanded a 9-man squad with the mission of holding a critical flank position. When overwhelming numbers of the enemy attacked under cover of withering artillery, mortar, and rocket fire, he withdrew his squad into a nearby house, determined to defend it to the last man. The enemy attacked again and again and were repulsed with heavy losses. Supported by direct tank fire, they finally gained entrance, but the intrepid sergeant refused to surrender although 5 of his men were wounded and 1 was killed. He boldly flung a can of flaming oil at the first wave of attackers, dispersing them, and fought doggedly from room to room, closing with the enemy in fierce hand-to-hand encounters. He hurled hand grenade for hand grenade, bayoneted 2 fanatical Germans who rushed a doorway he was defending and fought on with the enemy's weapons when his own ammunition was expended. The savage fight raged for 4 hours, and finally, when only 3 men of the defending squad were left unwounded, the enemy surrendered. Twenty-five prisoners were taken, 11 enemy dead and a great number of wounded were counted. Sgt. Turner's valiant stand will live on as a constant inspiration to his comrades his heroic, inspiring leadership, his determination and courageous devotion to duty exemplify the highest tradition of the military service (Center for Military History)

IN MEMORY OF

★ ★ ★ ★ ★

MEDAL OF HONOR RECIPIENT

DAY G. TURNER

RANK

STAFF SERGEANT, U.S. ARMY

UNIT

319TH INFANTRY REGIMENT, 80TH INFANTRY DIVISION

DATE OF DEATH

FEBRUARY 8, 1945

COMMEMORATED IN PERPETUITY AT

LUXEMBOURG AMERICAN CEMETERY

, LUXEMBOURG

"Time will not dim the glory of their deeds."

— GENERAL JOHN J. PERSHING

See monument on page about the village of Dahl

Francis J. Clark

Technical Sergeant Francis J. Clark originally from Salem, New York and was assigned to Company K, 109th Infantry Regiment, 28th Infantry Division. He survived the war and died in 1981 at the age of 69.

Clark received the Medal of Honor for action on 12 September 1944 near Kalborn, Luxembourg and on 17 September near Sevenig, Germany. His Medal of Honor citation reads:

He fought gallantly in Luxembourg and Germany. On 12 September 1944, Company K began fording the Our River near Kalborn, Luxembourg, to take high ground on the opposite bank. Covered by early morning fog, the 3d Platoon, in which T/Sgt. Clark was squad leader, successfully negotiated the crossing; but when the 2d Platoon reached the shore, withering automatic and small-arms fire ripped into it, eliminating the platoon leader and platoon sergeant and pinning down the troops in the open. From his comparatively safe position, T/Sgt. Clark crawled alone across a field through a hail of bullets to the stricken troops. He led the platoon to safety and then unhesitatingly returned into the fire-swept area to rescue a wounded soldier, carrying him to the American line while hostile gunners tried to cut him down. Later, he led his squad and men of the 2d Platoon in dangerous sorties against strong enemy positions to weaken them by lightning-like jabs. He assaulted an enemy machinegun with hand grenades, killing 2 Germans. He roamed the front and flanks, dashing toward hostile weapons, killing and wounding an undetermined number of the enemy, scattering German patrols and, eventually, forcing the withdrawal of a full company of Germans heavily armed with automatic weapons. On 17 September, near Sevenig, Germany, he advanced alone against an enemy machinegun, killed the gunner and forced the assistant to flee. The Germans counterattacked, and heavy casualties were suffered by Company K. Seeing that 2 platoons lacked leadership, T/Sgt. Clark took over their command and moved among the men to give encouragement. Although wounded on the morning of 18 September, he refused to be evacuated and took up a position in a pillbox when night came. Emerging at daybreak, he killed a German soldier setting up a machinegun not more than 5 yards away. When he located another enemy gun, he moved up unobserved and killed 2 Germans with rifle fire. Later that day he voluntarily braved small-arms fire to take food and water to members of an isolated platoon. T/Sgt. Clark's actions in assuming command when leadership was desperately needed, in launching attacks and beating off counterattacks, in aiding his stranded comrades, and in fearlessly facing powerful enemy fire, were strikingly heroic examples and put fighting heart into the hard-pressed men of Company K.

Source of photo: media.defense.gov

William D. McGee

Private William D. McGee is buried in the Luxembourg American Cemetery (Plot C, Row 7, Grave 13). Originally from Indianapolis Indiana, Pvt. McGee was assigned to Medical Detachment, 304th Infantry, 76th Infantry Division. He was awarded the Medal of Honor for his actions on March 18, 1945 near Mulheim, Germany where he was killed in action.

His Medal of Honor citation reads:

A medical aidman, he made a night crossing of the Moselle River with troops endeavoring to capture the town of Mulheim. The enemy had retreated in the sector where the assault boats landed, but had left the shore heavily strewn with antipersonnel mines. Two men of the first wave attempting to work their way forward detonated mines which wounded them seriously, leaving them bleeding and in great pain beyond the reach of their comrades. Entirely on his own initiative, Pvt. McGee entered the minefield, brought out one of the injured to comparative safety, and had returned to rescue the second victim when he stepped on a mine and was severely wounded in the resulting explosion. Although suffering intensely and bleeding profusely, he shouted orders that none of his comrades was to risk his life by entering the death-sown field to render first aid that might have saved his life. In making the supreme sacrifice, Pvt. McGee demonstrated a concern for the well-being of his fellow soldiers that transcended all considerations for his own safety and a gallantry in keeping with the highest traditions of the military service.

IN MEMORY OF
★ ★ ★ ★ ★
WILLIAM D. MCGEE

RANK
PRIVATE, U.S. ARMY

UNIT
304TH INFANTRY REGIMENT, 76TH INFANTRY DIVISION

DATE OF DEATH
MARCH 19, 1945

COMMEMORATED IN PERPETUITY AT
LUXEMBOURG AMERICAN CEMETERY
, LUXEMBOURG

"Time will not dim the glory of their deeds."
GENERAL JOHN J. PERSHING

WW II Museums in Luxembourg

There are five World War II museums in Luxembourg, and if you get the chance make time to visit all of them. Make sure to check out the museums websites for opening times.

National Museum of Military History Diekirch

10 Bamertal

L-9209 Diekirch

www.mnhm.net

Closed on Mondays

385th Bomb Group Perlé

Museum is dedicated to the 385th Bomb Group and documents aircraft crashes in Luxembourg during World War II.

6, Rue de l'Eglise

L-8826 Perlé

www.385bg.lu

museum385bg@pt.lu

Only open a few days a month so check website for opening days.

General Patton Memorial Museum Ettelbruck

5, rue Dr. Klein

L-9054 Ettelbruck

www.patton.lu

Closed on Mondays

Information Panel Ettelbruck (Patton Museum)

105 MM HOWITZER

The 105 mm howitzer was the standard U.S. light field howitzer during the Second World War.

Entering production in 1941, it quickly gained a reputation for accuracy and firepower.

The howitzer 105 mm mainly fired high explosive ammunition and had a range of 12,330 yards (11,270 m), making it particularly suitable for supporting infantry units.

During the Second World War, U.S. artillery regiments consisted of an HQ detachment, one 155 mm artillery battalion and three 105 mm artillery battalions. Both the 155 mm and 105 mm battalions had twelve guns each, divided into three batteries of four guns. That gave each regiment a total of twelve 155 mm howitzers and thirty-six 105 mm howitzers.

The 105 mm howitzer equipped as well the 313th, 314th and 905th Field Artillery Battalions of the 80th Infantry Division which liberated Ettelbruck in December 1944.

By the end of the Second World War, 8536 towed howitzers of this type had been built and post-war production continued at Rock Island Arsenal until 1953.

Caliber:	105 mm
Weight:	4,980 lb (2,260 kg)
Length:	19 ft 6 in (5.94 m)
Maximum firing range:	12,330 yards (11,270 m)
Rate of fire:	15 rounds per minute

105 MM HOWITZER

L'obusier de 105 mm était la pièce d'artillerie légère standard de l'armée américaine pendant la Seconde Guerre Mondiale.

Développé à partir de 1941, l'obusier gagnait rapidement une réputation pour sa précision et sa puissance de feu. L'arme tirait principalement des obus explosifs de calibre 105mm à une distance allant jusque 11,27 kilomètres, ce qui la rendait particulièrement adaptée pour l'appui des unités d'infanterie.

Pendant la Seconde Guerre Mondiale, un régiment d'artillerie américain était composé d'un détachement d'état-major, d'un bataillon d'artillerie lourde équipé d'obusiers de calibre 155 mm et de trois bataillons d'artillerie de 105 mm. Chaque bataillon était doté de 12 obusiers, répartis sur trois batteries disposant chacune de 4 pièces. Au total, un régiment disposait donc de 12 pièces de 155 mm et de 36 pièces de 105 mm.

L'obusier tracté de 105 mm équipait également les 313ème, 314ème et 905ème bataillons d'artillerie de la 80ème Division d'infanterie américaine qui libérait la ville d'Ettelbrück en décembre 1944.

Jusqu'à la fin de la guerre, 8536 obusier de 105 mm avaient été produits. Pendant la période d'après-guerre, la fabrication était poursuivie à l'arsenal de Rock Island jusque 1953.

Calibre:	105 mm
Poids:	2260 kg
Longueur:	5.94 m
Portée maximale:	11270 m
Cadence de tir:	15 coups par minute

105 MM HOWITZER

Die 105 mm-Haubitze war das meistverbreitete leichte Artilleriegeschütz der amerikanischen Armee im Zweiten Weltkrieg.

Nachdem die Produktion im Jahre 1941 angelaufen war, wurde das Geschütz schnell wegen seiner Treffsicherheit und der hohen Schusskadenz geschätzt. Die Haubitze verschoss hauptsächlich Explosivgeschosse des Kalibers 105 mm und hatte eine Reichweite bis 11,27 Kilometer. Hierdurch eignete sich die Waffe hervorragend für die Feuerunterstützung der Infanterieeinheiten.

Während des Zweiten Weltkrieges bestand ein amerikanisches Artillerieregiment aus einer Stabseinheit, einem schweren Artilleriebataillon mit 155 mm Geschützen und 3 Bataillonen mit 105 mm Haubitzen. Jedes Bataillon verfügte über 12 Geschütze, aufgeteilt auf 3 Batterien mit je 4 Haubitzen. Ein Artillerieregiment war also mit 12 schweren Geschützen des Kalibers 155 mm und 36 leichten Geschützen des Kalibers 105 mm ausgerüstet.

Die Artilleriebataillone (313th, 314th und 905th Field Artillery Battalion) der 80. US Infanteriedivision, die Ettelbruck im Dezember 1944 befreite, waren mit der 105 mm Haubitze ausgerüstet.

Bis Kriegsende waren 8536 Haubitzen des Kalibers 105 mm ausgeliefert worden, während der Nachkriegszeit wurde die Produktion jedoch im Arsenal von Rock Island bis 1953 fortgesetzt.

Kaliber:	105 mm
Gewicht:	2260 kg
Länge:	5.94 m
Maximale Reichweite:	11,270 m
Feuerkadenz:	15 Schuss pro Minute

Musee de la bataille des Ardennes Wiltz

35, rue de Château

L-9516 Wiltz

www.touristinfowiltz.lu

Check website for opening times

Musée de la bataille des Ardennes Clervaux

Château de Clervaux

L-9516 Clervaux

www.clervaux.lu/fr/musee-de-la-bataille-des-ardennes-1.html

Check website for opening times

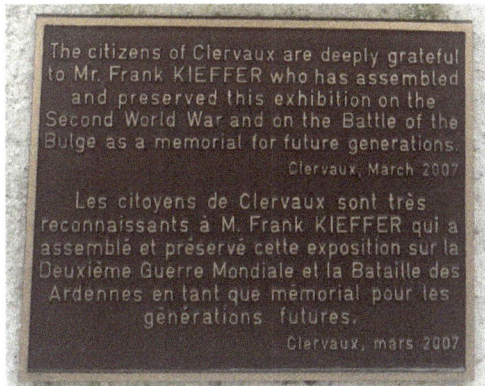

Remembrance Organizations

In addition to the museums in Luxembourg preserving the memory and dedicated to remembering World War II, there are also several organizations honoring the soldiers that fought to liberate Luxembourg. Here is a list of these organizations. Support is always welcome.

Cercle d'Etudes sur la Bataille des Adennes (CEBA)	www.facebook.com/CEBALuxembourg
US Veterans Friends	www.usvf.lu
Friends of Patton's 26th Infantry Division	www.fpyd.org
American War Memorials Overseas, Inc	www.uswarmemorials.org
Quadriga (historical vehicles collectors)	www.quadriga.lu
CVM Steinsel (military vehicles)	www.cvmsteesel.lu
The Ghost Army Legacy Project	www.ghostarmy.org
WWII Battlefield Research and Preservation Group G. D. Luxembourg	https://www.schoolandcollegelistings.com /LU/Luxembourg/145689668826196/WWII-Battlefield-Research-and-Preservation-Group

About the Author

Robert O. Walton, PhD

Dr. Bob Walton is a tenured Associate Professor in the Department of Decision Sciences, College of Business, with Embry-Riddle Aeronautical University-Worldwide (ERAU).

Bob received his undergraduate degree in geography from the University of North Carolina at Wilmington and holds a master of aeronautical science, master of business administration, master of logistics and supply chain management, and a PhD in business administration.

He retired from the US Army Reserves as a Major and has always enjoyed military history. Bob has lived in Luxembourg for 24 years and is a dual US - Luxembourg citizen. He has extensively toured the battlefields of Luxembourg and Europe.

Location Maps

The following maps are divided into quadrants of Luxembourg and provide a general location of the US monuments. The numbers on the map correspond to the page number in the text. Please note that these are only general locations, for more accurate locations please use the latitude/longitude on the monument page.

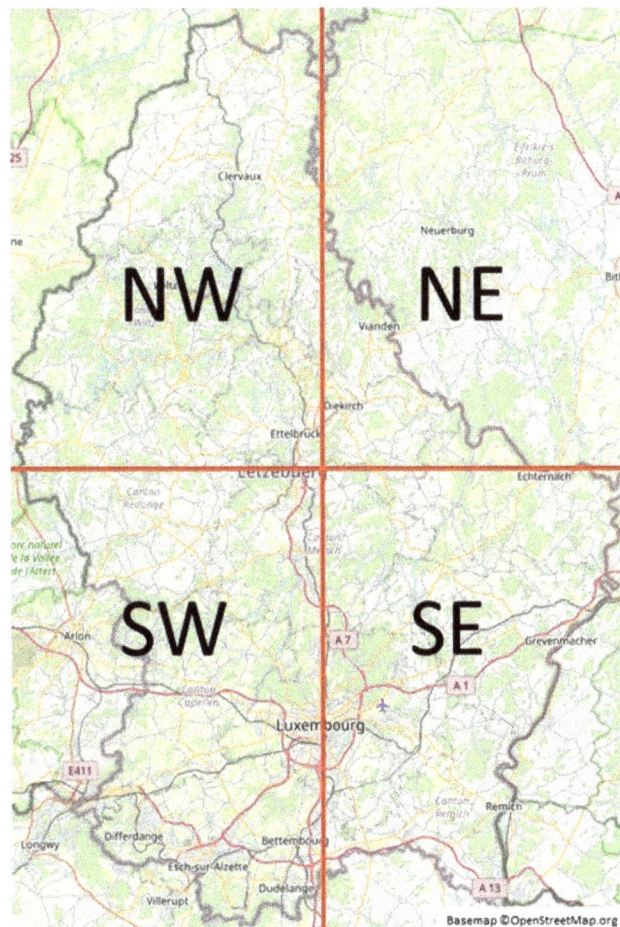

Basemap ©OpenStreetMap.org

Northwest Quadrant

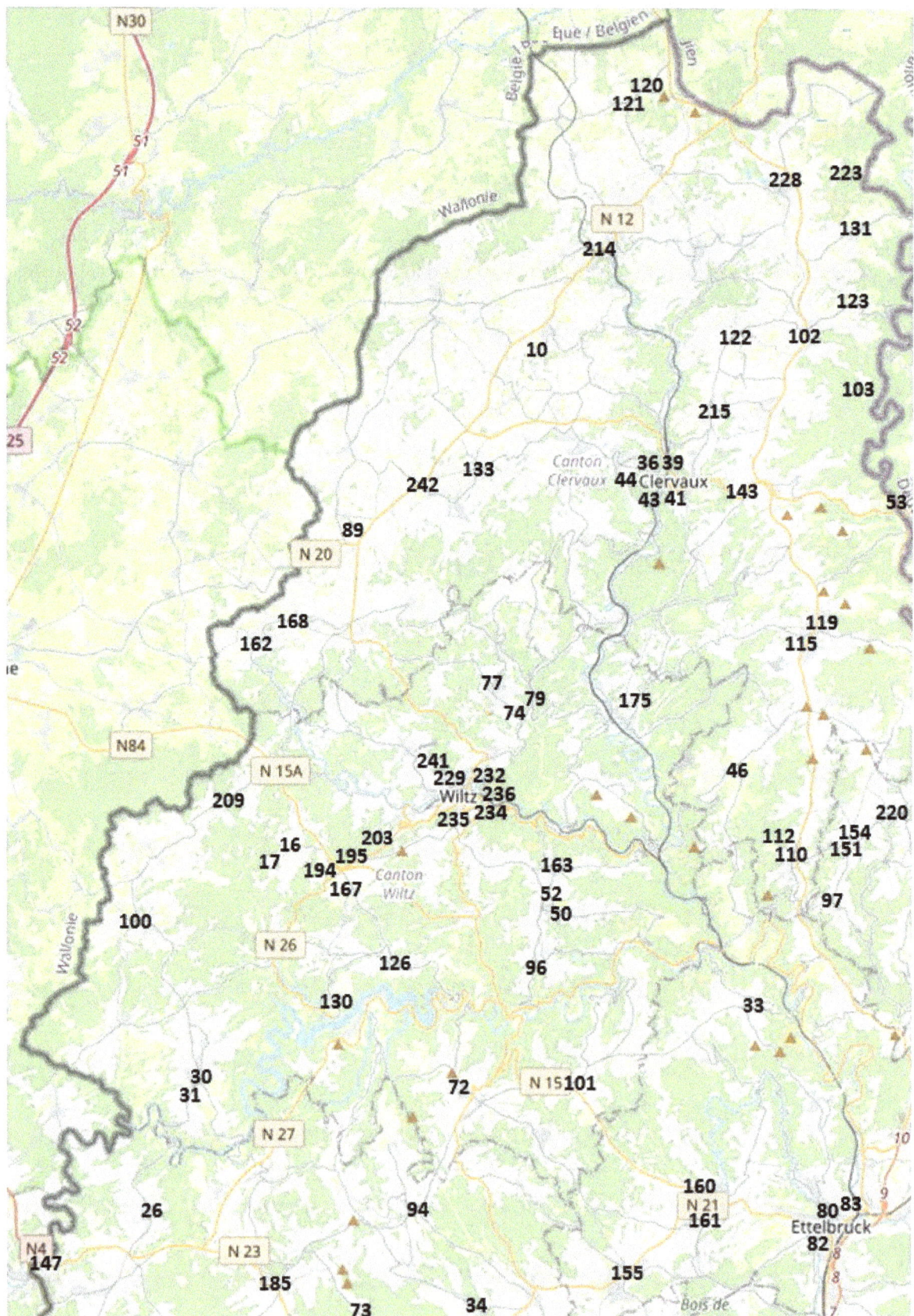

Basemap ©OpenStreetMap.org

Northeast Quadrant

Basemap ©OpenStreetMap.org

Southwest Quadrant

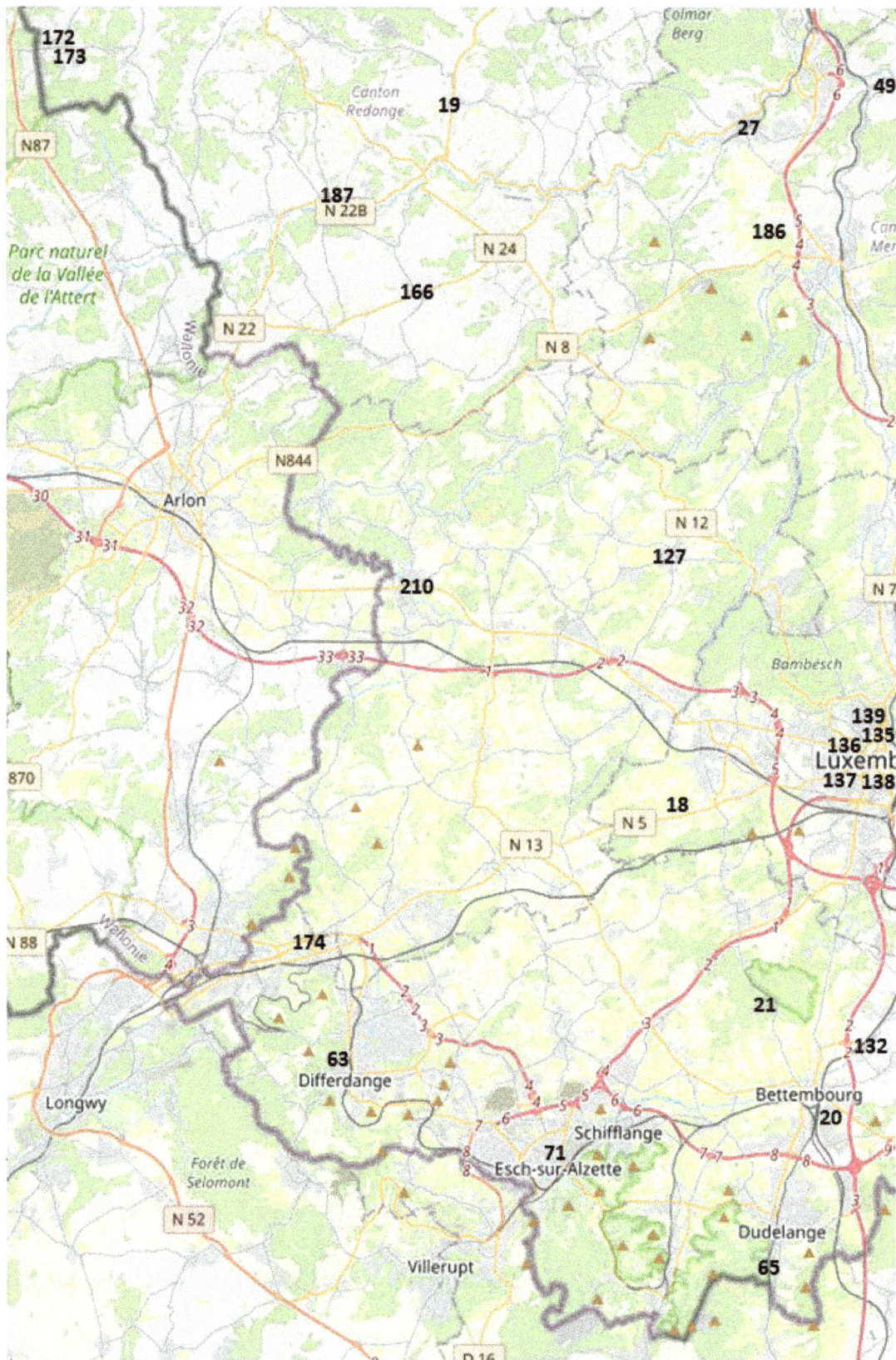

Basemap ©OpenStreetMap.org

Southeast Quadrant

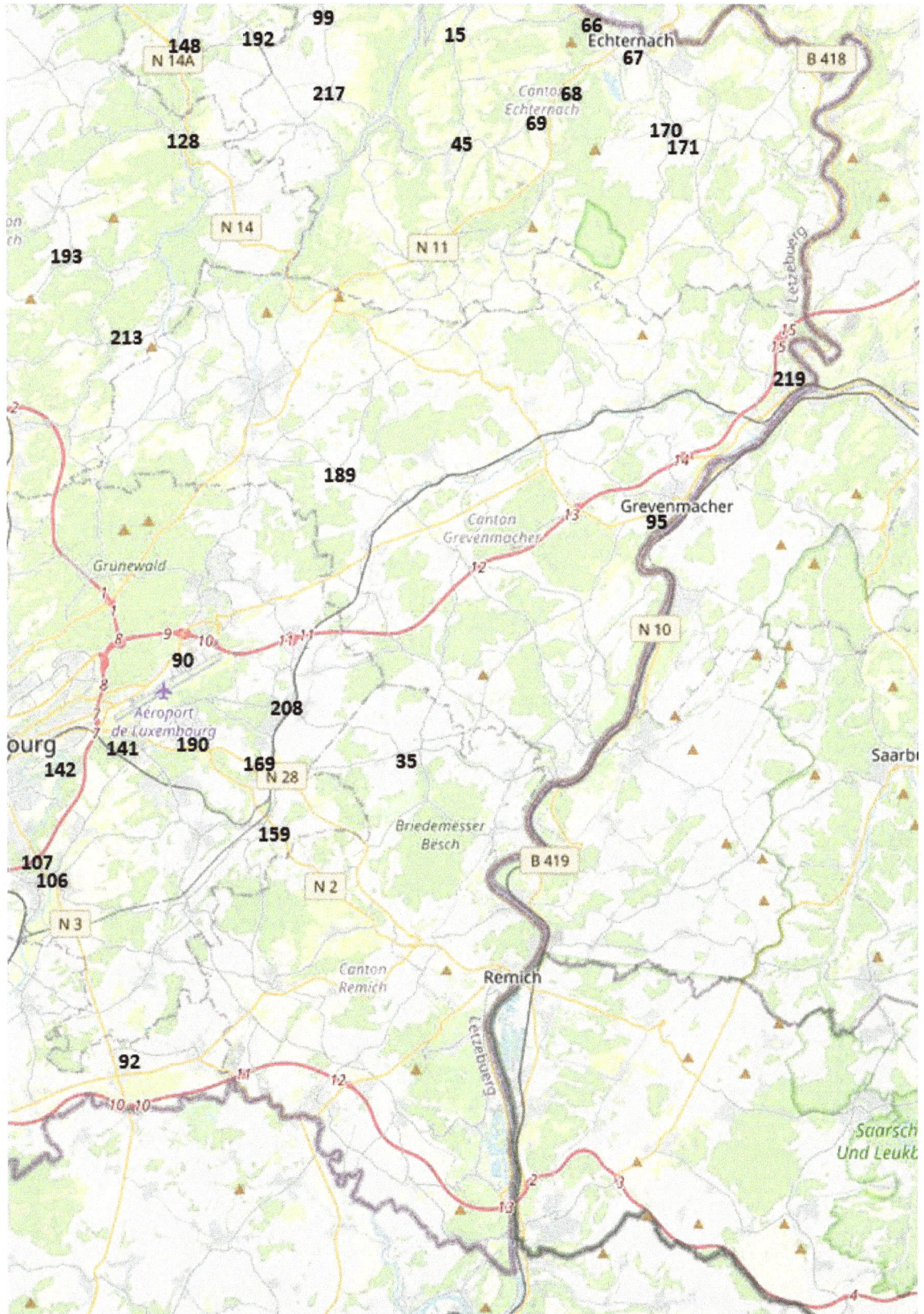

Basemap ©OpenStreetMap.org

Index

References

American Battle Monument Commission. (2019). Day G. Turner. Retrieved from
http://www.abmc.gov

American War Memorials Overseas, Inc. (n.d.). https://www.uswarmemorials.org/

Dupuy, T. N. (1994). Hitler's Last Gamble: The Battle of the Bulge, December 1944 – January 1945.
Harper Collins Publishers.

Center of Military History. (2019). Medal of Honor Recipients. Retrieved form
http://history.army.mil

HonorStates (n.d.). HonorStates.org

Luxembourg Tourism Office (2004). 60th Anniversary of the two liberations of Luxembourg.
Brochure.

MacDonald, C. B. (1985). A Time For Trumpets: The untold story of the Battle of the Bulge. Quill
publishing

MHOH (n.d.). Military Hall of Honor. www. Militaryhallofhonor.com

National Museum of Military History Diekirch (n.d.). Roster of US Units Engaged in Luxembourg:
September 1944 – February 1945

Unit History (n.d.) U.S. Army Center of Military History. www.history.army.mil